Springer Proceedings in Physics 65

Springer Proceedings in Physics

Managing Editor: H. K. V. Lotsch

44 *Optical Fiber Sensors*
Editors: H. J. Arditty, J. P. Dakin, and R. Th. Kersten

45 *Computer Simulation Studies in Condensed Matter Physics II: New Directions*
Editors: D. P. Landau, K. K. Mon, and H.-B. Schüttler

46 *Cellular Automata and Modeling of Complex Physical Systems*
Editors: P. Manneville, N. Boccara, G. Y. Vichniac, and R. Bidaux

47 *Number Theory and Physics*
Editors: J.-M. Luck, P. Moussa, and M. Waldschmidt

48 *Many-Atom Interactions in Solids*
Editors: R .M. Nieminen, M. J. Puska, and M. J. Manninen

49 *Ultrafast Phenomena in Spectroscopy*
Editors: E. Klose and B. Wilhelmi

50 *Magnetic Properties of Low-Dimensional Systems II: New Developments*
Editors: L. M. Falicov, F. Mejía-Lira, and J. L. Morán-López

51 *The Physics and Chemistry of Organic Superconductors*
Editors: G. Saito and S. Kagoshima

52 *Dynamics and Patterns in Complex Fluids: New Aspects of the Physics–Chemistry Interface*
Editors: A. Onuki and K. Kawasaki

53 *Computer Simulation Studies in Condensed Matter Physics III*
Editors: D. P. Landau, K. K. Mon, and H.-B. Schüttler

54 *Polycrystalline Semiconductors II*
Editors: J. H. Werner and H. P. Strunk

55 *Nonlinear Dynamics and Quantum Phenomena in Optical Systems*
Editors: R. Vilaseca and R. Corbalán

56 *Amorphous and Crystalline Silicon Carbide III, and Other Group IV – IV Materials*
Editors: G. L. Harris, M. G. Spencer, and C. Y.-W. Yang

57 *Evolutionary Trends in the Physical Sciences*
Editors: M. Suzuki and R. Kubo

58 *New Trends in Nuclear Collective Dynamics*
Editors: Y. Abe, H. Horiuchi, and K. Matsuyanagi

59 *Exotic Atoms in Condensed Matter*
Editors: G. Benedek and H. Schneuwly

60 *The Physics and Chemistry of Oxide Superconductors*
Editors: Y. Iye and H. Yasuoka

61 *Surface X-Ray and Neutron Scattering*
Editors: H. Zabel and I. K. Robinson

62 *Surface Science: Lectures on Basic Concepts and Applications*
Editors: F. A. Ponce and M. Cardona

63 *Coherent Raman Spectroscopy: Recent Advances*
Editors: G. Marowsky and V. V. Smirnov

64 *Superconducting Devices and Their Applications*
Editors: H. Koch and H. Lübbig

65 *Present and Future of High-Energy Physics*
Editors: K.-I. Aoki and M. Kobayashi

Volumes 1–43 are listed on the back inside cover

K.-I. Aoki M. Kobayashi (Eds.)

Present and Future of High-Energy Physics

Proceedings
of the 5th Nishinomiya-Yukawa
Memorial Symposium on Theoretical Physics,
Nishinomiya City, Japan, October 25–26, 1990

With 134 Figures

Springer-Verlag
Berlin Heidelberg New York
London Paris Tokyo
Hong Kong Barcelona
Budapest

Professor Dr. Ken-Ichi Aoki
Physics Department, Kanazawa University,
Kanazawa 920, Japan

Professor Dr. Makoto Kobayashi
Theory Division, KEK National Laboratory for High-Energy Physics,
Oho-machi, Tsukuba-gun, Ibaraki-ken 305, Japan

ISBN-13:978-3-642-84743-1 e-ISBN-13:978-3-642-84741-7
DOI: 10.1007/978-3-642-84741-7

Library of Congress Cataloging-in-Publication Data. Nishinomiya-Yukawa Memorial Symposium (5th: 1990: Nishinomiya-shi, Japan) Present and future of high-energy physics: proceedings of the 5th Nishinomiya-Yukawa Symposium on Theoretical Physics, Nishinomiya City, Japan, October 25-26, 1990 / K.-I. Aoki, M. Kobayashi (eds.). p. cm. – (Springer proceedings in physics; v. 65) Includes index.
ISBN-13:978-3-642-84743-1 1. Particles (Nuclear physics)–Congresses. I. Aoki, Ken-ichi. II. Kobayashi, M. (Makoto) III. Title. IV. Series. QC793.N57 1990 539.7'2–dc20 92-9578

Typesetting: Camera ready by authors

54/3140-543210 – Printed on acid-free paper

Preface

The fifth Nishinomiya-Yukawa Memorial Symposium on Theoretical Physics was held on October 25 and 26, 1990, in Nishinomiya City, Japan. The symposium was designed to present the forefront of current research and future prospects in high-energy physics, particularly for young physicists in this field.

We feel that the 20th century will be regarded as an era when particle physics was characterized by intimate collaboration between theory and experiment which led to a very rapid development of our knowledge. The structure of nature was unveiled at deeper and deeper levels, culminating in the standard model of elementary particles which describes all known interactions (except gravitation) in a simple and accurate manner. We are now targeting the central region of the standard model, where spontaneous electroweak symmetry breaking occurs. There may emerge a variety of elementary particles, interactions, and associated phenomena, including the diverse fermion mass spectrum, CP violation, the flavour mixing of quarks, various weak interactions and electromagnetism, and possibly as yet unseen supersymmetric partners.

As we head towards the 21st century, we shall explore this region and beyond. Experimentally there are proposals for accelerators to access new energy scales and to perform various precision experiments, whilst there is much theoretical activity in studying gauge field theories, non-perturbative methods, numerical simulations, and so on.

We hoped that this symposium would cover these topics. This volume contains 10 of the invited lectures given at the symposium. They all hint at the key to the rich physics of the future. To our regret, however, the important contribution of M.E. Peskin is not available.

The organizing committee of the symposium consisted of:

K.-I. Aoki (Yukawa Institute, Kyoto University)
K. Hikasa (KEK)
M. Kobayashi (KEK)
A. Maki (KEK)
S. Sugimoto (Osaka University)
E. Takasugi (Osaka University)

The symposium was organized under the auspices of the Education Board of Nishinomiya City and the Yukawa Instiute for Theoretical Physics (formerly known as the Research Institute for Fundamental Physics), Kyoto University.

Kyoto, Tsukuba
December 1991

Ken-Ichi Aoki
Makoto Kobayashi

Opening Address

I am very pleased to open the 5th Nishinomiya-Yukawa Commemorative Symposium. We have already held this Symposium four times, and I feel grateful that gradually many young physicists have decided to come and the number of persons supporting this project is continually increasing.

As the mayor of Nishinomiya I would like to express my gratitude for the great efforts made by both the members of the Steering Committee and the participants.

In 1986 many theoretical physicists constructed "The Birth Monument of Mesotron Theory" at Kurakuen in Nishinomiya. Taking this opportunity, Nishinomiya City decided to hold the "Nishinomiya-Yukawa Commemorative Symposium" in order to honour Dr. Yukawa's great work, to recognize Nishinomiya as the birth city of the Mesotron Theory and to awaken our citizens' interest in science.

As a memorial in its 5th year and on the occasion of the 65th anniversary of our municipality, we now hold this Symposium. There are sessions on high-energy physics, nuclear physics, a commemorative lecture by Professor Nishijima of Chuo University, and an exhibition of Dr. Yukawa's work.

Honouring the many research efforts, we would like to present two Nishinomiya-Yukawa Commemorative Prizes, adding one more this year. From daily life to space, the fundamentals of physics contribute to the development of human beings.

We, the citizens of Nishinomiya, are very happy and proud of this Nishinomiya-Yukawa Commemorative Symposium and the benefits which it brings to research in Physics. Thank you.

Yoneji Yagi
Mayor of Nishinomiya

Contents

Physics at the TRISTAN e^+e^- Collider

Y. Watanabe

Department of Physics, Faculty of Science, Tokyo Institute of Technology, 2-12-1, Oh-okayama, Meguro-ku, Tokyo 152, Japan

Abstract. The physics activity at TRISTAN is reviewed. Information on the axial and vector coupling constants is obtained through the studies of charge asymmetries and polarization in the annihilation processes of e^+e^- into fermion pairs. Further evidence is given on the $B^0\overline{B^0}$ mixing, and some information is obtained on $B^0_s\overline{B^0_s}$ mixing.

An evidence is given on the existence of the triple gluon vertex, a crucial feature of QCD. Mass limits obtained of new particle searches are discussed. Finally, the near future prospects at TRISTAN are reviewed.

1. Introduction

The TRISTAN e^+e^- collider started its operation in November 1986, and had served as an energy frontier machine for a little more than three years. The c.m.s. energy of the TRISTAN had been steadily increased; the start value was 50 GeV and it eventually reached to 64 GeV. This increase was in part achieved by an extensive use of superconducting RF cavities, which produced the accelerating gradient of more than 5 MV/m. In fact the TRISTAN is the first high energy machine that successfully accelerated beams with super RF cavities.

In 1989, SLC at SLAC and LEP at CERN have come into action at energies around Z^0 peak, and soon produced impressive amount of results. Thus, the role of TRISTAN had to be necessarily changed, and the phase 2 program has been started, where an effort is made to increase the luminosity as much as possible at an optimum c.m.s. energy of 58 GeV.

By the end of summer 1990, each of the three TRISTAN experiments, AMY, TOPAZ, and VENUS have accumulated the integrated luminosity of about 60 pb^{-1}, corresponding to event samples of about 7k multi-hadron,

700 $\mu^+\mu^-$ and 500 $\tau^+\tau^-$. Various studies have been made analyzing these event samples.

In this report, I try to give an overview of the physics accomplishments made atTRISTAN up to now, but only on selected topics[1], and then to give an outlook of the TRISTAN phase 2.

2. The energy range of TRISTAN

Before discussing about physics results from TRISTAN, it is useful for the reader to have an idea on the energy range covered by TRISTAN. Figure 1 shows the total hadronic cross sections measured at various accelerators as a function of the c.m.s energy $\sqrt{s}$. The $1/s$ decrease is apparent down to the TRISTAN energy range, followed by a big peak corresponding to the Z^0 resonance. Namely, the energy range of the TRISTAN is just at a local minimum in terms of cross section.

Cross sections are at the local minimum, but the interference effect between the photon and the Z^0 is near the maximum in the TRISTAN energy range, as shown in fig.2. Thus, detail studies can be made on electroweak effects at TRISTAN.

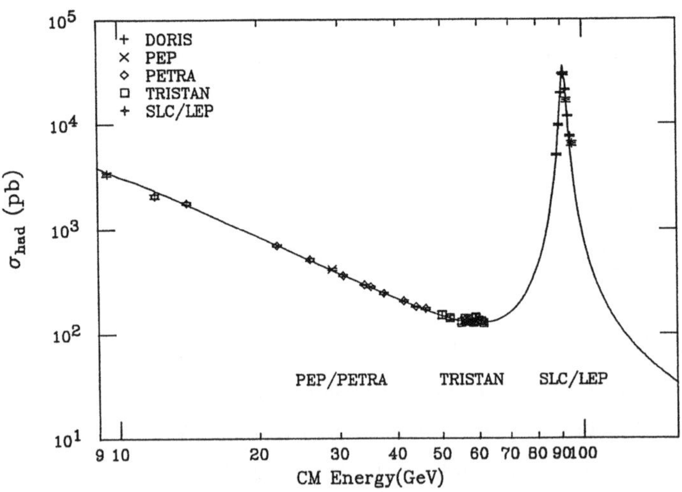

1. Total Hadronic cross section for $e^+e^- \to$ hadrons measured at DORIS, PEP, PETRA, TRISTAN and LEP plotted vs. $\sqrt{s}$ in log-log scale.

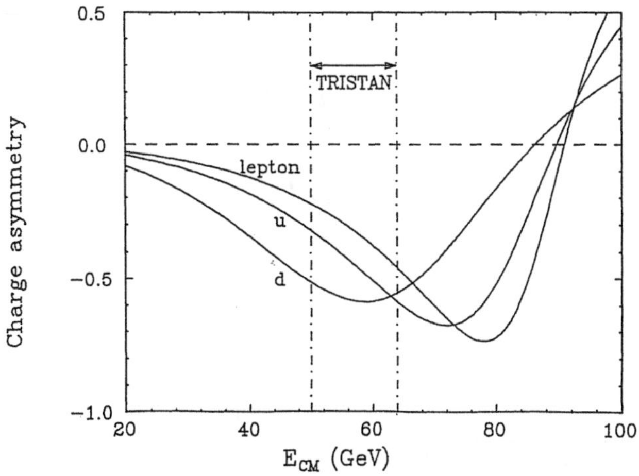

2. Charge asymmetries for the e^+e^- annihilation into lepton pairs, quark pairs of up-type and down-type expected of the standard model.

3. Study Electroweak Effects

3.1 Measurement of R

A fundamental quantity is the R-ratio, which is defined as

$$R \equiv \frac{\sigma_{hadron}}{\sigma_0}, \tag{1}$$

$$\sigma_0 \equiv \frac{4\pi\alpha^2}{3s} \approx \frac{87}{s(GeV^2)} nb. \tag{2}$$

Experimentally, the R-ratio can be obtained as,

$$R = \frac{1}{\sigma_0} \frac{N^{obs}_{hadron} - N_{BG}}{\eta(1+\delta)\int Ldt}, \tag{3}$$

where N^{obs} is the observed number of hadronic events, N_{BG} is the estimated number of background events, η is the detector acceptance, $1+\delta$ is the radiative correction factor, $\int Ldt$ is the integrated luminosity. For the $1+\delta$, we use the calculation by Fujimoto and Shimizu[2], which includes all the electroweak diagrams up to $O(\alpha^3)$.

In fig.3, the measured R-ratio is plotted as a function of $\sqrt{s}$. The solid curve is the prediction of the standard theory. We did not observe any step

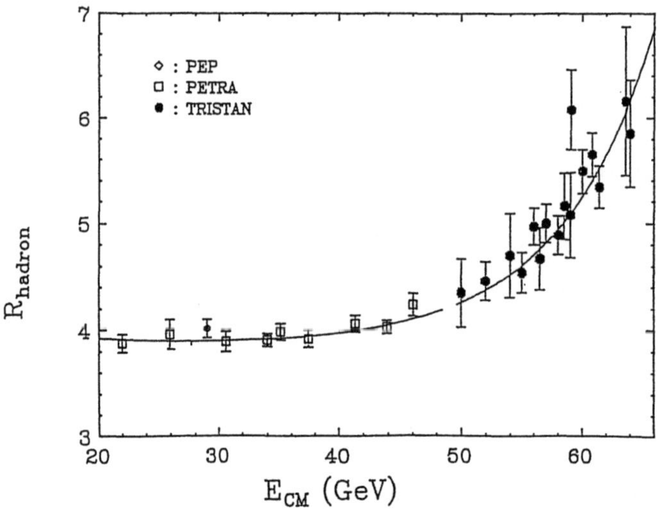

3. R-ratio measured at PEP, PETRA and TRISTAN plotted vs. E_{CM}

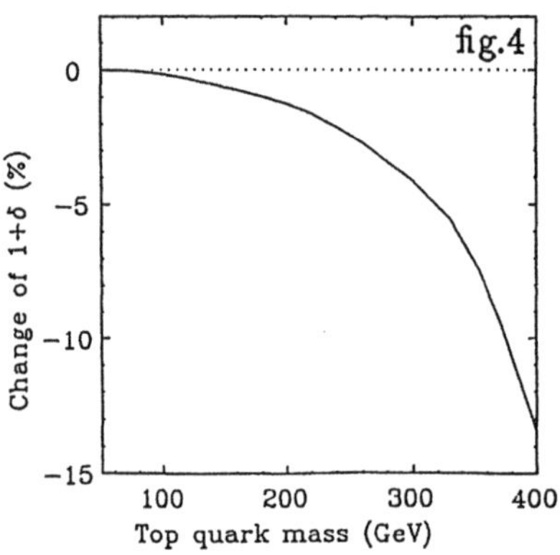

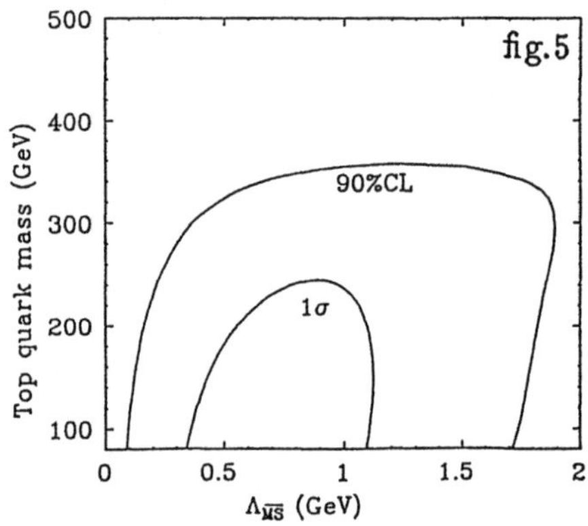

4. Dependence of the radiative correction factor $1+\delta$, on the top quark mass.

5. One sigma and 90% confidence level allowed region in the top quark mass(m_t) $\Lambda_{\overline{MS}}$ plane.

increase, which should indicate productions of a new flavor such as the top quark, the 4th generation quark or heavy lepton.

The values of R depend on the top quark mass m_t through the radiative correction factor as shown in fig.4. The data were fitted with the standard prediction of R, leaving m_t and $\Lambda_{\overline{MS}}$ as free parameters. The result is shown in fig.5. From this analysis, we set the upper limit to the top quark mass at 350 GeV at 90% confidence level.

3.2 Differential Cross Section for $e^+e^- \rightarrow f\overline{f}$

The differential cross section for e^+e^- annihilation into a fermion pair is given by

$$\frac{d\sigma}{d\Omega} = \frac{3}{16\pi}\sigma_0 N_c(R_f(1+\cos^2\theta) + B_f\cos\theta), \tag{4}$$

where σ_0 is given by eq.2, N_c is the color factor, which is 1 for leptons and 3 for quarks. The angle θ is defined to be the angle between the incident electron and the produced fermion (not anti-fermion). The $\cos\theta$ term comes from the $\gamma - Z^0$ interference. In the standard model, R_f and B_f are given by

$$\begin{aligned} R_f &= Q_f^2 - 2Q_f v_e v_f Re(g(s)) + (v_e^2 + a_e^2)(v_f^2 + a_f^2) \mid g(s) \mid^2, \\ B_f &= -4a_e a_f (Q_f Re(g(s)) - 2v_e v_f \mid g(s) \mid^2), \end{aligned} \tag{5}$$

where Q_f is the charge of the fermion, $g(s)$ is the Z^0 propagator factor,

$$g(s) = \frac{1}{16\sin^2\theta_W\cos^2\theta_W}\frac{s}{s - M_Z^2 + iM_Z\Gamma_Z}, \tag{6}$$

and v_f and a_f are vector and axial vector coupling constants, respectively, and are given in terms of the 3rd component of weak isospin, charge and the Weinberg angle as

$$\begin{aligned} v_f &= 2T_3 - 4Q_f\sin^2\theta_W, \\ a_f &= 2T_3. \end{aligned} \tag{7}$$

Differential cross sections were measured for the following processes;

$$e^+e^- \longrightarrow \mu^+\mu^-,\ \tau^+\tau^-,\ c\overline{c},\ b\overline{b}.$$

The identification of the final states, $\mu^+\mu^-$ and $\tau^+\tau^-$ is rather straight forward. The identification of $c(\overline{c})$ -quark final states utilizes the decay chain of

$$\begin{aligned} D^{*+} \longrightarrow\ & D^0\pi^+ \\ & \rightarrow k^-\pi^+,\ k^-\pi^+\pi^0,\ k^-\pi^+\pi^+\pi^0\ . \end{aligned} \tag{8}$$

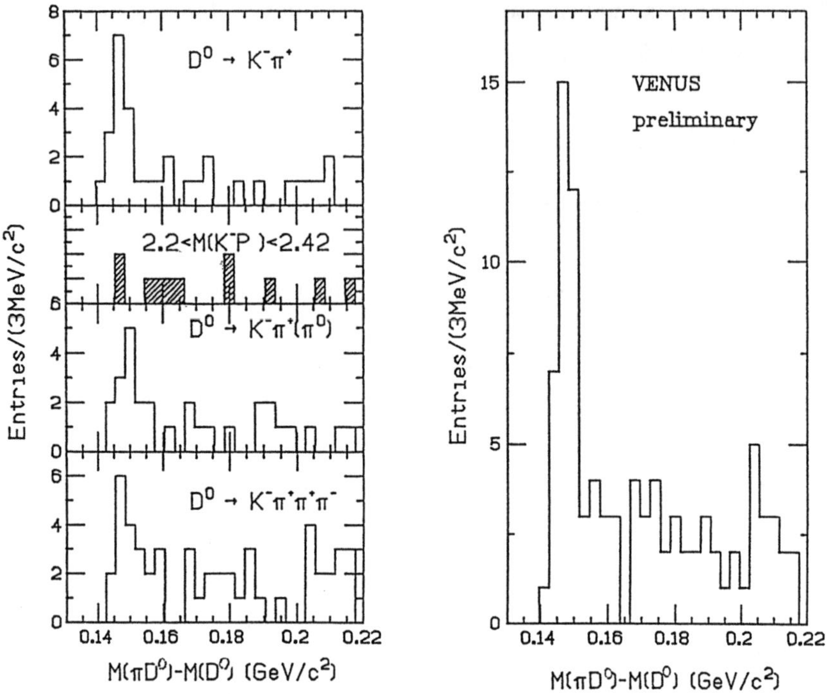

6. Mass difference of $D^{*\pm}$ and $D^0(\overline{D^0})$.

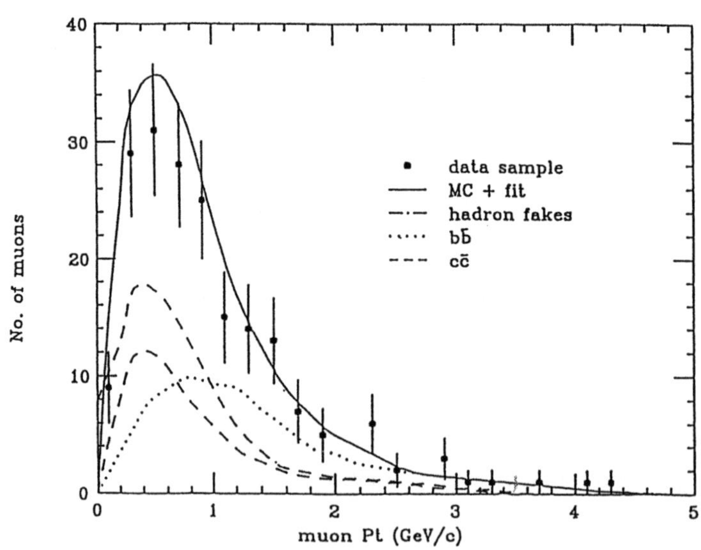

7. P_T distribution of leptons, relative to the jet axis. b-quarks are enriched by selecting high p_T leptons.

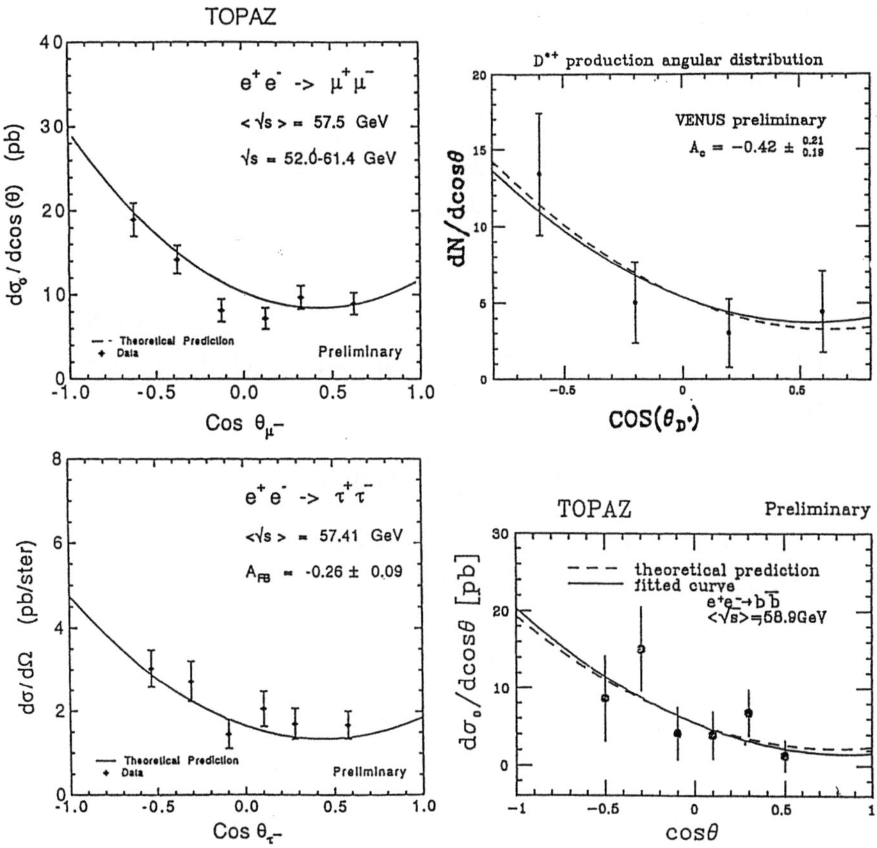

8. Differential cross section for $e^+e^- \to \mu^+\mu^-$, $\tau^+\tau^-$, $c\bar{c}$, and $b\bar{b}$, obtained by fitting the existing data.

A clear peak is visible in the mass difference of D^{*+} and D^0 as shown in fig.6.

The production of $b(\bar{b})$ -quark was identified through leptonic decays of b-quark. A b-quark sample can be enriched by finding large p_T leptons relative to the jet axis as shown in fig.7.

The measured cross sections for these processes are plotted in fig.8. The forward-backward asymmetry is clearly visible.

3.3 Charge Asymmetry

The forward-backward charge asymmetry is defined and comes out to be

$$
\begin{aligned}
A_f &\equiv \frac{\int_0^{\frac{\pi}{2}}\left(\frac{d\sigma(\theta)}{d\cos\theta}-\frac{d\sigma(\pi-\theta)}{d\cos\theta}\right)d\cos\theta}{\int_0^{\pi}\frac{d\sigma(\theta)}{d\cos\theta}d\cos\theta} \\
&= \frac{3}{8}\frac{B_f}{R_f}\ . \qquad (9)
\end{aligned}
$$

Fig.9 shows the observed asymmetry plotted as a function of energy. In the TRISTAN energy range, $Re(g(s))$ dominates, and A_f is, in a good approximation, expressed as,

$$
A_f \approx -\frac{3}{2}\frac{a_e a_f}{Q_f}Re(g(s)). \qquad (10)
$$

Thus, by measuring A_f, information on the axial coupling constant can be obtained. The following values were obtained by fitting all the existing data

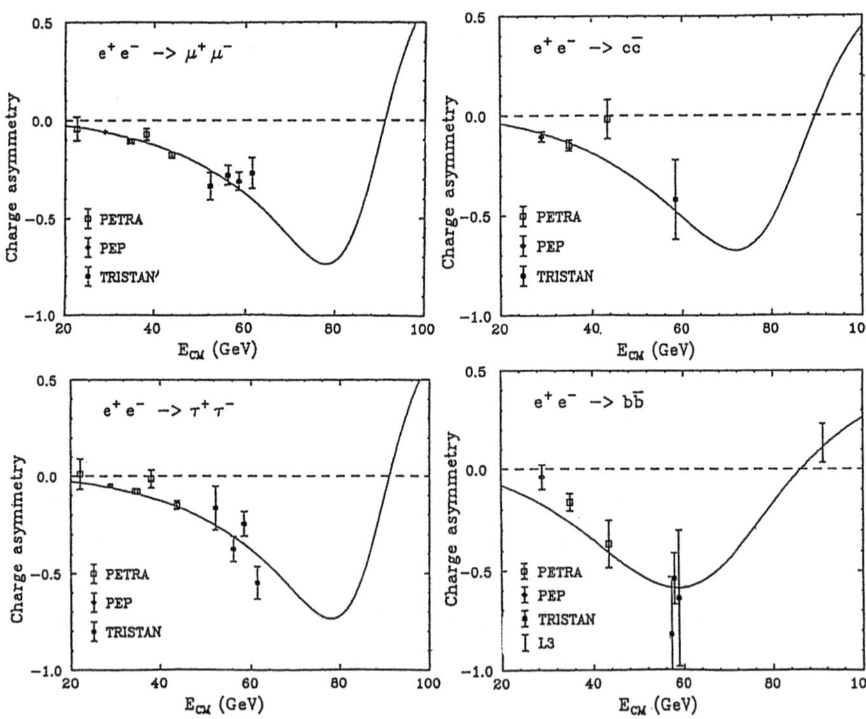

9. Measured charge asymmetry vs. $\sqrt{s}$ for $e^+e^- \to \mu^+\mu^-$, $\tau^+\tau^-$, $c\bar{c}$, and $b\bar{b}$.

on asymmetry measurements, and by assuming $a_e = -1$,

$$\begin{aligned} a_\mu &= -1.034 \pm 0.044 \\ a_\tau &= -0.908 \pm 0.064 \\ a_c &= +1.040 \pm 0.160 \\ a_b &= -0.061 \pm 0.140 \end{aligned} \tag{11}$$

These values are plotted in fig.10. The obtained values are consistent with the expected value of $a_f = \pm 1$, except for a_b, whose value deviates from the expectation by 2.8 σ.

3.4 $B^0\overline{B^0}$ mixing

In fact, the discrepancy of a_b away from the standard value of -1 is an evidence for the $B^0\overline{B^0}$ mixing. The argument goes as follows. The identification and the sign assignment of b-quarks are made through high p_T leptons, but some of the produced b-quarks dress up to B^0 mesons some of which turn into $\overline{B^0}$ mesons due to the $B^0\overline{B^0}$ mixing, thus giving rise to l^+ rather than l^-. Then the observed asymmetry is reduced as,

$$A_f^{obs} = (1-2\chi)A_f^{true}, \tag{12}$$

where χ is defined as,

$$\chi \equiv f_d \frac{r_d}{1+r_d} + f_s \frac{r_s}{1+r_s} = \frac{1}{2.3}\frac{r_d}{1+r_d} + \frac{0.3}{2.3}\frac{r_s}{1+r_s}, \tag{13}$$

for the pick-up probabilities of $u:d:s = 1:1:0.3$, and

$$r_d \equiv \frac{\Gamma(B_d \to l^+)}{\Gamma(B_d \to l^-)}, \quad r_s \equiv \frac{\Gamma(B_s \to l^+)}{\Gamma(B_s \to l^-)} \quad . \tag{14}$$

A fit to the data from TRISTAN together with those from UA1, MAC and Mark II[3] gave

$$\chi = 0.131 \pm 0.054 \quad . \tag{15}$$

Combining with the measurements by ARGUS and CLEO[4], a contour plot in r_d vs. r_s was obtained as shown in fig.11. From this, large r_s is favored, i.e. $B_s\overline{B_s}$ mixing may be maximal, as expected from the values of Kobayashi-Maskawa matrix elements.

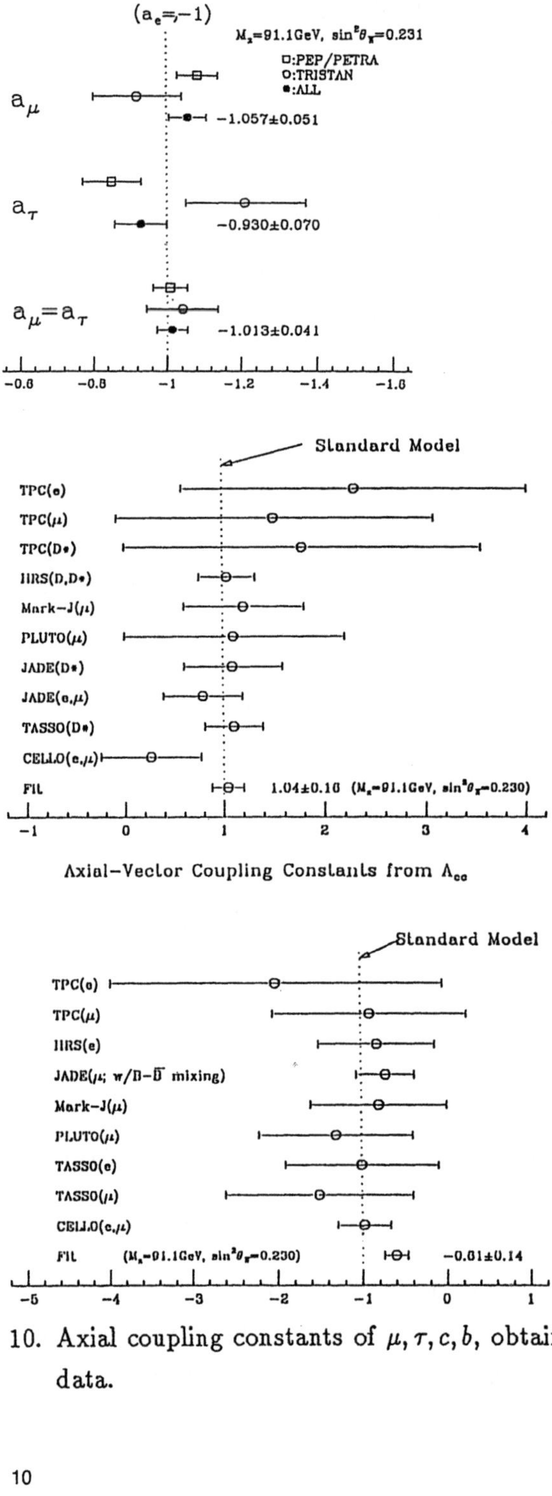

10. Axial coupling constants of μ, τ, c, b, obtained by fitting the existing data.

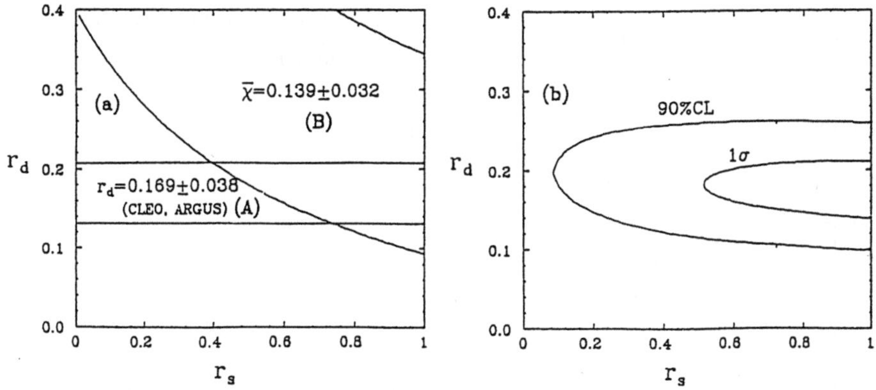

11. a)Allowed values for r_d and r_s. Region (A) is from r_d measurements by ARGUS and CLEO, and (B) is from χ.

b)One sigma and 90% confidence level allowed region for r_d and r_s from a fit to all the data.

3.5 Charge asymmetry of jets

The identification of quark flavors gives a clear set of data sample at the cost of statistics. When quark flavors are not identified, a large cancellation of asymmetries occur among flavors, but it is offset by large statistics. The overall asymmetry is given by

$$A_{jet} = f_d A_d - f_u A_u + f_s A_s - f_c A_c + f_b A_b, \tag{16}$$

where f_q is the proportion of $q\overline{q}$ production.

First, clear two jet events were selected by requiring the thrust to be greater than 0.8. The charge of each jet was identified by calculating the weighted mean of charges;

$$Q_{jet} \equiv \sum_{i=1}^{n} Q_i \eta_i^{\alpha}, \tag{17}$$

where Q_i is the charge of i-th particle in the jet, and η_i is the rapidity of the particle relative to the jet axis,

$$\eta_i = \frac{1}{2} \ln \frac{E^i + p_{\parallel}^i}{E^i - p_{\parallel}^i}, \tag{18}$$

and α is chosen to be 0.8, which was obtained by optimizing the charge

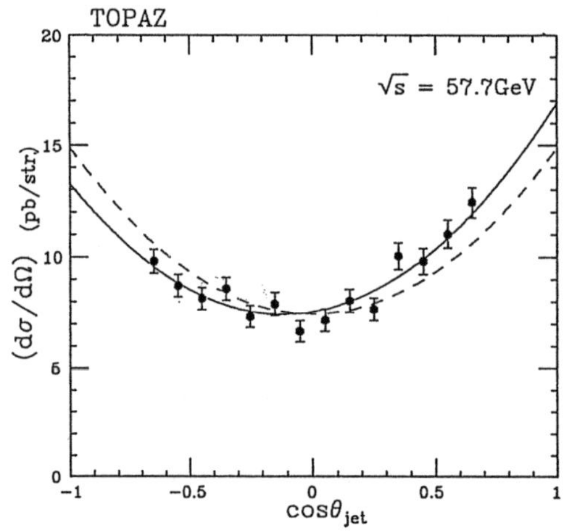

12. Differential cross section for the "charge identified" jet.

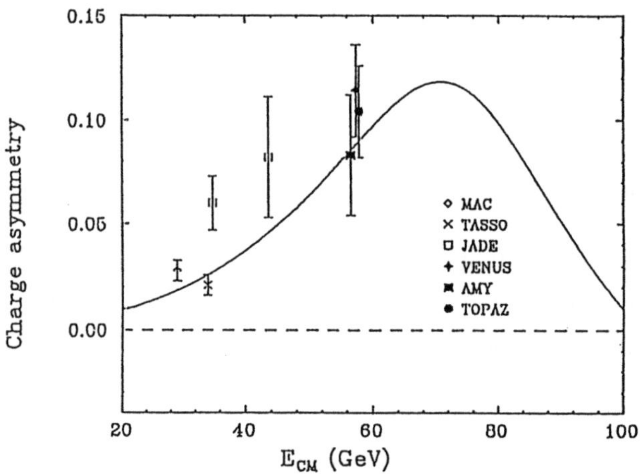

13. The charge asymmetry of jet plotted vs. $\sqrt{s}$

identification efficiency through Monte Carlo studies. The larger of the two jets in Q_{jet} is assigned as the positive jet. Correctness of the charge assignment probability was found to be about 67 %. The differential cross section thus obtained is shown in fig.12. An asymmetry is clearly visible, and the obtained value was

$$A_{jet} = 0.091 \pm 0.014 \tag{19}$$

This value agrees well with the standard model prediction as shown in fig.13.

3.6 Polarization of τ

According to the standard model of the electroweak interaction, final state fermions are longitudinally polarized, which can be observed in τ leptons produced in the reaction $e^+e^- \to \tau^+\tau^-$. The polarization is given, in a good approximation in the TRISTAN energy range, as

$$P_{\tau^-}(\theta) = -P_{\tau^+}(\theta) = Re(g(s))[v_e a_\tau + \frac{2cos\theta}{1+cos^2\theta} a_e v_\tau], \tag{20}$$

where $g(s)$ is given by eq.6. Thus measuring polarization, one can obtain information on the vector coupling constants. The polarization of τ can be measured through the momentum distribution of decay products. For τ decay into muons or electrons, the momentum distribution is given by

$$\frac{df(x)}{dx} = a(x) + Pb(x), \tag{21}$$

where

$$x = (p - p_{min})/(p_{max} - p_{min}) \tag{22}$$

$$a(x) = \frac{1}{3}(5 - 9x^2 + 4x^3) \tag{23}$$

$$b(x) = \frac{1}{3}(1 - 9x^2 + 8x^3) \ . \tag{24}$$

For $\tau \to \pi\nu$ and $\rho\nu$ decays, the π and ρ momentum spectra are given by

$$\frac{df(x)}{dx} = 1 + 2\alpha P(x - \frac{1}{2}) \tag{25}$$

(26)

$$\alpha = \begin{cases} 1 & \text{for } \pi \text{ mode} \\ (m_\tau^2 - 2m_\rho^2)/(m_\tau^2 + 2m_\rho^2) = 0.46 & \text{for } \rho \text{ mode} \end{cases} \tag{27}$$

The measured x distributions for all four decay modes are shown in fig.14 By fitting all these distributions simultaneously, vector coupling constants were derived, assuming $a_e = a_\tau = -1$. The obtained values are

$$v_\tau = -0.95 \pm 0.71, \tag{28}$$

and plotted in fig.15 together with previous measurements. Even though the error bars are still large, the measurement improved the accuracy of v_τ significantly over the previous measurements.

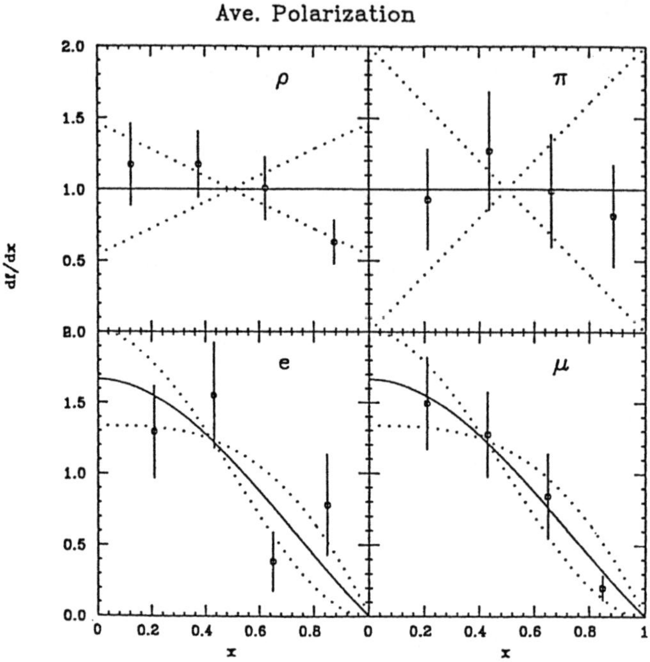

14. Normalized momentum (x) distribution of e, μ, π and ρ from τ decay.

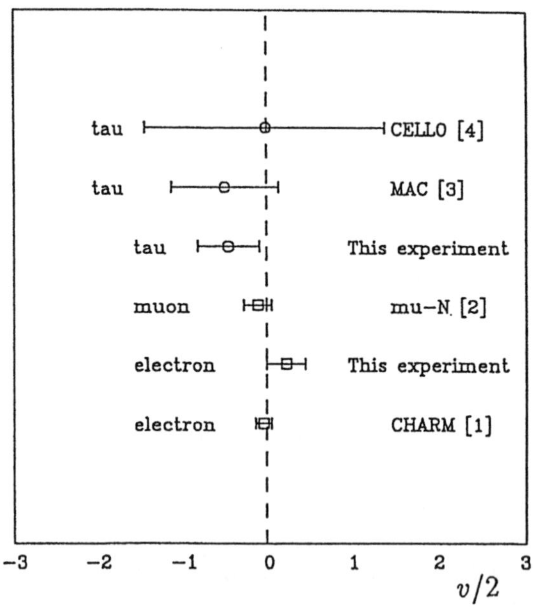

15. Vector coupling constant of τ so far measured.

4. Study of QCD

Various aspects of QCD (Quantum Chromodynamics) have been studied through the analyses of hadronic events. I discuss only a few among them here.

4.1 s dependence of α_s

One of the important consequence of QCD is the asymptotic freedom, where the strong coupling constant α_s gets smaller at higher Q^2. Experimentally, this can be investigated by plotting R_3, the 3-jet fraction of hadronic events as a function of energy. In the second order QCD, R_3 is given by

$$R_3 \equiv \frac{N_{3-jet}}{N_{total}} = c_1(y_{cut})\alpha_s(s) + c_2(y_{cut})\alpha_s^2(s), \tag{29}$$

where y_{cut} is the cut-off in merging particles into a jet. Jet clustering algorithm is rather simple, and goes essentially as follows.

1. For any pair of particles i and j, calculate $y_{ij} = m_{ij}^2/E_{vis}^2$.
2. Find the pair i and j, which gives minimum y_{ij} and merge into a single particle if $y_{ij} < y_{cut}$.
3. Repeat 1 and 2 until no more merging is possible.

A typical result of jet clustering is shown in fig.16. When y_{cut} is increased, R_2 increases, and all the others decrease as expected. At fixed y_{cut}, any

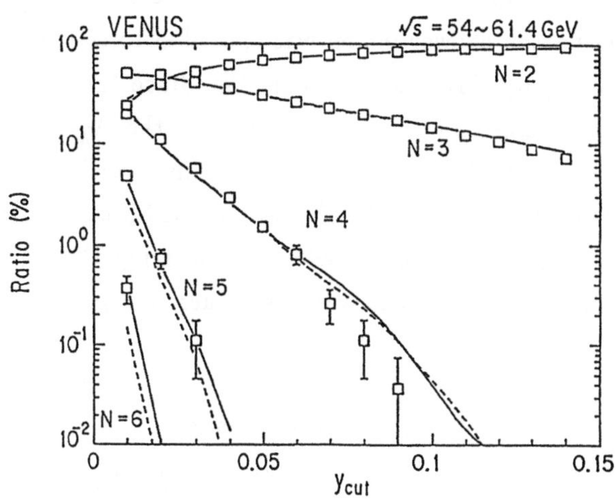

16. Number of jets vs. y_{cut}.

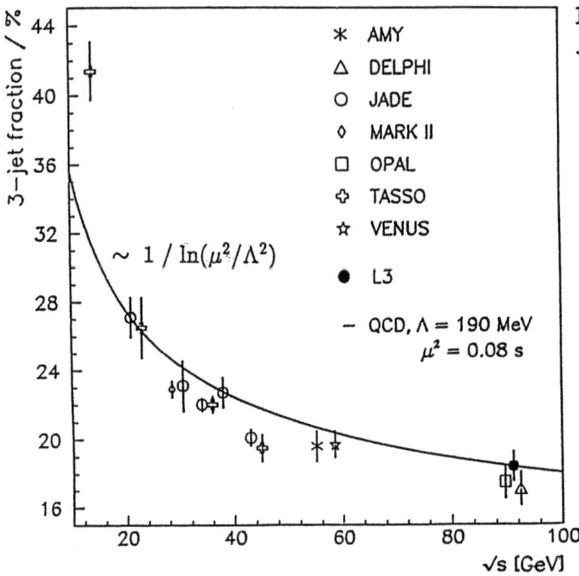

17. The 3-jet fraction vs. $\sqrt{s}$. Note $R_3 \propto \alpha_s$.

s dependence of R_3 directly comes from that of α_s. As seen from fig.17, where recent LEP results are also included, α_s clearly runs and decreases with energy.

4.2 Determination of $\Lambda_{\overline{MS}}$ using NLL MC

In order to extract the information on $\Lambda_{\overline{MS}}$, theoretical model has to be well defined. Recently, the parton shower model was extended to the next-to-leading log order (NLL-PS) by Kato and Munehisa[5]. Since the renormalization scheme can be specified in this approximation, the $\Lambda_{\overline{MS}}$ can be properly defined in the model. The NLL-PS was combined with a few fragmentation models, such as LUND or EPOCS to form a Monte Carlo generators. The determination of $\Lambda_{\overline{MS}}$ was made by comparing the 3-jet fraction, with Monte Carlo predictions with various values of $\Lambda_{\overline{MS}}$. The determined values are

$$\begin{aligned} \Lambda_{\overline{MS}} = \quad & 240 \pm 70 \pm 60 MeV \quad (AMY; Lund\ frag.) \\ & 254^{+55}_{-47} \pm 56 MeV \quad (VENUS; EPOCS\ frag.) \\ & 232^{+23}_{-79} (stat. + sys.) \quad (TOPAZ; Lund\ frag.) \end{aligned} \tag{30}$$

4.3 Triple gluon vertex

Even though QCD, the theory of strong interaction, is well established, one of its crucial feature, the existence of triple gluon vertex had not been experimentally proven. The AMY group was the first to give an evidence[6]. Other two groups at TRISTAN and experiments at LEP recently provided further evidences.

The study can be made by examining 4-jet events. Theoretically, 4-jet events come from the diagrams shown in fig.18, where fractions of the contribution expected are also shown for the case of QCD and for that of abelian theory. Note of course that the triple gluon vertex is the unique feature of nonabelian group theory.

The way to distinguish the existence of triple gluon vertex is based on the following observation. The plane made by the two gluons that originate from the triple gluon vertex tends to be parallel to that made by the final two quarks. This is because of the helicity and the angular momentum conservation, as well explained in ref.7. So θ_{BZ}, a variable sensitive for this investigation, is defined to be the angle between the two planes.

Experimentally, clear 4-jet events were selected first. For each event, jets were numbered such that

$$E_1 > E_2 > E_3 > E_4 \quad . \tag{31}$$

Another variable θ_{NR} is the angle between $\vec{p_1} - \vec{p_2}$ and $\vec{p_3} - \vec{p_4}$. Fig.19 shows the results. Clearly QCD is favored, and abelian theory is ruled out.

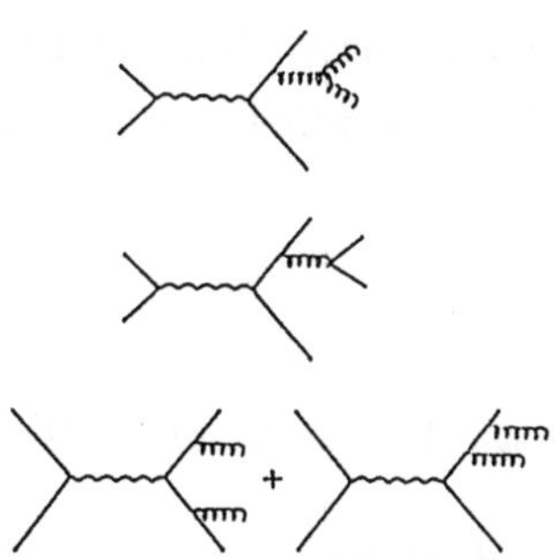

	QCD	Abelian Model
(a)	28.4%	65.5%
(b)	5.4%	34.5%
(c)	66.2%	0%

18. Feynman diagrams that give rise to 4 jets, and their fraction expected of QCD and of an abelian model.

• AMY

$\sqrt{s}$=50 - 57GeV, 139 four-jet events

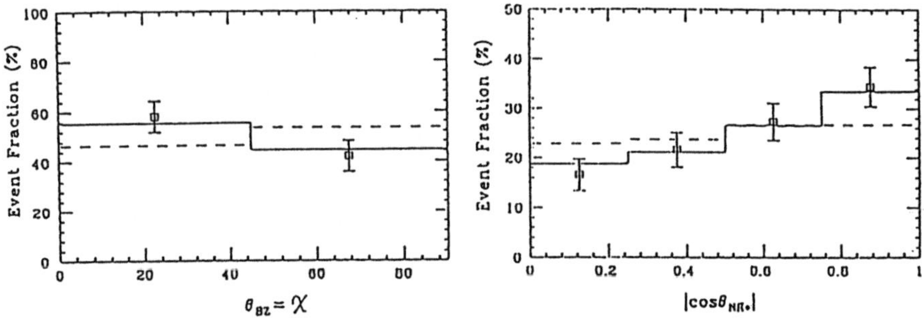

• VENUS

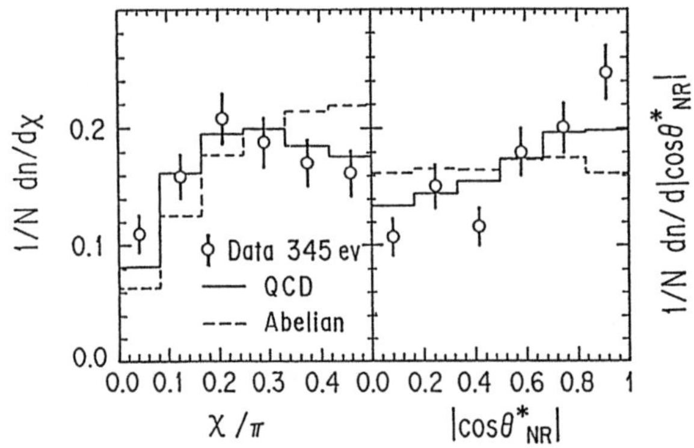

19. Distribution of $\chi \equiv \theta_{BZ}$ and of $\cos\theta_{NR}$.

5. Search for new particles

Various new particles were searched for at TRISTAN, and mass limits were obtained. Now most of the limits have been superseded by the results from LEP as shown in fig.20. There are certain new particles that are difficult to search or have not been searched at LEP-I. I mention a few of these particles here.

5.1 Search for an extra Z boson

The neutral current Lagrangian with one extra Z boson can be written as

$$L_{NC} = -e(J^{\mu}_{EM}A_{\mu} + J^{\mu}_{1}Z^{0}_{1\mu} + J^{\mu}_{2}Z^{0}_{2\mu}) \tag{32}$$

where A and Z^0_1 are the photon and Z fields of the standard $SU(2)_L \times$

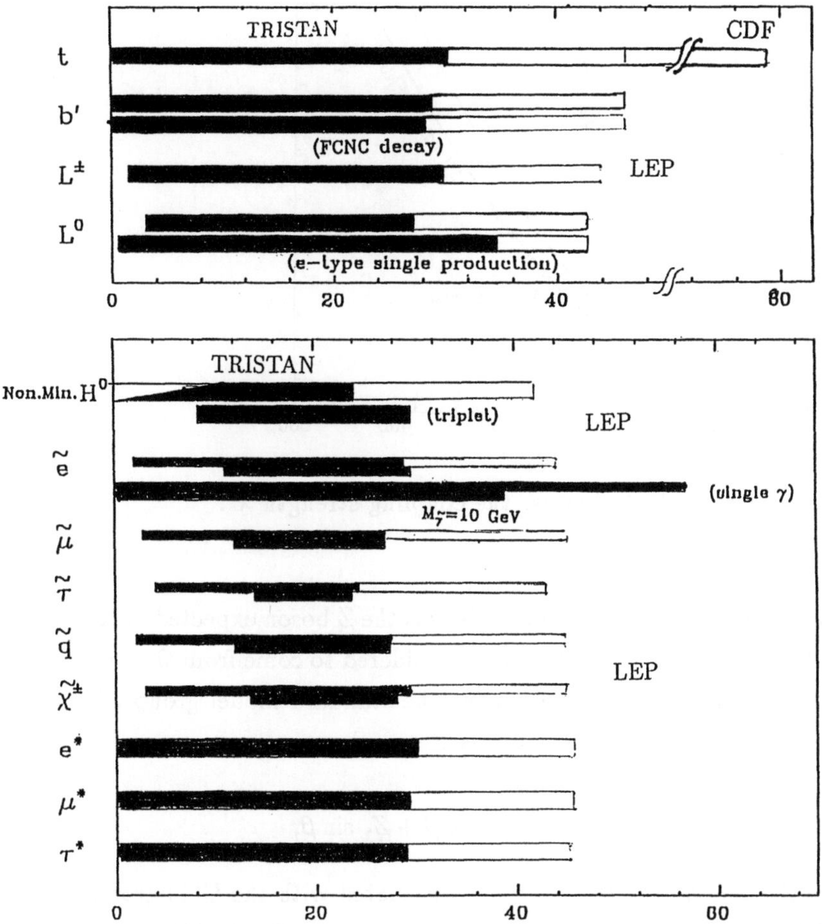

20. Mass limits for various new particles. Solid bars are from TRISTAN, and open ones are from CDF and LEP.

$U(1)_Y$ model and Z_2^0 is the extra Z boson associated with the extra U(1) symmetry. Deviation from the standard theory prediction occurs through the interference between the extra Z and the known gauge bosons.

For the extra Z boson, two cases were considered. One is a simple extension of the known Z boson. Namely, the coupling of quarks and leptons to the extra Z boson is assumed to be the same as that to the standard Z boson, except for an overall factor λ. The result of analysis on this assumption is shown in fig.21, where the mass limit obtained vs. λ is plotted. For $\lambda = 1$, $mass(Z') > 426GeV$ was obtained.

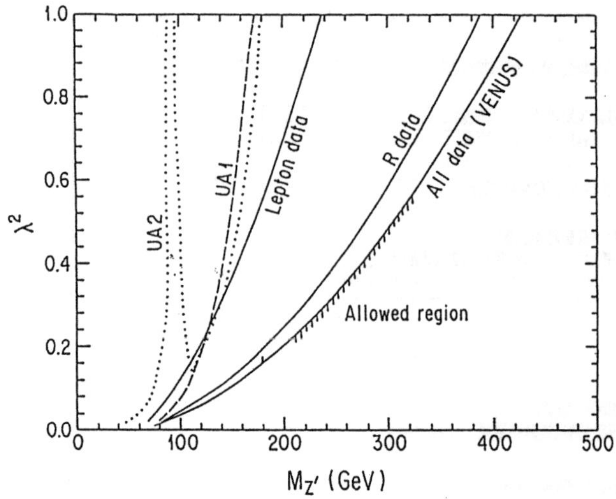

21. Mass limit of Z' boson vs. the coupling strength λ^2.

Another very interesting possibility is the Z boson expected in the E_6 inspired model. The extra Z boson is considered to come from the breakdown $E_6 \to G \times U(1)_\beta$, where G contains the standard model group $SU(3)_C \times SU(2)_L \times U(1)_Y$. The new Z boson Z_β can be expressed as

$$Z_\beta = Z_\psi \cos\beta + Z_\chi \sin\beta, \tag{33}$$

where Z_ψ and Z_χ are the Z bosons associated with the U(1) groups of the breakdown $E_6 \to SO(10) \times U(1)_\psi$ and $SO(10) \to SU(5) \times U(1)_\chi$[8]. Special values of β correspond to particular models of Z boson, and the coupling constants to quarks and leptons are given as a function of β. Fig.22 shows the expected deviations from the standard model prediction for the case of Z boson corresponding to $\beta = \tan^{-1}\sqrt{3/5}$. In this case, the coupling constants turned out to take values such that there should be some excess for multi-hadronic cross section and some deficit for leptonic ones in the TRISTAN energy range as seen in the figure. Fig.23 shows the result of mass limit to the Z_β boson as a function of β[9].

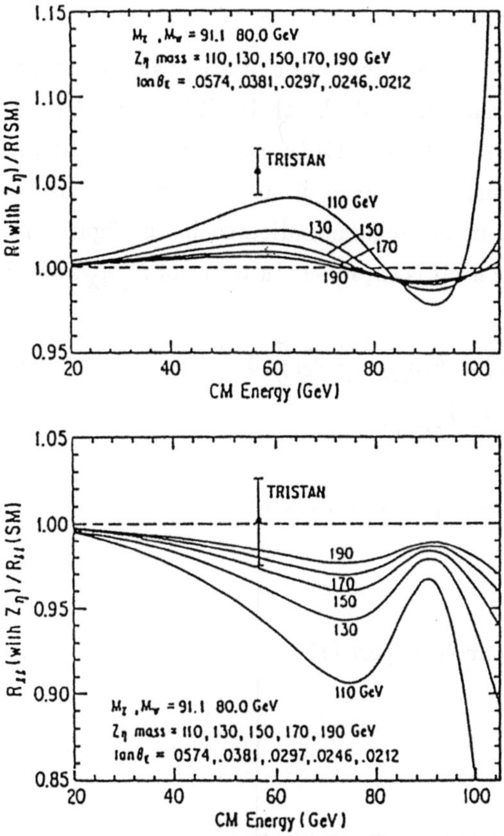

22. Total cross sections for $e^+e^- \to l^+l^-$ and $q\overline{q}$ vs. energy for various masses of Z_β boson.

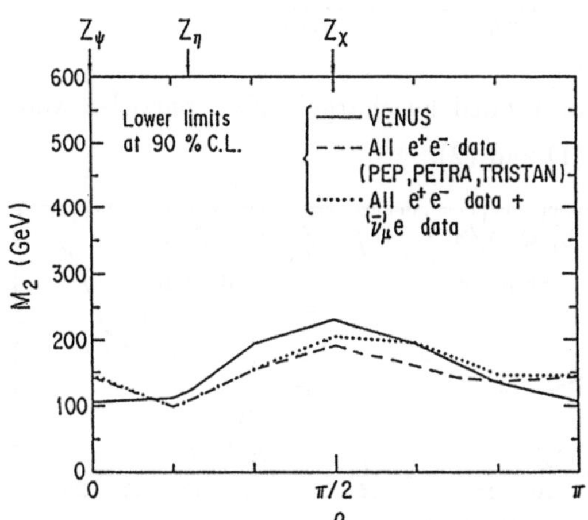

23. Mass limit to the Z_β boson vs. the mixing angle β.

5.2 Heavy stable charged particles

The TOPAZ group searched for inclusive production of heavy stable charged particles through dE/dx measurements made by their TPC (Time Projection Chamber)[11]. Fig.24 shows a plot of dE/dx vs. momentum divided by the charge. Clear bands corresponding to ordinary particles are visible. New particles were searched for in the regions indicated in the figure. No new particles were found and the limit is expressed in terms of $F_{Q\overline{Q}X}$, which is defined as

$$\sigma(e^+e^- \to Q\overline{Q}X) = F_{Q\overline{Q}X}\frac{4\pi\alpha}{s}e_Q^2\frac{\beta}{2}(3-\beta^2) \tag{34}$$

Fig.25 shows the obtained limit to $F_{Q\overline{Q}X}$.

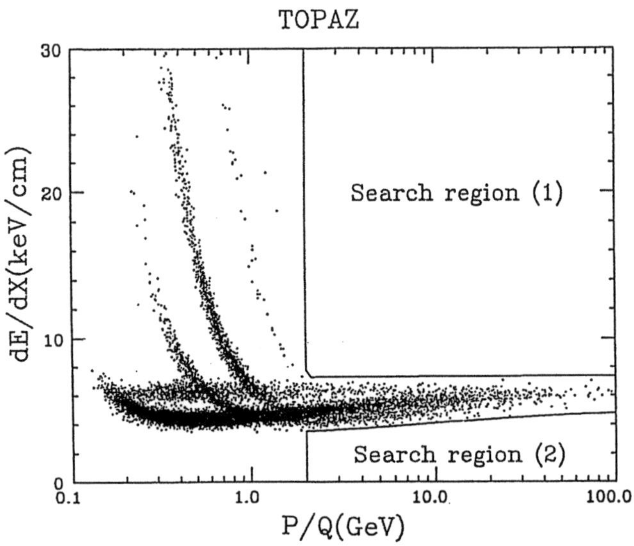

24. dE/dx vs. p/Q(momentum divided by charge). New particles were searched for in the region (1) and (2).

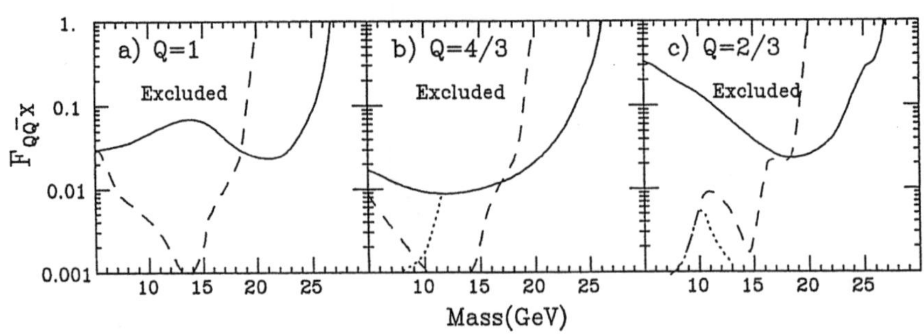

25. Limit to $F_{Q\overline{Q}X}$ vs. mass of a stable particle.

6. Future prospects

With the start of LEP and SLC in 1989, the role of TRISTAN as an energy frontier machine had ended, and the operation of TRISTAN phase 2 has been started. Namely, the luminosity is maximized at an optimum energy of 58 GeV. During the present shut-down, 4 pairs of superconducting mini-β quadrupole magnets are being installed at the four intersection regions, which should result in a factor two increase in the luminosity. The expected luminosity for each experiment is then about 1 pb^{-1}/day, which allows us to obtain a little more than 100 hadronic events a day.

Detectors have been upgraded to expand their physics capabilities. They include:

1. installation of a vertex chamber (AMY, TOPAZ and VENUS)
   This is to tag b quarks and τ leptons to allow detail studies of their properties, such as the life times, coupling constants etc.

2. Hermetic calorimetry down to small angles (TOPAZ and VENUS)
   One of the physics topics that cannot be studied at LEP-I is the search for SUSY particles by looking at events of single γ with missing mo-

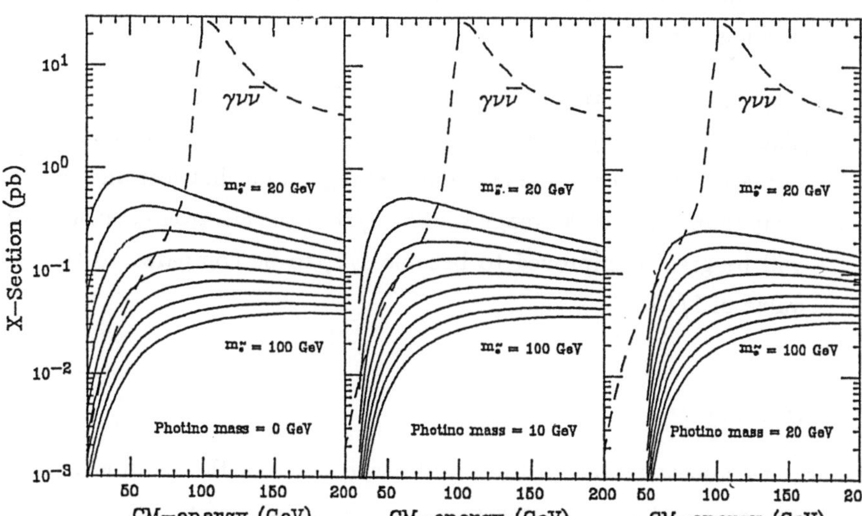

26. The cross section for $e^+e^- \to \gamma\tilde{\gamma}\tilde{\gamma}$ and $\gamma\nu\overline{\nu}$ as a function of c.m.s. energy for various masses of scalar electron.

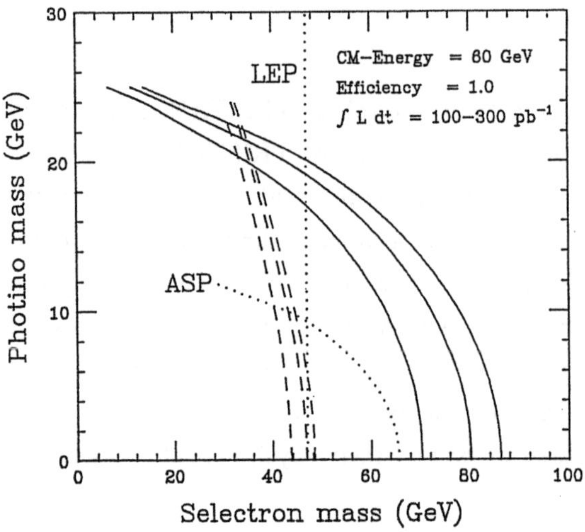

27. Limit we expect to obtain in the plane of the photino mass and the scalar electron mass at given values of the integrated luminosity.

mentum. Fig.26 shows the expected cross sections as a function of the c.m.s. energy for radiative production of photino pairs, $e^+e^- \rightarrow \gamma\tilde{\gamma}\tilde{\gamma}$. Plotted in the same figure is the major background to this process $e^+e^- \rightarrow \gamma\nu\bar{\nu}$, which clearly dominates at Z^0 region. Fig.27 shows the mass limits we expect to obtain at certain values of integrated luminosities.

3. Capability of electron identification (VENUS)
   The VENUS group is currently installing a transition radiation detector in the reserved space in the barrel region, i.e. between the central drift chamber and the magnet. The S/N of the electron identification will be greatly improved to become 1/1, to be compared to the old value of 1/10.

7. Summary

Charge asymmetries due to the interference between the photon and the Z^0 were measured for the processes $e^+e^- \rightarrow \mu^+\mu^-$, $\tau^+\tau^-$, $c\bar{c}$, $b\bar{b}$ and 2-jets. Axial coupling constants were derived from the measured asymmetries. They all agree well with the prediction of the standard model. For the case of $b\bar{b}$ production, further evidence for $B^0\overline{B^0}$ mixing was obtained, and the maximal mixing for $B_s^0\overline{B_s^0}$ was preferred according to an overall fit to the existing data. The polarization of τ leptons was measured and its vector coupling constant was derived with an error bar still large but considerably improved over the previous measurements.

Various studies were made on QCD. It is clear now that α_s is indeed running. The value of $\Lambda_{\overline{MS}}$ was estimated to be about 250MeV, comparing data with NLL-PS Monte Carlo predictions made with various values of $\Lambda_{\overline{MS}}$. Perhaps, one of the most important results from the TRISTAN e^+e^- collider would be the first clear confirmation of the existence of the triple gluon vertex, a crucial feature of QCD.

New particles were searched for and mass limits were obtained, but most of them were superseded by those from SLC/LEP. Some of the limits are still viable. Mass of an additional Z boson with the same coupling constants as the standard Z boson was limited to be greater than 426GeV at 95%CL. Mass limit for E_6 inspired Z boson was also obtained as a function of the mixing angle β.

The phase-2 program of TRISTAN has been started, where luminosity is expected to reach 1 pb^{-1}/day with the help of a pair of superconducting mini-β quadrupole magnets placed at each interaction region. Detectors have been upgraded to maximize the physics outputs at the TRISTAN phase 2 operation.

References

1. The topics discussed in this report were mostly based on the following material.

   K.Abe, "e^+e^- physics below Z^0", KEK-Preprint 90-136, to be appeared in Proc. of the 25th International Conference on High Energy Physics, Singapore, 1990.

   M.Yamauchi, "Electroweak physics at TRISTAN", KEK-Preprint 90-135, to be appeared ibid.

   R.Itoh, "Recent results from the three TRISTAN experiments", KEK-Preprint 90-128

   K.Fujii, "e^+e^- physics below Z^0",KEK-Preprint 89-182.

2. J.Fujimoto and Y.Shimizu, Mod. Phys. Lett. A3, 581(1988)

3. UA1 collab. C.Albajar et al., Phys. lett. 186B, 247(1987)
MAC collab. H.R.Band et al., Phys. lett. 200B, 221(1988)
Mark II collab. A.J.Weir et al., Phys. lett. 240B, 289(1990)

4. ARGUS collab. M.V.Danilov, Proc. of the 1989 International Symposium on Lepton Photon Interaction at High Energies, Stanford(1989).
CLEO collab. D.L.Kreinick, ibid.

5. K.Kato and T.Munehisa, CERN-TH-5719-90(1990)

6. AMY collab. L.H.Park et al., Phys. Rev. lett. 62, 1713(1989)

7. M.Bengtsson, Z. Phys. C42, 75(1989); further references therein.

8. K.Hagiwara et al., Phys. Rev. D41, 815(1990); further references therein.

9. VENUS collab. K.Abe et al., KEK-Preprint 90-34, to be published in Phys. Lett. B.

10. TOPAZ collab. I.Adachi et al., KEK-Preprint 90-84, to be published in Phys. Lett. B.

High-Energy Antiproton-Proton Collisions – The Collider Detector at Fermilab

G.W. Brandenburg

Harvard University, High Energy Physics Laboratory,
42 Oxford Street, Cambridge, MA 02138, USA

Abstract. Recent results from experiments at antiproton-proton colliders are presented with emphasis on those from the Collider Detector at Fermilab (CDF). The data on hadron jet, W, and Z production are all in excellent agreement with the standard model. The lower limit for the top quark mass is found to be somewhat above the W mass. No evidence is found for W' or Z' states or for supersymmetric particles.

1. Introduction

Antiproton–proton collider experiments have provided very exciting data during the 1980's. From the initial discovery of the W and Z bosons at CERN through the continuing search for the top quark at Fermilab, there has been a wealth of information from these experiments – thus far all in excellent agreement with the Standard Model. In the decade to come these experiments together with those at LEP will continue to apply stringent tests within the Standard Model and to look for exciting new effects beyond it.

In this talk I will take a brief look at the accelerators and experiments that have made this work possible and then summarize the most recent results. I will concentrate primarily on results from my own experiment, CDF, reviewing the following major topics: Jet/QCD studies (section 4), W/Z physics (section 5), and finally the top quark search (section 6). In each of the physics sections I will try to project what can be expected from future data – of course that which is unexpected may prove to be the most exciting!

2. Hadron Colliders – the Tevatron

The key element which has made the study of high energy antiproton–proton collisions possible was the development of the hadron collider. The first such machine was the CERN ISR, but the development of the SPS collider together with its antiproton source by Van de Meer, Rubbia, and others opened the current epoch of experimentation. This machine was followed by the Fermilab Tevatron collider, which used similar techniques for the production and cooling of antiprotons. However, Fermilab was able to utilize the superconducting Tevatron as a storage ring allowing higher energies than the SPS ring with its conventional magnets. I will describe the main features of the Fermilab system below.

The first step is the production of antiprotons. This is done using 120 GeV protons from the old Fermilab main ring, incident on a lithium target. Antiprotons are filtered from the resulting debris and are channeled into an 8 GeV pair of rings called the Debuncher/Accumulator. Here the antiprotons are stochastically "cooled" such that the

Springer Proceedings in Physics, Vol. 65 Present and Future of High-Energy Physics
Editors: K.-I. Aoki and M. Kobayashi

phase space they occupy will easily fit within the available phase space of the larger rings. It takes several hours to accumulate in excess of 10^{11} antiprotons, at which point they can be transferred in six bunches to the Tevatron ring.

Once the six bunches of antiprotons have been inserted into the Tevatron, it is a relatively straightforward matter to also insert six comparable bunches of protons directly from the old main ring. With the antiprotons and protons inserted and rotating in opposite directions, the Tevatron is gently raised from its injection energy of 150 GeV to its maximum energy of 900 GeV. The two sets of six bunches pass each other at twelve points around the circumference of the ring, however, the beams are only focussed to a collision point where an experiment has been installed. Previously CDF has been the only major experiment taking data (at the "B0" collision point), while in the upcoming run CDF will be joined by the new D0 experiment.

In the previous Tevatron collider run lasting from July 1988 to July 1989 the machine delivered a total integrated luminosity of 9 pb^{-1} at a collision energy of $\sqrt{s}$ = 1800 GeV, and the CDF experiment was able to log approximately half of this to tape. (Delivered luminosity is stated in units of inverse cross section – when multiplied times the cross section for a process one obtains the expected number of associated events.) Fermilab plans to run the Tevatron collider at approximately two year intervals through the 1990's making improvements each time. Some of the projected improvements are: install electrostatic separators to reduce beam–beam interactions at unused crossing points, increase the number of bunches (20–40), raise the Tevatron energy to 1000 GeV, and replace the old main ring with a more efficient Main Injector in a separate tunnel. With these improvements the delivered luminosity can be expected to increase by a factor of roughly five in each successive run. Thus we can expect between 25 and 50 pb^{-1} in the CDF/D0 run starting next summer.

It is important to note that increased luminosity at a hadron collider is practically speaking almost as valuable as increased energy. The most interesting processes require very high energy parton–parton interactions. Because of the rapid decrease of the proton structure function with increasing parton momentum fraction, larger luminosity results in a measureable rate for ever higher energy parton–parton collisions.

3. The Experiments – CDF

The experiments which have been designed to study high energy hadron collisions have to take advantage of two important aspects of the most interesting interactions. First these processes tend to involve large masses and momentum transfers, so the experiments should emphasize measurement of outgoing particles at large angles. Second the interesting massive objects tend to decay weakly, so it is important to be able to identify leptons. Thus the most important components are tracking to identify the trajectories of muons and electrons, segmented calorimetry to measure the energy of electrons and quark jets, and complete calorimetric coverage to infer the presence of neutrinos.

Fig. 1. shows the major detectors that have been or will be used to study antiproton–proton collisions. The original two CERN detectors complemented each other in that both had the major features pointed out above, but UA1 emphasized particle tracking in a dipole field while UA2 emphasized segmented calorimetry for electron and jet measurement. The Collider Detector at Fermilab (CDF) has both excellent particle tracking in a solenoidal field and very good calorimetry. However, the new Fermilab detector, D0, will be highlighted by a liquid argon calorimeter which will provide even better energy measurement. The UA1, UA2 program at CERN is now winding down, but CDF and D0 will begin major new data taking runs at Fermilab in the summer of 1991.

Because I will be focussing primarily on results from the CDF detector I will describe its components in somewhat more detail [1]. The sections of the detector are logically divided into different pseudorapidity regions, where this familiar approximation for the longitudinal rapidity variable is given by $\eta = \ln(\tan(\theta/2))$. The innermost element is a vertex time–projection chamber which is used to accurately locate the event vertex along the beam line and to ensure that there are not multiple interactions. Surrounding the vertex chamber is an 84 layer cylindrical drift chamber which extends out to a radius of 1.3 m and covers the interval $-1.0 < \eta < +1.0$. The tracking chambers are enclosed in a superconducting magnet with an axial field of 1.5 Tesla, resulting in the accurate measurement of the momentum vectors for large angle charged particle tracks. The solenoid and the tracking chambers are surrounded by calorimetry: the region $|\eta| < 1.1$ is covered by segmented scintillator plate calorimeter, while for larger η values extending to $|\eta| = 4.2$ there is coverage by gas proportional tube calorimeters. Finally a set of drift chambers for identifying penetrating muon tracks are located outside the calorimeter in the region $|\eta| < 0.6$. The "central region" of the detector ($|\eta| < 1$), which includes particle tracking, calorimetry, and almost complete muon chamber coverage, is used as a starting place for most of the physics analyses. A schematic view of the CDF Detector is shown in Fig. 2.

Because the rates for uninteresting processes tend to be astronomically large, it is necessary to have an elaborate, very fast trigger scheme to sort out the most interesting events for further analysis. Typical event signatures that can be used in this trigger are clusters of calorimeter energy representing electrons or parton jets, penetrating tracks from muons, and a large value of the "missing transverse energy". (Transverse energy is defined in analogy to transverse momentum as $E_T = E \sin(\theta)$, where E is the energy of a calorimeter cluster. Assuming that the vector sum of the outgoing E_T for all particles is approximately zero, the missing E_T is defined as the negative of the total observed E_T summed over all calorimeter segments.)

In CDF the trigger is subdivided into four logical levels. An event must pass the lower levels to reach the higher levels. The lower levels are simple topologically and hence very fast, while the higher levels are more complicated and introduce "dead time" into the experiment. The highest level consists of a complete event reconstruction on a "farm" of on–line microprocessors. Table 1 summarizes the CDF trigger system and the size of the interaction cross section that was accepted by each level in the last run. The net result of this trigger scheme was to reduce a primary event rate of 80 khz down to a final data stream of approximately 2 events per second. It should be noted that the final data set always includes random samplings of the more prolific processes as well as all of the most exotic triggers.

Just as the the accelerators are upgraded between major runs, the detectors must keep pace with their own improvements. Some of these are dictated by the demands placed on the system by the projected increases in luminosity. Other improvements are designed to enhance the ability of the detectors to study various processes. For the next Tevatron run a

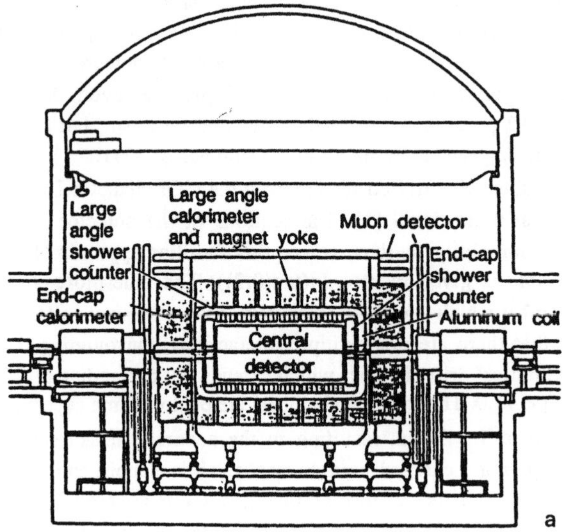
Large angle
calorimeter
and magnet yoke
Muon detector
Large
angle
shower
counter
End-cap
calorimeter
End-cap
shower
counter
Aluminum coil
Central
detector
a

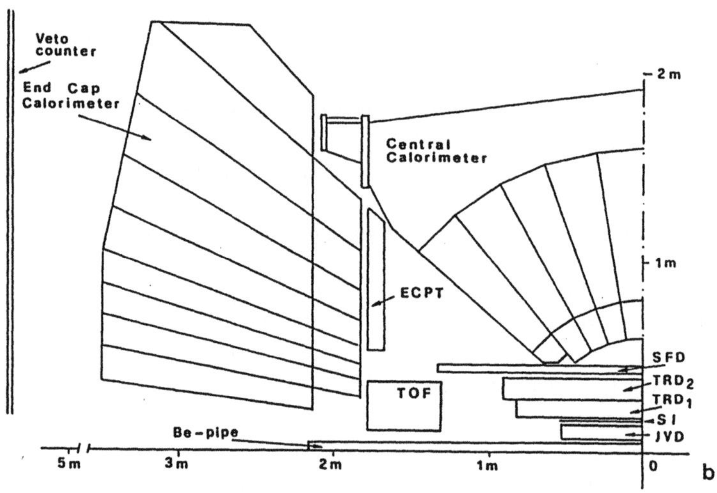
Veto
counter
End Cap
Calorimeter
Central
Calorimeter
ECPT
TOF
SFD
TRD2
TRD1
SI
JVD
Be-pipe
2 m
1 m
5 m
3 m
2 m
1 m
0
b

Figure 1

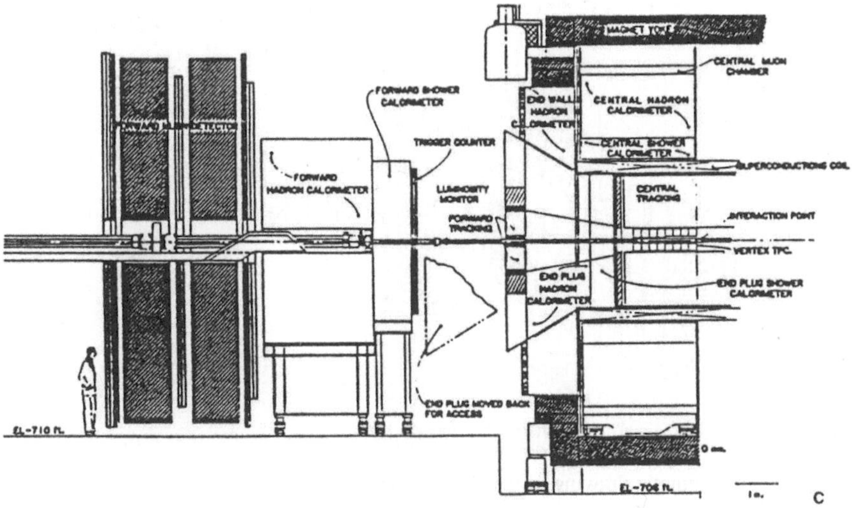

c

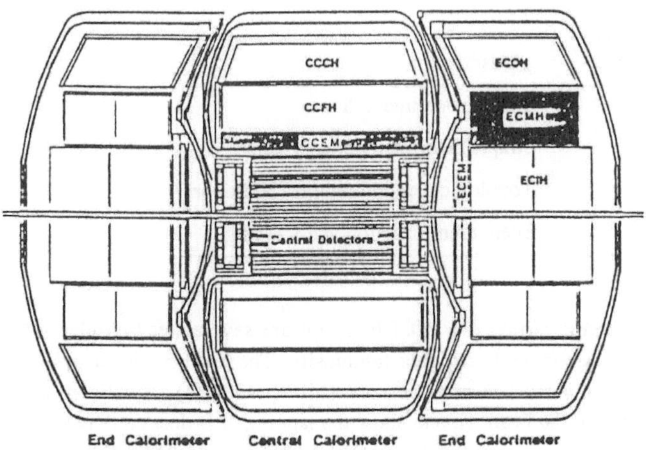

d

Figure 1 The four major $\bar{p}$–p collider experiments: UA1 (a) and UA2 (b) at CERN, CDF (c) and D0 (d) at Fermilab.

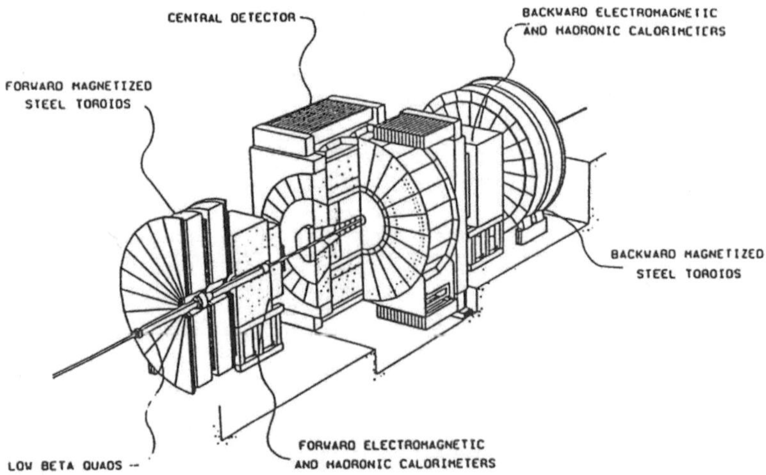

Figure 2 A schematic drawing of the CDF detector.

Table 1 CDF Trigger Levels

Level	Cross Section	Description
0	45 mb	Beam–beam interactions
1	1 mb	Simple calorimeter energy sums
2	3 μb	Topological combinations (30 flavors)
3	1 μb	Event reconstruction and tracking

number of upgrades are being carried out at CDF. There are several readout electronics improvements designed to accommodate higher luminosity. The level 3 trigger system is being rebuilt with more powerful microprocessors (MIPS R3000's). A silicon vertex detector is being added around the beam pipe to allow the tagging of B decays. A preradiator layer is being added in front of the central calorimeters to aid in the identification of direct photons. Finally the central muon system is being enhanced with additional steel to eliminate hadron "punch–through" and with additional chambers to extend its angular coverage. The latter improvements will be very useful for particle searches as well as for electroweak studies. In future runs CDF will further extend its muon coverage and will replace its gas tube calorimeters with faster scintillator tile versions.

4. Jets Physics - QCD

It is assumed that free quarks and gluons never emerge from a hadronic interaction, and that instead one sees collimated "jets" of particles which result from the hadronization of the partons. Such jets were originally observed in electron–positron collider experiments as a statistical elongation of the overall angular distribution of the outgoing particles. Now in high energy hadron collider experiments jets appear as highly collimated clusters of

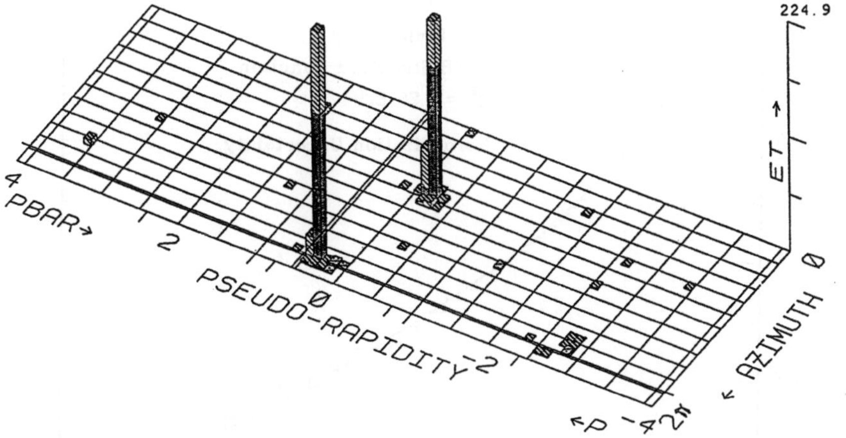

Figure 3 A lego plot of two jet event from CDF.

particles which stand out distinctly from one another. They can be most dramatically observed in a two dimensional histogram ("lego plot") of the transverse energy deposited in the calorimeters, where the two axes are the pseudorapidity, η, and the azimuth angle, ϕ. In Fig. 3 a lego plot for a high energy elastic parton–parton scatter is shown. The jets representing the two outgoing partons must be back-to-back in azimuth, but may occur at any pseudorapidity values as the incoming partons have their longitudinal momenta statistically distributed according to the structure functions.

As can be seen from the figure it is a relatively easy matter to select the jets by defining a small area on the surface of the lego plot containing most of the jet energy. It is important to note that the product $dE_T^2\, d\eta\, d\phi$ is the element of invariant phase space, and as a consequence jets will have a symmetric E_T profile in the η–ϕ plane. In fact selecting the energy inside a circular contour centered on a jet is similar to selecting only those jet fragments with with η_{jet} above a certain limit, where η_{jet} is calculated with respect to the jet axis instead of the beam axis. For example our CDF jet selection algorithm uses a circular area of radius, $R_{\eta\phi} = 0.7$, which corresponds to an approximate η_{jet} lower limit of 1.0. Thus at most a few low energy fragments of a jet will be missed by the algorithm. Nonetheless is is necessary to correct the resulting jet energy for such losses as well as for losses due to geometric gaps in the calorimeter and for the overall calibration of the calorimeter energy scale.

The most recent CDF inclusive cross section is shown as a function of E_T in Fig. 4 for jets in the central region as selected by the above algorithm. The data falls almost seven orders of magnitude from $E_T \approx 20$ GeV to $E_T \approx 400$ GeV. It is shown in comparison with the Next to Leading Order QCD calculation of Ellis, Kunst and Soper [2], where the QCD prediction has been absolutely normalized to the data. The agreement between theory and experiment is remarkable.

The same data are shown in Fig. 5, but this time compared instead to a lowest order QCD calculation plus a contact term representing possible quark compositeness. The contact term is proportional to $1/\Lambda^2$, where is Λ is the compositeness energy scale. At the

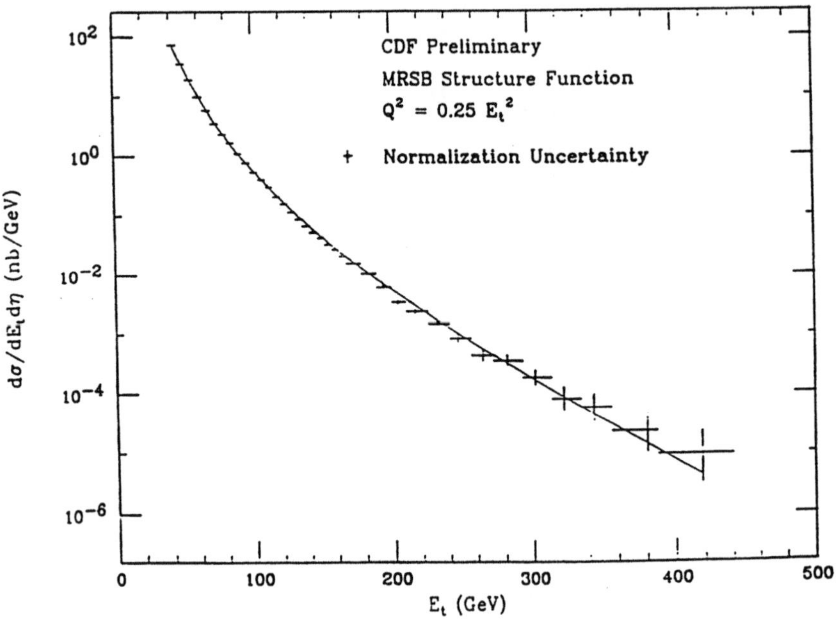

Figure 4 The inclusive jet differential cross section as a function of E_T from CDF ($\sqrt{s}$ = 1800 GeV). The curve is the NLO QCD prediction of Ellis, Kunst and Soper.

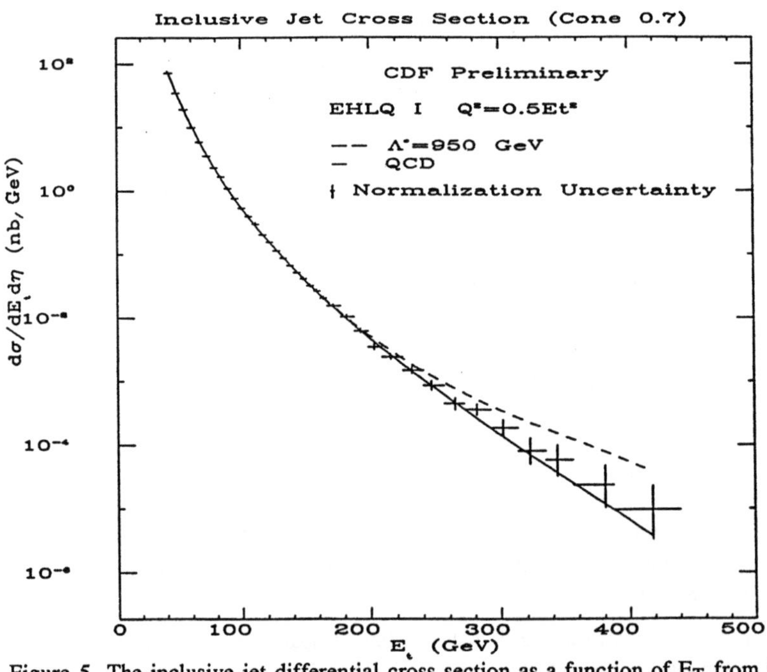

Figure 5 The inclusive jet differential cross section as a function of E_T from CDF ($\sqrt{s}$ = 1800 GeV). The curves are lowest order QCD plus quark compositeness terms.

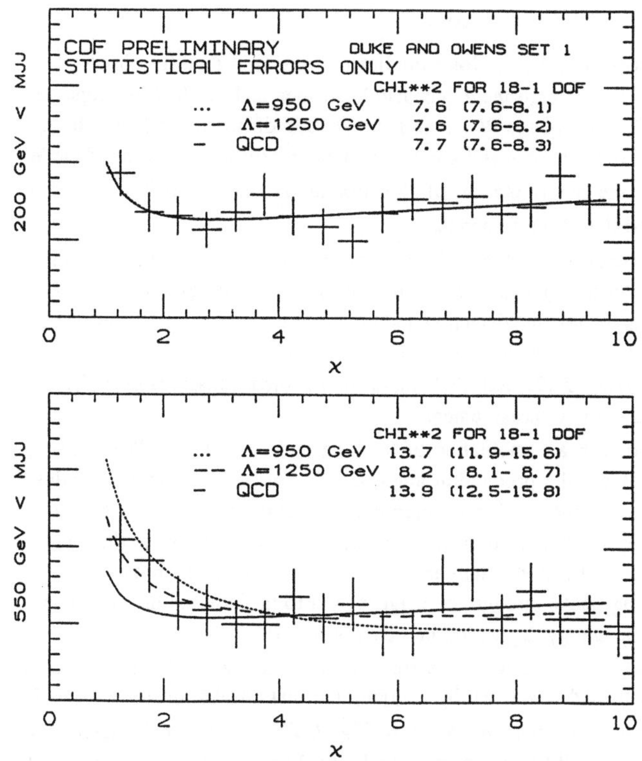

Figure 6 The dijet angular distribution from CDF as a function of $\chi = (1+\cos\theta^*)/(1-\cos\theta^*)$.

95% confidence level only values of Λ greater than 950 GeV are consistent with the data. The corresponding distance scale is less than $2 \cdot 10^{-17}$ cm or 0.0002 fermi. It is interested to note that the length and distance scales addressed by Yukawa some fifty years ago differed from these by a factor of roughly 5000. In the next CDF run it should be possible to extend the measurement of the inclusive jet spectrum beyond $E_T = 500$ GeV and to extend the compositeness scale limit beyond 1200 GeV.

The primary component of the inclusive jet cross section are the dijet events which represent parton–parton elastic scattering. The angular distribution for these events has also been measured and is displayed in Fig. 6 as a function of $\chi = (1+\cos\theta^*)/(1-\cos\theta^*)$, where θ^* is the Collins–Soper approximation of the center–of–mass scattering angle [3]. The χ distribution would be approximately flat for $\chi > 2$ in the case of simple Rutherford scattering, but is modified by higher order corrections. The figure is divided into a two parts by the dijet mass; the high dijet mass part is the one where one would expect to see deviations from QCD. The curves shown are for QCD plus a contact term as in Fig. 5, and the results obtained here are consistent with those from the inclusive spectrum.

There are studies also underway of the CDF data for multijet events. Events with as many as six jets have been observed, and thus far these data have proven to be completely consistent with the predictions of QCD.

5. Electroweak Physics - W,Z Production

The intermediate gauge bosons, although long predicted, were first directly observed by UA1 and UA2 at CERN [4]. Although the Z has also been seen and has had its properties precisely measured at LEP, study of the W remains the domain of hadron colliders. In this section I will review the results on the Z and present the most recent measurement of the W mass. I will also discuss the determination of the mixing parameter, $\sin^2\theta_W$, and the measurement of other properties of the W and Z.

Although the leptonic decays of the W and Z represent only a small fraction of their total width (1/6 for the W, 1/20 for the Z), these decays are the simplest topologies to detect. The key to their measurement is the reliable detection and precise measurement of electrons and muons. It is also important to have a "hermetic" calorimeter in order to infer the possible presence of a neutrino. UA1, UA2, and CDF have all devoted considerable effort to understanding and calibrating these measurements.

Electrons are measured with a combination of tracking and calorimetry. The tracking provides a measurement of the the production angle and is used to discriminate against backgrounds such as π^0's and photons. The primary energy measurement for the electrons comes from the calorimetry, which also provides rejection of charged hadrons. In the case of muons tracking provides both the momentum and angle measurement. Muon backgrounds are eliminated by connecting tracks through the magnet steel to the external drift chambers and by requiring a minimum energy deposition in the calorimeters.

At CDF the calibration of the momentum measurement in the central tracking chamber is done with cosmic ray muons and is verified by studying the mass peaks for the ψ/J and Υ. The dimuon peaks for these two states are shown in Fig. 7. Fits to both peaks agree well with PDG averages. In fact as will be seen below, the best calibration of the tracking in the future will be obtained by comparing our Z mass value to the average value from LEP.

The electron energy measurement of the calorimetry is first calibrated using test beam electrons. This calibration is fine tuned in situ using electrons from W's and comparing their momentum from tracking with their calorimeter energy. Although the tracking has

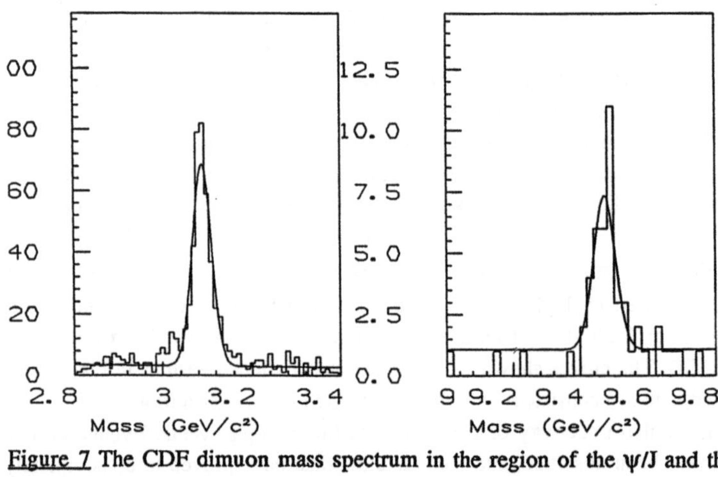

Figure 7 The CDF dimuon mass spectrum in the region of the ψ/J and the Υ. The fitted masses are 3.097 ± 0.001 and 9.469 ± 0.010 GeV respectively.

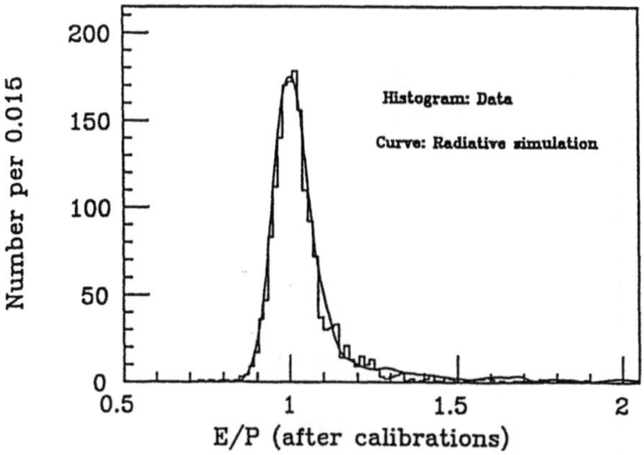

Figure 8 The ratio of calorimter energy to track momentum for electrons from W decays. The curve includes radiative corrections to the momentum.

already been calibrated, there is a high probability that an electron will radiate some of its energy as it passes through the inner tracking chambers. The calorimeter will see the radiated photon and thus measures the total original energy of the electron. In Fig. 8 the ratio of E/p for W electrons is compared to a calculation which takes into account the radiative corrections to the momentum. This comparison is used to set the energy scale factor for the calorimetry. Because of the smaller statistics of the Z sample, this method results in a smaller scale uncertainty than a comparison of our Z signal to LEP.

The Z with its decay into two charged leptons is the easiest signal to measure. The CDF mass peaks for both the electron and muon channels [5] are shown in Fig. 9. The locations of the two peaks are in excellent agreement with each other, and the combined CDF result of $M_Z = 90.9 \pm 0.4$ GeV is in good agreement with the latest LEP average [6] of $M_Z = 91.18 \pm 0.03$ GeV. In future runs as our Z sample increases in size the latter comparison will become CDF's primary energy scale calibration.

Although the leptonic decays of the W are about an order of magnitude more prolific than those of the Z, the W decays cannot be completely reconstructed because of the missing neutrino. The signature of the W is a single charged lepton with large E_T that is not balanced by any other visible E_T. This is illustrated in Fig. 10, where the E_T of the electron is plotted against the missing E_T for the W candidates in the electron channel. The events are roughly concentrated on a 45° line indicating that the electron E_T balances the missing E_T. Since the longitudinal momentum of the neutrino is not determined it is not possible to calculate the W mass directly. However, using only the transverse components of energy one can calculate a quantity called the transverse mass, $M_T = 2\, E_{Te}\, E_{T\nu}\, (1\text{-}\cos\Delta\phi)$, where $\Delta\phi$ is the azimuth angle separation of the electron and neutrino. This variable has a Jacobian peak at the W mass, and is relatively insensitive to the transverse momentum of the W itself. The M_T plots for the electron and muon W samples from CDF are shown in Fig. 11. The data samples shown here require a charged lepton with $E_T > 25$ GeV, missing energy with $E_T > 25$ GeV, no jet clusters with $E_T > 7$ GeV, and no extra tracks with $p_T > 15$ GeV. The latter two cuts are intended to improve the mass resolution of the sample.

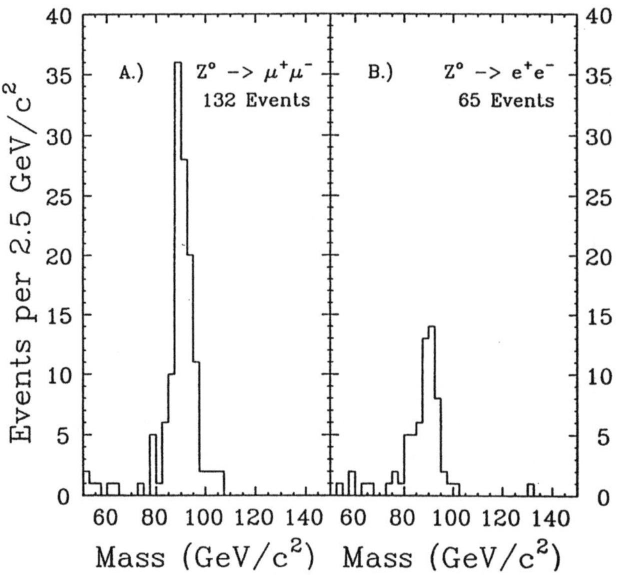

Figure 9 CDF Z mass peaks for dimuons (left) and dielectrons (right).

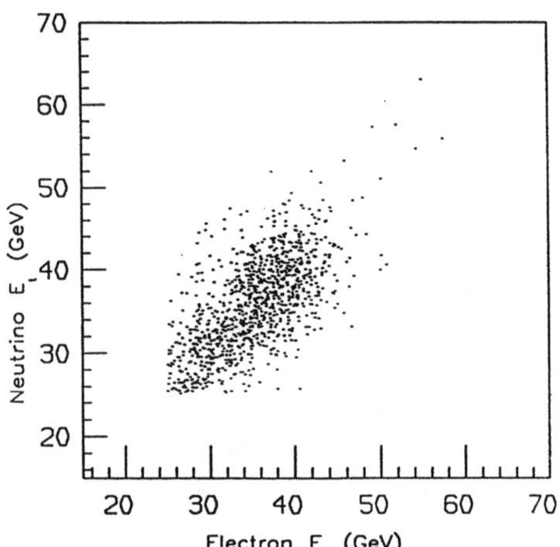

Figure 10 A scatter plot of missing E_T versus electron E_T for W candidates.

In order to determine the W mass from the data shown in Fig. 11 it is necessary to generate monte carlo simulations of the M_T spectrum for various values of M_W [7]. This monte carlo assumes the standard W production and decay dynamics and is done with a range of quark structure functions. The assumptions about experimental resolution come from both the test beam results and from other data samples. For example the "minimum bias" events taken with a total cross section trigger are used to determine the missing E_T resolution. Finally the W p_T distribution is calibrated against the Z sample. Although fits were attempted with the W width left as a second free parameter, the best results were obtained by fixing this to the nominal value of 2.1 GeV. The results of these fits for the electron and muon channels have comparable errors and are completely consistent with each

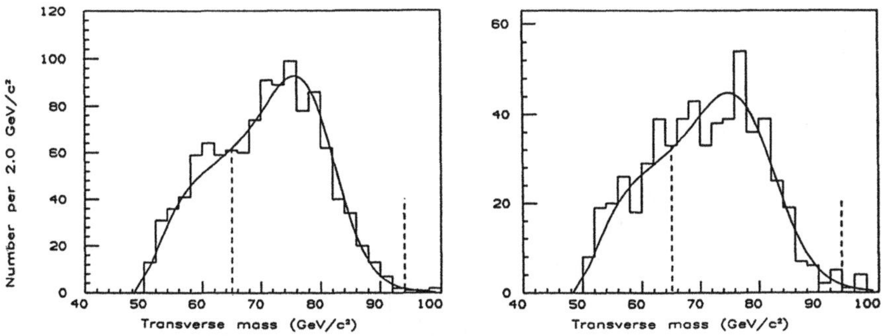

Figure 11 Transverse mass plots for the eν data (left) and μν data (right).

other yielding a combined result of $M_W = 79.9 \pm 0.4$ GeV, where the error is a combination of statitistical and systematic uncertainties. This is to be compared with the UA2 result [8] from electrons of $M_W = 80.5 \pm 0.5$ GeV. A large part of the quoted uncertainty in the CDF result depends on the statistics of the data and will benefit directly from the increased luminosity in future runs. For example the uncertainty in the energy scale will benefit from the increased size of the Z sample. In the upcoming CDF run the integrated luminosity should increase by a factor of five which will make it possible to decrease the uncertainty in M_W to about 0.2 GeV.

The measurement of the W mass can be combined together with the Z mass to obtain the weak mixing parameter, $\sin^2\theta_W = 1 - M_W^2/M_Z^2$. Combining the two CDF masses we get $\sin^2\theta_W = .231 \pm .008$, and combining the CDF W mass with the LEP Z mass we get .2317 ± .0075. This particular definition of $\sin^2\theta_W$ is not dependent on the top quark mass through radiative corrections, but can be combined with other derivations to obtain an upper limit on the top quark [9]. Depending on assumptions these upper limits range from 170 to 230 GeV. With the improvement in the W mass uncertainty in the next CDF run the accuracy of the $\sin^2\theta_W$ determination will improve by approximately a factor of two.

Although I have concentrated on the leptonic decays of the W and Z, the UA2 collaboration have taken advantage of their excellent calorimetry to observe the $\bar{q}$–q → dijet decay mode of these bosons [10]. Their data is shown is Fig. 12, where part a) shows the dijet mass spectrum enhanced by a multiplicative factor of M_{jj}^6. This factor makes it possible to see a small rise around 80 GeV. Part b) shows the same data divided by a smooth background estimate and now clearly shows a peak. This peak is then fitted to a pair of Gaussians with the ratio of the two peaks constrained to the nominal value of M_W/M_Z. The fit recovers the correct W mass, namely $M_W = 78.9 \pm 1.5$ GeV. This remarkable result shows how difficult it is to find mass peaks in dijet and multijet channels. Not only is the background from QCD processes overwhelming, but the mass resolution even under the best of circumstances is barely adequate.

Turning to the production properties of the W and Z, in Fig. 13 the total cross sections are shown for UA2 and CDF energies [11]. The cross sections are in good agreement with a QCD prediction. The predominant uncertainty in the cross sections is in the integrated luminosity. An alternative way of presenting the results shown in Fig. 13 is to calculate the ratio, R, of the cross sections for the leptonic modes [12]. In this ratio the luminosity uncertainties cancel. Furthermore if one accepts the theoretical values for the boson total

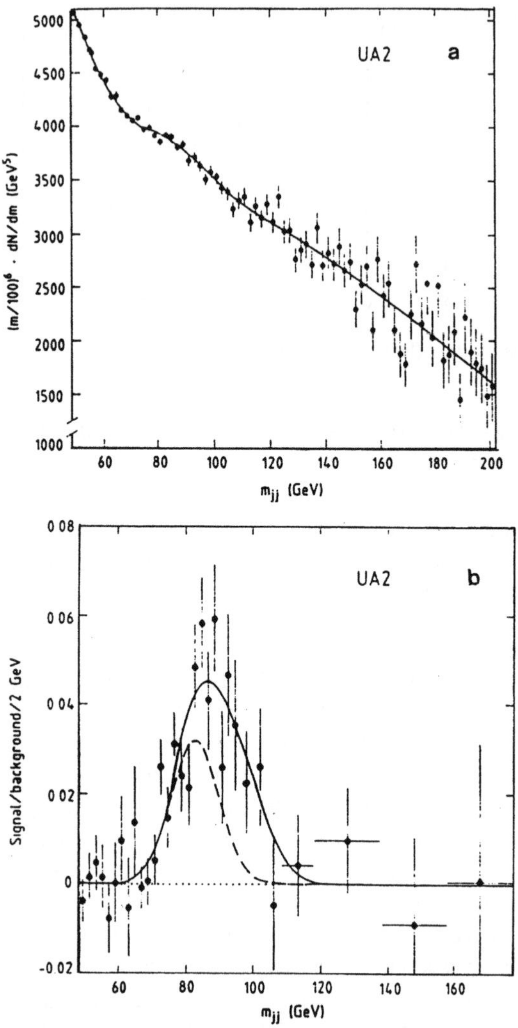

Figure 12 UA2 dijet mass data. a) figure has been enhanced by an M_{jj}^6 factor. b) figure has been divided by smooth background curve.

cross sections and leptonic widths then R depends only the ratio of the W and Z widths. Since LEP has measured the Z width very accurately, R becomes an indirect measurement of the W width. The CDF results for these quantities are $R = 10.2 \pm 0.9$ and $\Lambda_W = 2.19 \pm 0.20$ GeV.

The differential cross sections for W and Z production have also been measured and found to be in good agreement with QCD calculations. The CDF data is shown together with a QCD calculation by Reno and Arnold in Fig. 14. These data are obtained by relaxing the requirement that there be no additional jet activity in the events.

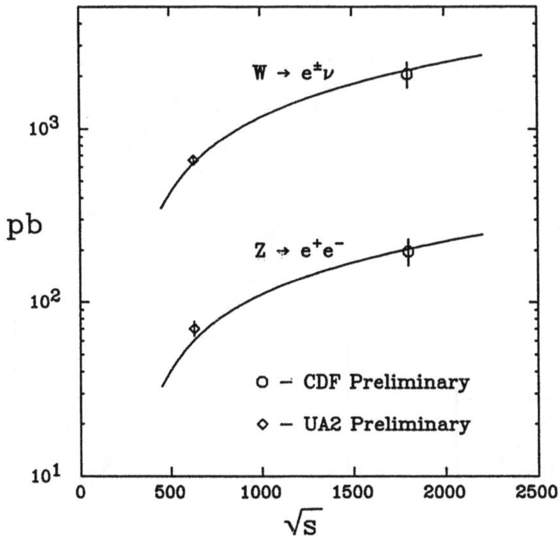

Figure 13 W and Z total leptonic cross sections from CDF and UA2.

Finally we take a look at the angular distribution of the Z decay, in particular the decay asymmetry. Since the Z is slightly left-handed, its decay distribution is expected to be asymmetric in an amount proportional to $\sin^2\theta_W$. There is also a small asymmetry which results from the interference of the Z with the Drell–Yan background. Fig. 15 shows the distribution of the Collins–Soper angle for the Z electron sample after corrections have been made for acceptance and efficiency. The curve that is shown is the result of a maximum likelihood fit yielding $\sin^2\theta_W = .229 \pm .016$. Because of the approximate charge symmetry of the experiment the systematic uncertainty in this result is very small and the quoted error is primarily statistical. This definition of $\sin^2\theta_W$ does depend on the top quark mass through radiative corrections. In Fig. 16 this result is transformed to the M_W/M_Z definition as a function of M_t. The M_W/M_Z results from CDF and UA2 are also shown. With the current level of statistics it is not yet possible to set a meaningful upper limit on the M_t, but this will improve in future runs. The W decay asymmetry, which is dependent on the structure functions and not on $\sin^2\theta_W$, is also currently being studied.

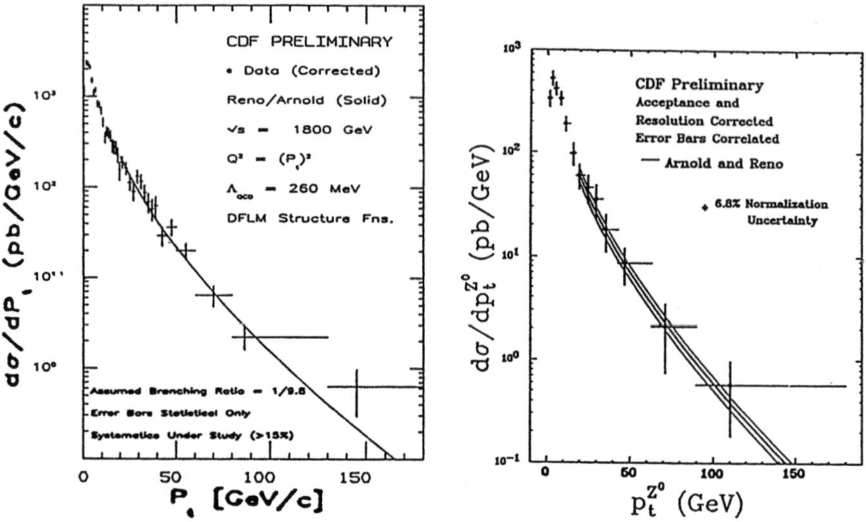

Figure 14 Differential cross sections for the W (left) and the Z (right) as a function of p_T.

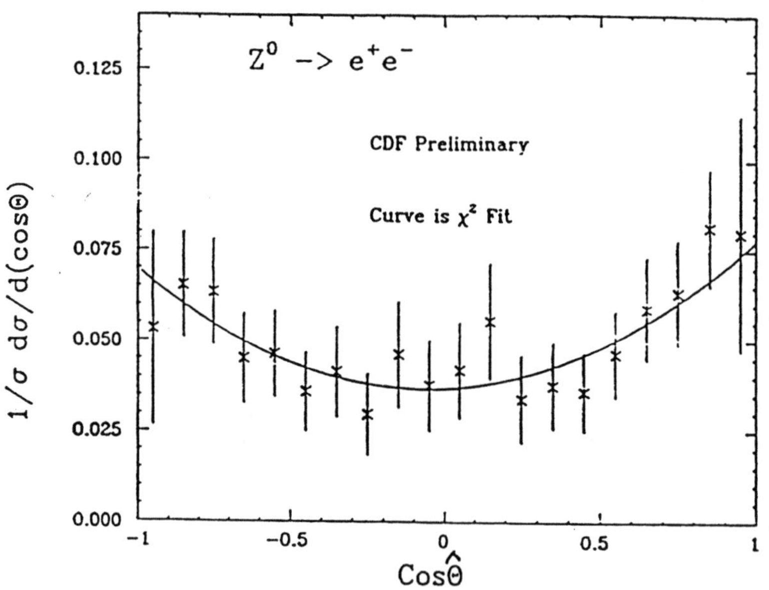

Figure 15 Z angular distribution as a function of the cosine of the center-of-mass angle (Collins–Soper representation).

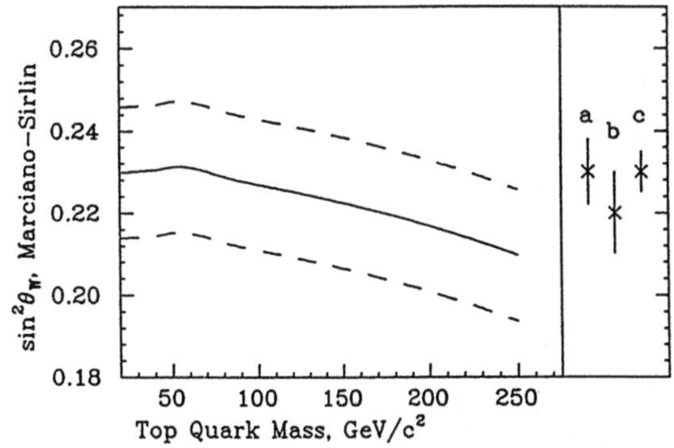

Figure 16 $\sin^2\theta_W$ as determined from the Z asymmetry versus top quark mass. The points on the right are $\sin^2\theta_W$ from M_W/M_Z: a) CDF, b) UA2, c) combined.

6. Heavy Flavors – Top Quark Search

In the previous runs of both CDF at the Tevatron collider and UA2 at the SPS collider the most exciting topic has been the search for the top quark. The absence of a toponium signal at Tristan had set a lower limit of 30 GeV and the earlier CERN collider results had raised this to 40 GeV. As mentioned earlier the interpretation of $\sin^2\theta_W$ also led to an upper limit somewhere between 170 and 230 GeV. From the data taken in the last CDF run it has been possible to raise the top mass lower limit above the W mass as I will discuss below. Similar, but somewhat lower limits have also been set by UA2.

At the energy of the Tevatron collider, 1800 GeV, the dominant mode for top production will be the production of a top antiquark–quark pair. Both top quarks will then decay to a W boson, which may be virtual, and a b (or $\bar{b}$) quark. For top quark masses comparable to or below the W mass, the kinematics of the top decay dictate that the W will carry away most of the energy. Hence it is difficult to detect the b quarks or their fragments and instead the top search must focus on the two W boson remnants. This means there are three possible topologies to be considered: a) both W's decay to quark jets, b) one W decays to jets and the other decays leptonically, and c) both W's decay leptonically. Possibility a) is very difficult as we noted in the previous section because of the large QCD background to the four jet system and because of the poor dijet mass resolution. For both b) and c) the identification of electrons and muons is the key issue just as in the last section .

As was previously discussed electrons are selected using both tracking and calorimetry information. Both must be completely consistent with an electron signal and consistent with each other. In addition cuts are made to eliminate γ and π° conversions, and E_T threshholds are imposed to ensure trigger efficiency. Finally an "isolation" cut is made which eliminates events where there are other energy depositions near the electron (any extra E_T in a surrounding cone with $R_{\eta\phi}$ = 0.4 must be less than 5 GeV). This cut eliminates electrons from b quark production and decay, where the electron is accompanied by a nearby jet.

Likewise muons are selected by correlating a good track with either a track stub in the chambers outside the magnet steel or with a calorimeter cell with minimum energy deposition. A p_T threshhold ensures trigger efficiency as with the electrons, and a similar

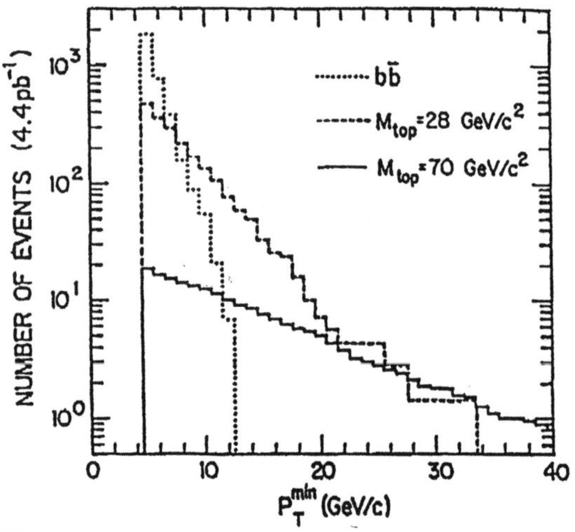

Figure 17 The number of events predicted by ISAJET to have both the electron E_T and the muon p_T greater than p_T^{min}. The curves are for $\bar{b}$–b and $\bar{t}$–t production as indicated.

isolation cut is imposed. In the muon case not only does the isolation cut remove muons from b's, but it also reduces the probability of finding a secondary muon which results from the decay of one of the hadrons in a jet.

The first topology that I will discuss is the double semi–leptonic case, where both W's from the $\bar{t}$–t pair decay leptonically. Here the simplest signature is the production of two oppositely charged leptons. Although the branching ratio is small (2/81),the cleanest combination is to have one electron and one muon [13], which has neither any Drell–Yan background nor Z signals to cope with. The idea is to select events with an oppositely charged electron–muon pair with both leptons above the same E_T (or p_T) threshhold. Fig. 17 shows the predicted number of events above any threshhold between 5 and 40 GeV for three different processes: direct $\bar{b}$–b production, $\bar{t}$–t production with M_t = 28 GeV, and $\bar{t}$–t with M_t = 70 GeV. The predictions were generated with ISAJET and assumed the integrated luminosity of the last CDF run. It can be seen that by setting the common threshhold at 15 GeV the $\bar{b}$–b background is completely eliminated, but reasonable efficiency for top masses up to 70 GeV is maintained.

A scatter plot of the electron E_T versus the muon p_T from the last CDF run is shown in Fig. 18. One event is observed above the dual 15 GeV threshholds. This event, although consistent with the interpretation as a $\bar{t}$–t decay, is located well above the threshholds, which is highly unusual with steeply falling spectra. Nonetheless if we assume that one event is observed then a 95% confidence level lower limit on the top mass is determined to be 72 GeV. This limit takes the efficiencies for detecting and reconstructing the two leptons fully into account.

The double semi–leptonic top search can be extended by including the dielectron and and dimuon channels after removing the Z signal and accounting for the Drell–Yan background. The Z is removed by a mass window cut from 75 to 105 GeV; in addition it is required that the missing E_T be greater than 20 GeV and the charged leptons not be back to back in azimuth.

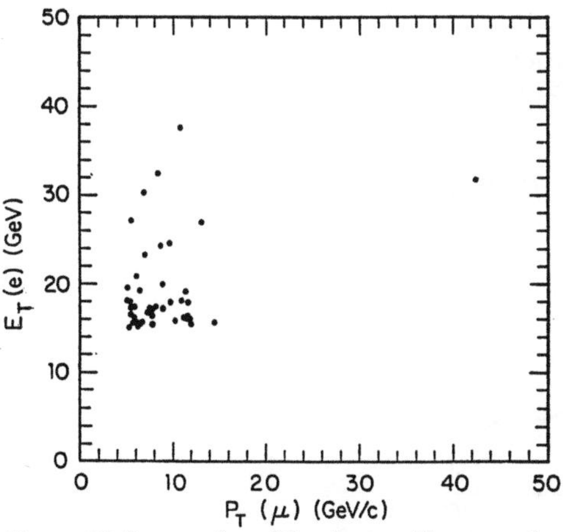

Figure 18 Scatter plot of the electron E_T versus the muon p_T for the CDF data.

After these cuts are made, the addition of these channels increases the acceptance for the top signal by about 50%. No new events are observed to pass the cuts, and as a result the lower limit is raised to 84 GeV (95% C.L.). This extension to the published CDF limit is still preliminary, but it is important to note that it requires the top quark mass to be significantly above the W mass. Hence the W in the top quark decay will be on the mass shell, which has direct consequences for the search methodology in the future.

We now turn to the single semi–leptonic channel where the second t (or $\bar{t}$) quark decays into two quark jets. The branching ratio for this channel is higher than for the double semi-leptonic, but there is now a major background, namely QCD production of the W plus two jets. The initial studies of this channel used the transverse mass variable to look for enhancements below the W jacobian peak [14]. The data sample required that there be an electron and missing energy both with $E_T > 20$ GeV, plus two (or more) jets both with $E_T > 10$ GeV. The published 95% C.L. lower limits obtained by this method were 69 GeV for UA2 and 77 GeV for CDF.

Now that it appears certain that the top mass is above M_W, the transverse mass is no longer a useful tool to investigate the single semi-leptonic events. Another possibility is to look for a W mass peak in the accompanying dijets, but as we have noted in section 5 this is extremely difficult and does not provide a unique $\bar{t}$–t signature. Therefore it is necessary to find evidence of the b or $\bar{b}$ quark fragments. A first attempt at this has been done by searching for a soft muon resulting from the b decay. In addition the single semi-leptonic sample has been expanded to include primary muons as well as electrons. The extra soft muon is required to have $p_T > 1.7$ GeV, and must be separtated from the nearest W jet by $R_{\eta\phi} > 0.6$ to eliminate punch–through. No events are observed with these extensions to the analysis. When this result is combined with the double semi–leptonic results the 95% C.L. lower limit on the top mass is raised to 89 GeV. This preliminary result is shown in Fig. 19, where the upper limit on the $\bar{t}$–t cross section as measured by CDF is plotted for each of the successive analysis steps described above. The intersection of each of these curves with the lower limit of the of the QCD prediction for $\bar{t}$–t production gives the corresponding top mass limit.

In the next CDF run if we are unlucky enough to find no additional events with the same analyses as described above, then the measured cross section upper limits shown in Fig. 19

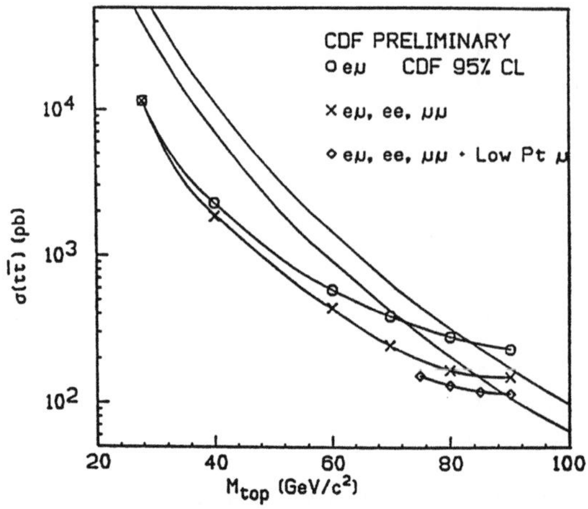

Figure 19 The preliminary CDF upper limits on the $\bar{t}$–t cross section as a function of the top quark mass for the sequential analyses described in the text. The pair of smooth curves give the limits of the theoretical prediction.

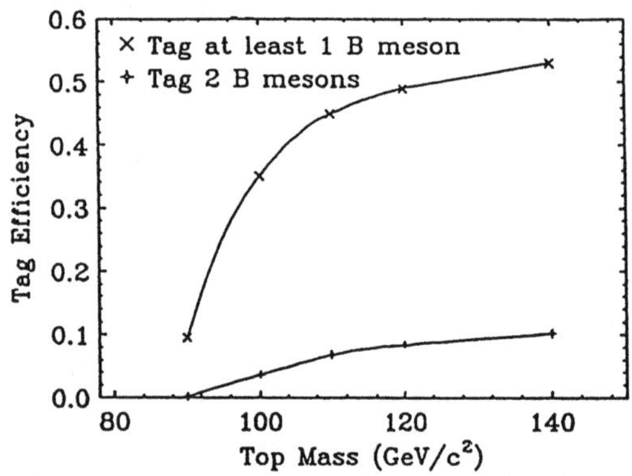

Figure 20 The predicted efficiency for tagging B mesons with the new silicon vertex detector in the next CDF run.

will be lowered in inverse proportion to the increase in integrated luminosity. For example if we receive a fivefold increase in luminosity, the extrapolation of these curves would indicate a M_t lower limit around 120 GeV. However, we also plan to improve the analyses, in particular by taking advantage of the increased energy of the b and $\bar{b}$ fragments with larger M_t. With the addition of a silicon vertex detector it should be possible to directly tag many of the b decays. An estimate of the efficiency of the b tagging as a function of M_t is shown in Fig. 20. Furthermore the extended muon coverage in the next run will improve the acceptance for both primary and secondary muons. With these improvements we expect to push the top mass limit well above 120 GeV.

Having mentioned b tagging it is important to add that CDF will also utilize the new silicon vertex detector for extensive b–physics studies in the next run. In the data from the previous run Υ, Υ', and Υ" peaks have been observed in the dimuon channel. We are also studying the inclusive decays of B's into ψ/J + X and other states as well as $\bar{B}B$ mixing.

7. Beyond the Standard Model

Although several searches for unorthodox states have taken place I will just mention two. First we have searched the dilepton mass spectra above the W and Z for higher mass gauge bosons. Aside from a couple of very high mass events, there is no evidence for any peak. Assuming standard couplings for such states, lower mass limits are found to be $M_{W'} > 380$ GeV and $M_{Z'} > 400$ GeV (95% C.L.).

Second, multijet events with large missing E_T are a possible signature of supersymmetric squarks and gluinos decaying to stable, undetected photinos plus hadrons. CDF has a sample of 98 events with missing $E_T > 40$ GeV which, however, can be fully explained by W and Z production and QCD multijets [15]. The resulting squark and gluino mass limits are shown in Fig. 21. The limits are worse for the case where the gluino is lighter than the squark because the final decay to a photino state is three–body and not two–body.

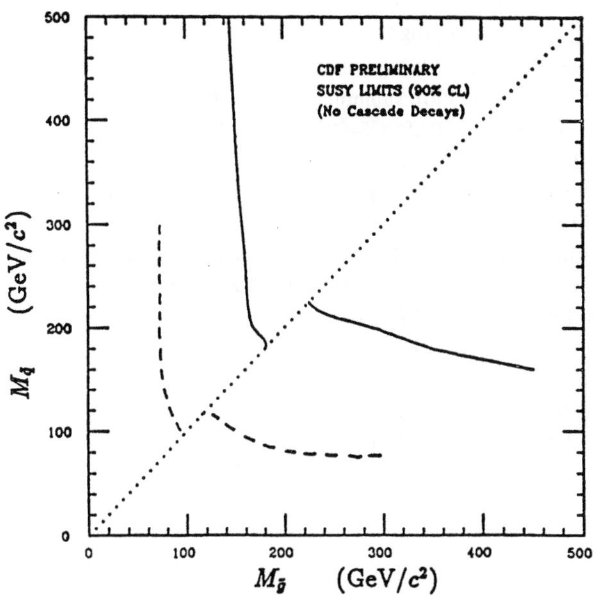

Figure 21 The preliminary mass limits for squarks (vert.) versus gluinos (horiz.) as obtained by CDF. The dashed lines give the previous published limits.

I would like to thank the organizers of the 5th Nishinomiya Yukawa–Memorial Symposium for their kind invitation and hospitality. I would also like to thank my CDF colleagues for their help with the preparation of this talk.

References

1. CDF Collaboration, NIM **A271**, 387 (1988).
2. S.D. Ellis, A. Kunst and D. Soper, Phys. Rev. Lett. **62**, 726 (1989).
3. CDF Collaboration, Phys. Rev. Lett. **62**, 3020 (1989).
4. UA1 Collaboration, Phys Lett. **B122**, 103 (1983);
   UA2 Collaboration, Phys Lett. **B122**, 476 (1983);
   UA1 Collaboration, Phys Lett. **B126**, 398 (1983);
   UA2 Collaboration, Phys Lett. **B129**, 130 (1983).
5. CDF Collaboration, Phys. Rev. Lett. **63**, 720 (1989).
6. B. Adeva *et al.*, Phys. Lett. **B231**, 509 (1989);
   D. Decamp *et al.*, Phys. Lett. **B231**, 519 (1989);
   M. Akrawy *et al.*, Phys. Lett. **B231**, 530 (1989);
   P. Aarnio *et al.*, Phys. Lett. **B231**, 509 (1989).
7. CDF Collaboration, Phys. Rev. Lett. **65**, 2243 (1990).
8. UA2 Collaboration, CERN EP/90-22 (1990), submitted to Phys. Lett. **B**.
9. P. Langacker, to be published in Proc. of PASCOS-90, Boston, 1990.
10. UA2 Collaboration, Phys Lett. **B186**, 452 (1987);
    UA2 Collaboration, CERN EP 89-178 (1989), to be published in Proc. of 8th Topical Workshop on Proton Antiproton Collider Physics, Castiglione, 1989.
11. CDF Collaboration, FERMILAB-PUB-89/245 (1989), submitted to Phys. Rev. Lett.;
    UA2 Collaboration, CERN EP/90-20 (1990), submitted to Phys. Lett. **B**.
12. CDF Collaboration, Phys. Rev. Lett. **64**, 152 (1990).
13. CDF Collaboration, Phys. Rev. Lett. **64**, 147 (1990).
14. CDF Collaboration, Phys. Rev. Lett. **64**, 142 (1990);
    UA2 Collaboration, Zeit. Phys. **C46**, 179 (1990).
15. CDF Collaboration, Phys. Rev. Lett. **62**, 1825 (1989).

Tests of the Standard Model at LEP

T. Mashimo

International Center for Elementary Particle Physics, Faculty of Science, University of Tokyo, 7-3-1 Hongo, Bunkyo-ku, Tokyo 113, Japan

Abstract. During the first year of LEP operation various tests of the standard model have been carried out by four main LEP experiments. The results are in good agreement with the predictions of the minimal standard model with three light neutrino generations. I review the measurements of the parameters of Z^0 resonance and a few results of QCD studies. Direct searches for new particles and new phenomena are covered by my fellow speaker Prof. D. Treille.

1 Status of LEP

The Large Electron Positron Collider (LEP) at CERN started its operation in July 1989 and was in operation for 3 months for physics from September to December 1989 and for 5 months from March to August in 1990. Since the start of the operation the luminosity of the beam has been steadily improved. So far the maximum luminosity of $\sim 6 \times 10^{30}\ cm^{-2}s^{-1}$. has been reached. A typical luminosity in 1990 was $\sim 3 \times 10^{30}\ cm^{-2}s^{-1}$. This is to be compared with the design luminosity $\sim 16 \times 10^{30}\ cm^{-2}s^{-1}$. The luminosity has been limited by so-called the beam-beam effect. For 1990 the scheduled physics time was 102 days, while usable physics coast was in total 42 days, thus the LEP running efficiency for the experiments was roughly 40%.

During the first year data have been recorded at several center-of-mass energies in scan of the Z^0 resonance. In 1990 roughly half of the total luminosity was devoted to the Z^0 peak energy. The total number of multi-hadronic events which are good and usable for physics analyses is $\sim$ 80K in 1989 and $\sim$ 500K in 1990 when summed over the four main LEP experiments (ALEPH, DELPHI, L3 and OPAL).

At the end of August 1990 the natural transverse beam polarization was measured by a polarized laser beam and the observed value of $\sim$ 10% was close to the predicted one. It is hoped that in 1991 measurement will be done using the depolarization frequency technique for precise determination of the absolute beam energy.

2 Electroweak physics results

2.1 Line-shape and asymmetry measurements

One of the main tasks of the LEP experiments has been the measurement of the energy dependence of the cross sections for ($e^+e^- \rightarrow$ fermion pair) around the Z^0 pole, the so-called Z^0 line shape. All four experiments have published [1] [2] [3] [4] or presented the results on the hadronic and leptonic line shapes and the forward-backward charge asymmetries of lepton-pair final states.

Table 1: Number of multi-hadronic events selected, the systematic error of the selection, and experimental and theoretical systematic errors on luminosity measurement.

	Number of events	Systematic error (%)	Luminosity syst. error experiment (%)	theory (%)
ALEPH	83588	0.6	1.1	0.7
DELPHI	68225	1.1	1.7	1.0
L3	61675	0.7	1.1	0.7
OPAL	112237	0.8	1.2	1.0

Table 2: Number of leptonic events selected and the systematic error of the selection. For the e^+e^- channel, the systematic error attributed to the t-channel calculation is listed separately.

	ALEPH	DELPHI	L3	OPAL
e^+e^-				
# of events	2500	1389	1991	3270
syst. error	1.0%	1.3%	0.7%	1.0%
t-ch. error	0.5%	1.0%	0.7%	0.5%
$\mu^+\mu^-$				
# of events	2100	1618	1244	4642
syst. error	0.9%	1.9%	1.3%	0.9%
$\tau^+\tau^-$				
# of events	2300	1016	1169	3412
syst. error	1.0%	2.7%	3.0%	1.9%
l^+l^-				
# of events	6700	3187	—	—
syst. error	0.6%	1.3%		

The results shown here are the latest available ones, that is, essentially the same as those presented at the 25th International Conference on High Energy Physics which took place at Singapore in August 1990 [5] [6] [7] [8] [9]. The analyses are based on the data samples from 1989 runs and only part of 1990 runs. The results should be regarded as preliminary. The numbers of multi-hadronic events and lepton events used for the analyses are listed in table 1 and 2 (from ref. [10]) together with the systematic errors on the event selections and the errors assigned to the luminosity measurements. Individual experiments measure the luminosity by counting the number of small angle Bhabha events recorded by their forward detectors. The theoretical systematic errors arise from radiative corrections to the Bhabha cross section. ALEPH and DELPHI also made inclusive lepton pair selections without distinguishing the final-state flavour. This results in smaller systematic errors and the samples can be used for the analyses in which lepton universality is assumed.

The four LEP experiments are now finalizing their results using all the data samples taken in 1989 and 1990. The final results are expected to appear toward the end of 1990. Deeper understanding of the detector performance will decrease the systematic uncertainties. The theoretical error in the luminosity measurements might also be reduced in the near future when a new Monte Carlo program is available.

Examples of the line shape and asymmetry measurements are shown in figure 1 and figure 2.

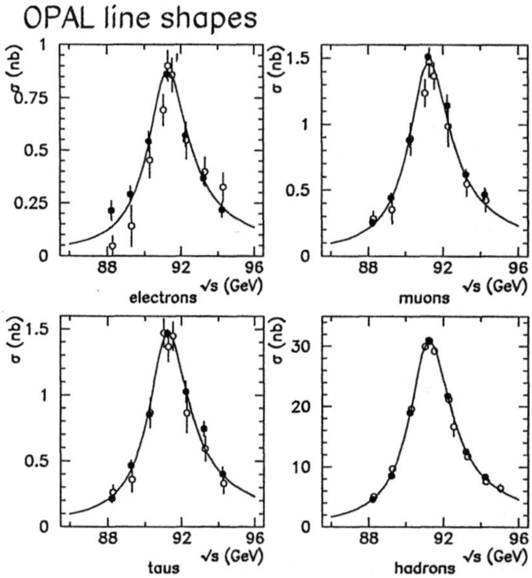

Figure 1: Cross sections measured by OPAL as functions of center-of-mass energy for $e^+e^- \to e^+e^-$, $e^+e^- \to \mu^+\mu^-$, $e^+e^- \to \tau^+\tau^-$ and $e^+e^- \to$ hadrons. The $e^+e^- \to e^+e^-$ cross section is integrated over $-0.7 < \cos\theta_{e^-} < 0.7$ and corrected for t-channel contributions. The $\mu^+\mu^-$, $\tau^+\tau^-$ and hadronic cross sections are corrected to full acceptance. The open and the full circles show the data from 1989 and 1990 runs, respectively. The solid lines are the results of the fit to the combined e^+e^-, $\mu^+\mu^-$, $\tau^+\tau^-$ and hadronic data (cross sections and asymmetries) assuming lepton universality.

All experiments have followed the basic strategies suggested in ref. [11], but the details of the methods of analyses for extracting Z^0 parameters are different from one experiment to another with respect to:

- Choice of programs:
  ALIBABA [12], Borrelli / Cahn formulae [13][14], Burgers' ZAPP program [15] (updated to program MIZA [16]), EXPOSTAR [17], Greco formulae implemented in Caffo-Remiddi program [18], ZFITTER [19], ZSHAPE [20]

- Treatment of t-channel and $s-t$ interference for $e^+e^- \to e^+e^-$ channel:
  These contributions are in general subtracted from the measured cross section, resulting in an additional systematic error for this channel.

- Combination of data samples:

  σ_{had} ,
  σ_e , σ_μ , σ_τ ,
  σ_{lept} (ALEPH and DELPHI) ,
  A_{FB}^{ee} , $A_{\text{FB}}^{\mu\mu}$, $A_{\text{FB}}^{\tau\tau}$,
  $A_{\text{FB}}^{\text{lept}}$ (ALEPH and DELPHI) ,
  $A_{\text{FB}}^{q\bar{q}}$ (forward-backward charge-asymmetry of hadron jets, ALEPH),
  P_τ (τ polarization, ALEPH)

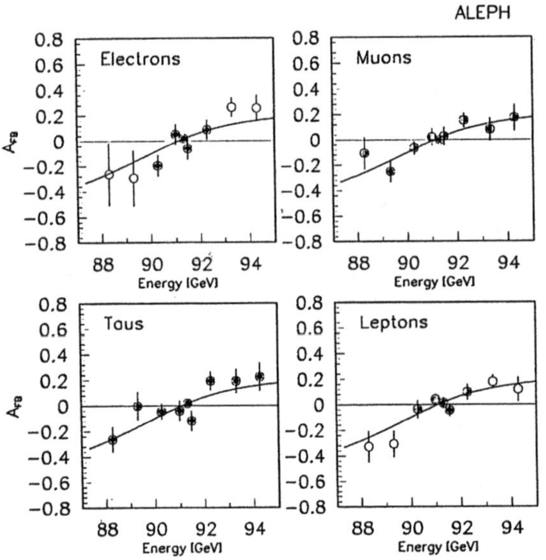

Figure 2: Forward-backward asymmetry for e^-, μ^-, τ^-, and l^- in lepton-pair events measured by ALEPH as a function of the center-of-mass energy. The full lines are the results of the fit. Points with open circles are not used in the fit in order to avoid larger systematic uncertainties coming from the "t-channel subtraction".

- Choice of parameter set extracted simultaneously in one fit: M_Z , Γ_Z , Γ_{ee} , $\Gamma_{\mu\mu}$, $\Gamma_{\tau\tau}$, $\Gamma_{l^+l^-}$, Γ_{had} , σ^{pole}_{had} , Γ_{inv} , N_ν , R_Z , $\hat{a}_l^2$, $\hat{v}_l^2$, $\hat{a}_l^2\hat{v}_l^2$, ρ , $\sin^2\bar{\theta}_W$, m_t , ...

I do not describe the details of their analyses. Instead I only show the summary of the results in the subsequent figures [21]. The combined results of the four experiments are also given in the figures. The procedure for averaging the results is the same as those adopted in other recent reviews (e.g. [9]): A common error is estimated and the individual experimental measurements in the figures are shown with the common error already subtracted. A weighted average of the four values is then calculated (shown as "Average" in the figure), using the remaining errors to determine the weights. Finally the common error is added back in quadrature to the error from the weighted average.

At the bottom of the figures the values predicted by the standard model are shown except M_Z, which is considered as an input to the standard model. They have been calculated by using the program ZFITTER [19] with the measured value of M_Z = 91.177 GeV/c^2. Unknown top mass is varied in the range $50 < m_t < 230$ GeV and the curves are shown for two (extreme) values of Higgs boson mass, 50 GeV and 1000 GeV, each for 3 different values of the strong coupling constants: α_s = 0.10, 0.12, and 0.14.

Figure 3 shows the results of the measurements of Z^0 mass M_Z, its total width Γ_Z, the hadronic pole cross section σ^{pole}_{had} and the hadronic width Γ_{had}. The uncertainty of the M_Z is now dominated by that of the absolute beam energy of LEP. In center-of-mass energy, E_{cm}, the uncertainty amounts to 0.030 GeV. During the 1990 the systematic error has been reduced somewhat, but in view of large uncertainty in this error the four

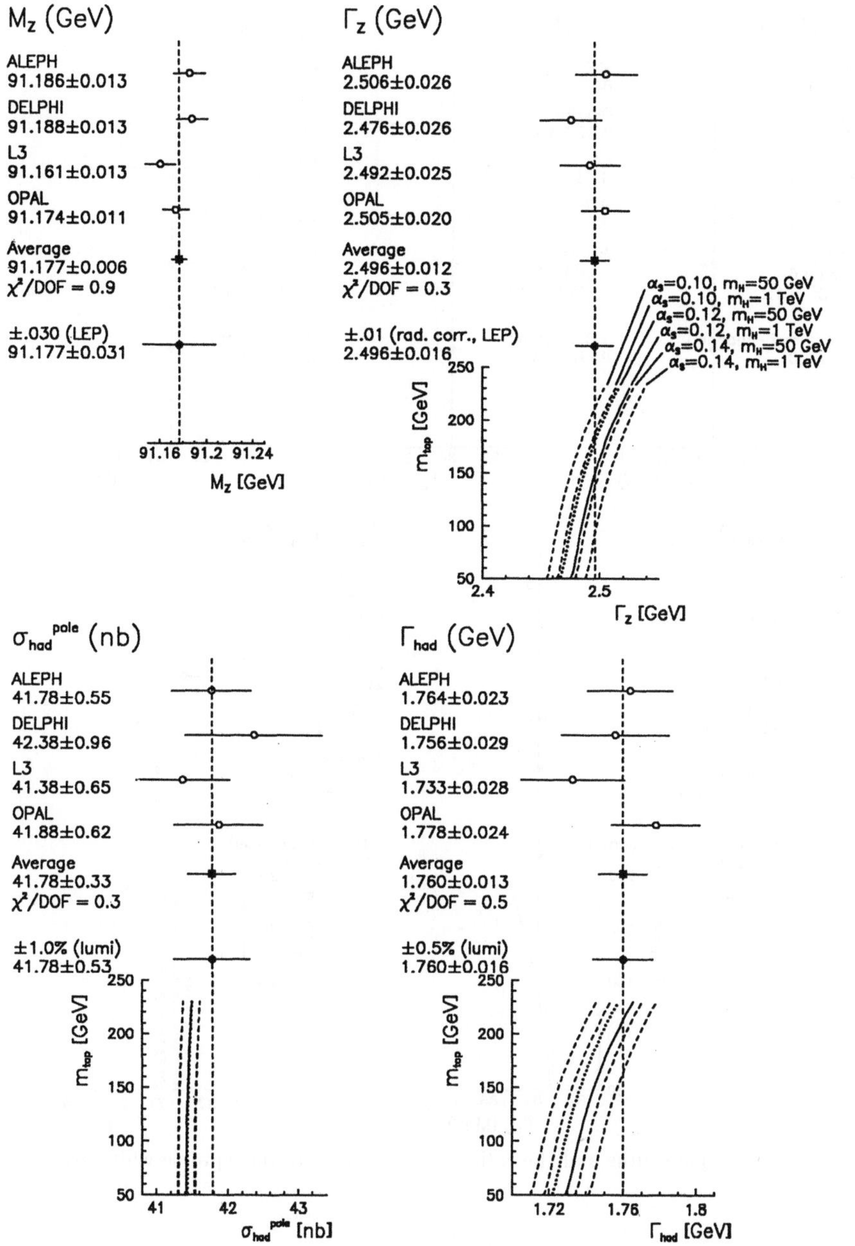

Figure 3: Z^0 mass M_Z, its total width Γ_Z, the hadronic pole cross section σ_{had}^{pole} and the hadronic width Γ_{had}.

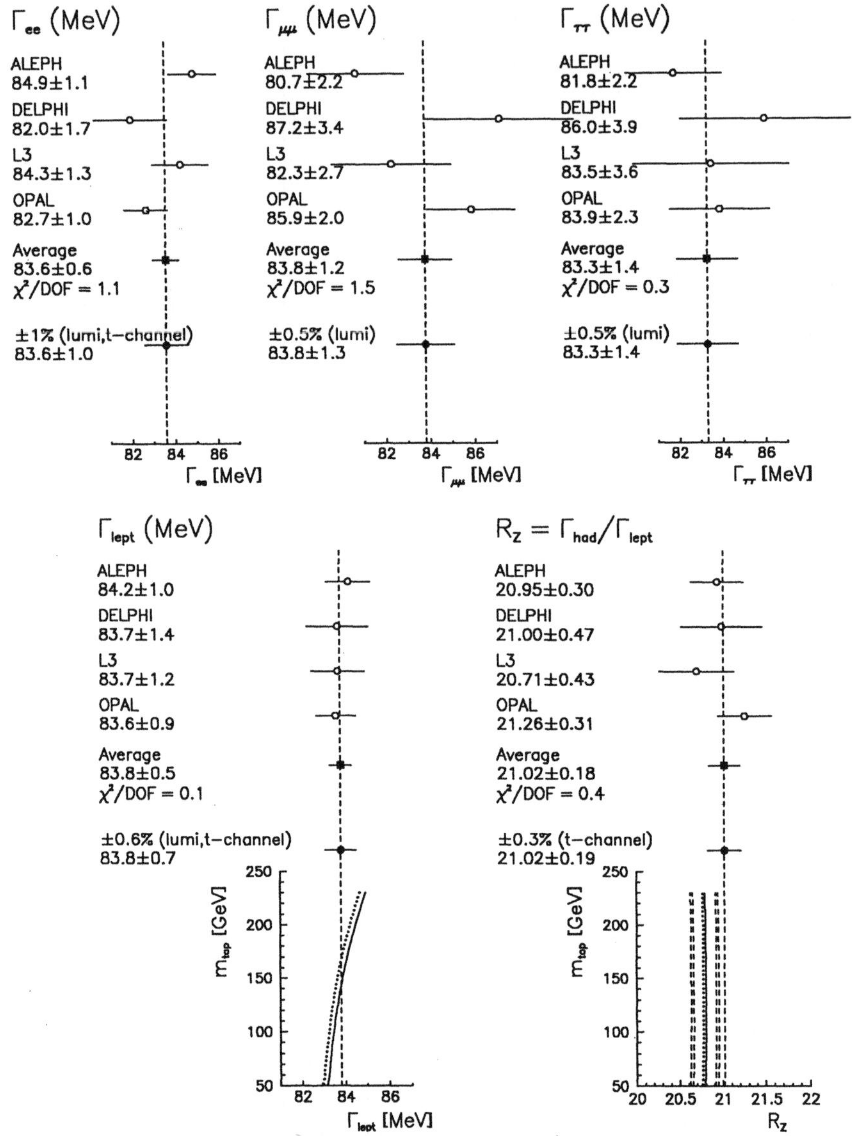

Figure 4: Leptonic widths and the ratio of the hadronic to the leptonic width R_Z.

LEP experiments recently agreed to quote the error as ±0.02 GeV, i.e. quote only one significant digit in the forthcoming publications. As mentioned before it is hoped that in 1991 transverse polarization will be used to measure the absolute energy and the systematic error will be reduced to below 10 MeV level in E_{cm}.

Figure 4 shows the lepton partial widths Γ_{ee}, $\Gamma_{\mu\mu}$, $\Gamma_{\tau\tau}$ and $\Gamma_{l^+l^-}$, where for the $\Gamma_{l^+l^-}$ the lepton universality $\Gamma_{l^+l^-} = \Gamma_{ee} = \Gamma_{\mu\mu} = \Gamma_{\tau\tau}$ is assumed.

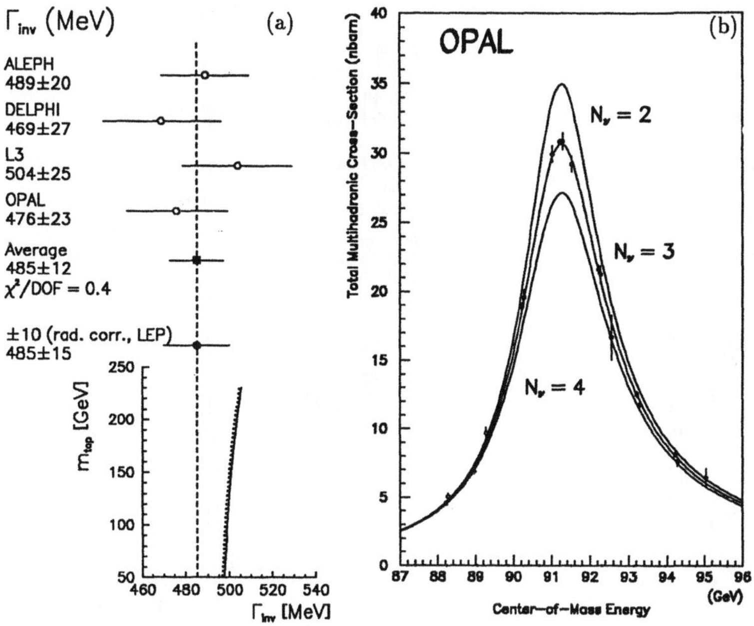

Figure 5: (a) The invisible width $\Gamma_{\rm inv}$
(b) Cross sections for $e^+e^- \to$ hadrons as a function of center-of-mass energy around Z^0 pole measured by OPAL. The open circles show the 1989 data and the full circles show the 1990 data. The curves show the standard model predictions for $N_\nu = 2$, 3 and 4.

Also shown in the figure is the ratio of the hadronic and leptonic widths $R_Z \equiv \Gamma_{\rm had}/\Gamma_{l^+l^-}$. The theoretical prediction for the ratio is much less dependent on the unknown top quark mass. It has also the advantage that the experimental systematic errors such as that of absolute luminosity largely cancel.

Figure 5 (a) shows the Z^0 invisible width $\Gamma_{\rm inv}$, defined as

$$\Gamma_{\rm inv} = \Gamma_Z - (\Gamma_{\rm had} + 3\Gamma_{l^+l^-})\,.$$

Assuming that $\Gamma_{\rm inv}$ can be entirely attributed to neutrinos and that the partial widths of different kinds of neutrinos are the same, the $\Gamma_{\rm inv}$ can be written as

$$\Gamma_{\rm inv} = N_\nu \cdot \Gamma_{\nu\nu},$$

where N_ν is the number of neutrino families to which the Z^0 couples. The N_ν can be expressed also as

$$N_\nu = \frac{\Gamma_{\rm inv}}{\Gamma_{\nu\nu}} = \frac{\Gamma_{l^+l^-}}{\Gamma_{\nu\nu}}\left[\sqrt{\frac{12\pi R_Z}{\sigma^{\rm pole}_{\rm had} M_Z^2}} - R_Z - 3\right]\,.$$

The formula has the advantage that the theoretical predictions of the ratios $\Gamma_{l^+l^-}/\Gamma_{\nu\nu}$ and $R_Z = \Gamma_{\rm had}/\Gamma_{l^+l^-}$ are much less dependent on the top and Higgs boson masses. Using the standard model prediction $\Gamma_{l^+l^-}/\Gamma_{\nu\nu} = 0.5010 \pm 0.0005$ and the LEP average values of R_Z and $\sigma^{\rm pole}_{\rm had}$, the following result has been obtained

$$N_\nu = 2.90 \pm 0.10$$

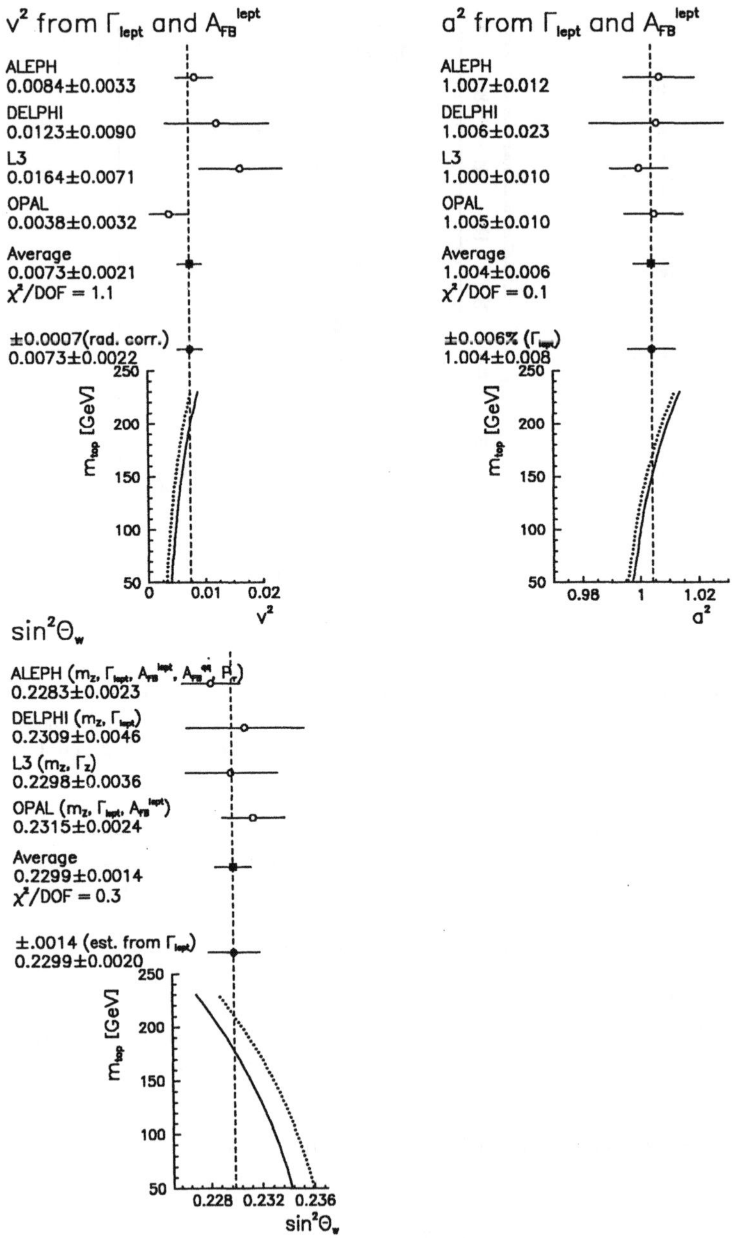

Figure 6: Squares of the effective vector and axial vector coupling constants, $\hat{v}_l^2$ and $\hat{a}_l^2$, and the effective weak mixing angle $\sin^2\overline{\theta}_W$.

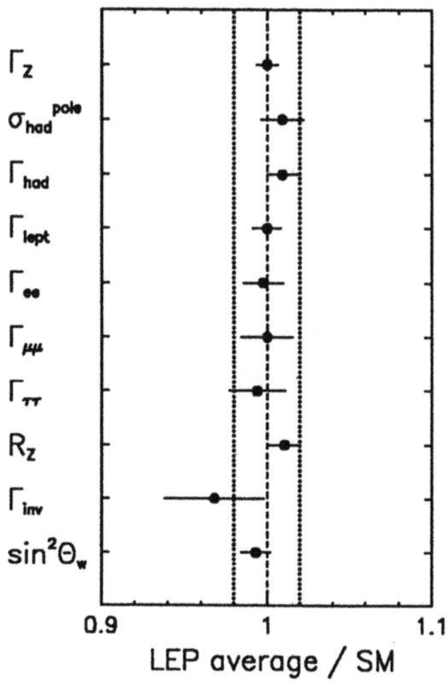

Figure 7: Ratios of the LEP average data to the standard model predictions.

for the number of light neutrino families, an important result which also constrains various dark matter candidates. Figure 5 (b) [22] demonstrates the sensitivity of the measurement for excluding possibilities of $N_\nu \neq 3$.

Figure 6 shows the squares of the effective vector and axial vector coupling constants, $\hat{v}_l^2$ and $\hat{a}_l^2$, and the effective weak mixing angle $\sin^2\overline{\theta}_W$, extracted from cross section and asymmetry measurements.

To summarize, figure 7 shows the comparisons between the average data and the standard model predictions, where in the latter the m_t, m_H and α_s are fixed as follows:

$$m_t = 150 \text{ GeV}, m_H = 100 \text{ GeV}, \alpha_s = 0.12 .$$

As you see, LEP experiments have already tested the standard model to ≤ 1 % level in many observables.

The various measurements of Z^0 parameters can be used to constrain the mass of the yet undiscovered top quark, if the results are interpreted within the minimal standard model. ALEPH quoted [5]:

$$m_t = 231 \pm 57 \pm 3_{M_Z} \pm 1_{\alpha_s} \pm 20_{m_H} \text{ GeV},$$

where the subscripts refer to the quantities of the origin of the uncertainties. OPAL showed [8] the following limit:

$$m_t = 154^{+55} \text{ GeV (no lower limit)},$$

where an uncertainty in α_s of ± 0.02 and a fixed Higgs mass of 100 GeV are assumed. The results of LEP experiments can be combined with the measurements of the ratio of W and Z^0 masses, M_W/M_Z, from hadron collider experiments CDF [23] and UA2 [24], and neutrino experiments CDHS [25] and CHARM [26] to obtain more stringent limits (see figure 8 from ref.[27]). The following mass limit was quoted as the combined result at the Singapore Conference [9]:

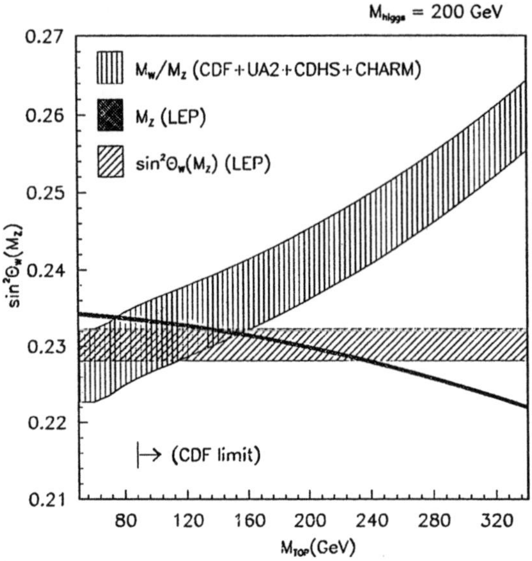

Figure 8: Regions in the $\sin^2\theta_W(M_Z^2)$ ($= \sin^2\overline{\theta}_W$ in the current context) versus the top mass constrained by measurements of M_Z, lepton-pair cross sections and M_W/M_Z. The Higgs boson mass of 200 GeV is assumed.

$$\begin{aligned} m_t &= 137 \pm 33 \pm 3_{M_Z} \pm 20_{m_H} \text{ GeV} \\ &= 137 \pm 40 \text{ GeV} \ . \end{aligned}$$

At present the contribution from LEP to tightening the combined limit mainly comes from the precise measurement of the M_Z. With much more statistics in the future the measurements of the line shapes and asymmetries will be a dominating factor for the top mass limit.

2.2 Quark partial widths

Not only the couplings of leptons but also the couplings of quarks to Z^0 have been studied by LEP experiments. Primary quarks of particular flavours have to be identified in multi-hadronic events. For tagging b quarks the leptons from semi-leptonic b decays are most often used. Expected signatures of the prompt leptons are high momentum and high transverse momentum with respect to the jet axis because of the hard fragmentation and large mass of the b quark. The same is true, though to a less extent, for the leptons from c decays. An example of the p_T distribution is shown in fig.9. The data at high p_T are dominated by $b \to \mu$ decays.

Based on the studies of prompt leptons, ALEPH [28], L3 [29] and OPAL [30] quote the following values on $\Gamma_{b\bar{b}}/\Gamma_{\rm had}$ before correcting for the branching ratio for the $b \to \mu$ decay:

		$\mathrm{BR}(b \to l\nu X) \cdot \Gamma_{b\bar{b}}/\Gamma_{\rm had}$
ALEPH	e	$0.0217 \pm 0.0019(\text{stat.}) \pm 0.0010(\text{syst.})$
ALEPH	μ	$0.0238 \pm 0.0028(\text{stat.}) \pm 0.0012(\text{syst.})$
L3	μ	$0.0248 \pm 0.0014(\text{stat.} + \text{syst.})$
OPAL	μ	$0.0206 \pm 0.0010(\text{stat.}) \pm 0.0018(\text{syst.})$

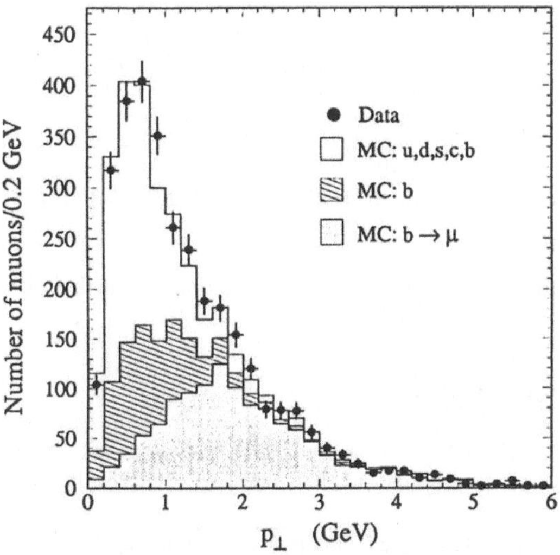

Figure 9: The measured distribution of the muon transverse momentum p_T with respect to the nearest jet. A cut on the momentum $p > 4\text{GeV}$ has been applied. The contributions of the various processes are indicated as expectations from Monte Carlo study. Muons from b quarks can originate also from the processes like $b \to c \to \mu$ and $b \to \tau \to \mu$.

ALEPH fitted the full p and p_T spectra of electrons and determined $\Gamma_{b\bar{b}}$ and $\Gamma_{c\bar{c}}$ simultaneously:

ALEPH	e	$\mathrm{BR}(b \to l\nu X) \cdot \Gamma_{b\bar{b}}/\Gamma_{\mathrm{had}} = 0.0219 \pm 0.0017(\mathrm{stat.}) \pm 0.0010(\mathrm{syst.})$
ALEPH	e	$\mathrm{BR}(c \to l\nu X) \cdot \Gamma_{c\bar{c}}/\Gamma_{\mathrm{had}} = 0.0133 \pm 0.0040(\mathrm{stat.})\ ^{+0.0038}_{-0.0031}(\mathrm{syst.})$

DELPHI have used [31] a different technique based on the "boosted sphericity product" [32] for separating the fraction of $b\bar{b}$ decays in the two jet events. It utilizes the fact that the jets induced by b quarks tend to be fatter due to the large rest mass of the B hadrons. Each of the two jets is boosted along the sphericity axis towards its hypothetical B hadron rest frame and the sphericity S is calculated in each hemisphere in these boosted frames. The distribution of the boosted sphericity product $S_1 \times S_2$ was used to derive the fraction of $b\bar{b}$ decays.

The following table summarizes the results of $\Gamma_{b\bar{b}}/\Gamma_{\mathrm{had}}$ with the values of the branching ratio $b \to l$ used:

		$\mathrm{BR}(b \to l\nu X)$	$\Gamma_{b\bar{b}}/\Gamma_{\mathrm{had}}$
ALEPH	e	0.102 ± 0.010	$0.215 \pm 0.017(\mathrm{stat.}) \pm 0.024(\mathrm{syst.} + \mathrm{br.})$
DELPHI	$S_1 \times S_2$	—	$0.209 \pm 0.030(\mathrm{stat.}) \pm 0.031(\mathrm{syst.})$
L3	μ	0.118 ± 0.011	$0.210 \pm 0.012(\mathrm{stat.} + \mathrm{syst.}) \pm 0.019(\mathrm{br.})$
OPAL	μ	0.10 ± 0.01	$0.206 \pm 0.010(\mathrm{stat.}) \pm 0.018(\mathrm{syst.}) \pm 0.020(\mathrm{br.})$

All values are in good agreement with the standard model prediction of 0.217.

DELPHI presented a result [33] on $\Gamma_{c\bar{c}}/\Gamma_{\mathrm{had}}$ derived from the observation of the pion coming from the decay $D^{*+} \to D^0\pi^+$. The charged pion in this decay is almost at rest

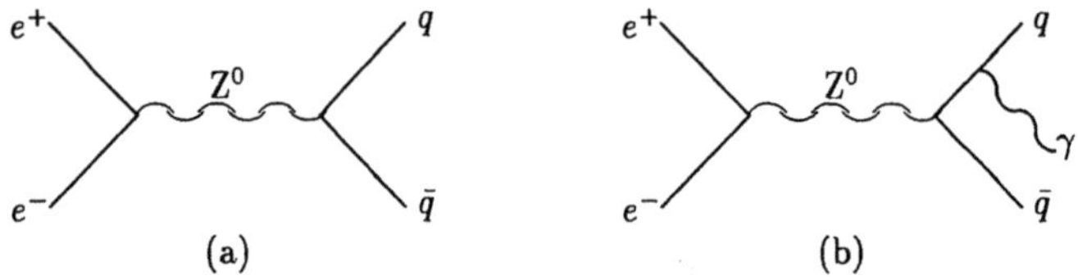

Figure 10: Z^0 decays into a quark pair (b) with or (a) without a final state radiation.

in the D^* rest frame, hence it has a low p_T relative to the jet direction. By fitting the p_T^2 spectrum, DELPHI derived

$$\boxed{\text{DELPHI} \quad \Gamma_{c\bar{c}}/\Gamma_{\text{had}} = 0.162 \pm 0.030(\text{stat.}) \pm 0.050(\text{syst.}),}$$

assuming the probability $c\bar{c} \to D^{*\pm} + X$, $D^{*+} \to D^0\pi^+$ is 0.31 ± 0.05 as measured by CLEO [34]. A similar analysis was reported also by ALEPH [28] and their result is:

$$\boxed{\text{ALEPH} \quad \text{BR}(c \to \pi D^0) \cdot \Gamma_{c\bar{c}}/\Gamma_{\text{had}} = 0.0290 \pm 0.0035(\text{stat.}) \pm 0.0023(\text{syst.}).}$$

Using $\text{BR}(c \to l\nu\text{X}) = 9.0 \pm 1.3\%$, their result on $\text{BR}(c \to l\nu\text{X}) \cdot \Gamma_{c\bar{c}}/\Gamma_{\text{had}}$ mentioned before was expressed as

$$\boxed{\text{ALEPH} \quad \Gamma_{c\bar{c}}/\Gamma_{\text{had}} = 0.148 \pm 0.044(\text{stat.}) \,^{+0.045}_{-0.038}(\text{syst.})\,,}$$

in agreement with the standard model value of 0.171. As for L3, they quoted [29]:

$$\boxed{\text{L3} \quad \Gamma_{c\bar{c}} = 221 \,^{+48}_{-42} \text{ MeV}}$$

derived from a fit to the p and p_T distributions of muons with $\Gamma_{c\bar{c}}$ and $\Gamma_{b\bar{b}}$ as free parameters.

OPAL measured the partial widths of "u"-type quarks (u and c: charge $\frac{2}{3}e$) and "d"-type quarks (d, s and b: charge $\frac{1}{3}e$) using multi-hadronic events with final state photon radiations [35]. The principle of the method is that the rate for the process of figure 10 (a) is proportional to $c_q \equiv a_q^2 + v_q^2$, while the rate for the reaction of figure 10 (b) is $\propto Q_q^2 c_q$, where Q_q is the electric charge of the quark.

Assuming the universality: $c_u = c_c$ and $c_d = c_s = c_b$, the above argument leads to the following two relations:

$$\Gamma_{\text{had}} \quad \propto \quad \sum_q c_q = 3\,c_d + 2\,c_u \,,$$

$$\Gamma_{Z^0 \to q\bar{q}\gamma} \quad \propto \quad \sum_q Q_q^2 c_q = \frac{e^2}{9}(3\,c_d + 8\,c_u)\,.$$

By measuring the two quantities, Γ_{had} and number of $q\bar{q}\gamma$ events, one can solve for c_d and c_u, or equivalently for the partial widths $\Gamma_{u\bar{u}}$ and $\Gamma_{d\bar{d}}$ using the relation:

$$\Gamma_{q\bar{q}} = \frac{c_q}{3c_d + 2c_u} \cdot \Gamma_{\text{had}}.$$

Looking for isolated energetic (> 10GeV) photons in about 77000 multi-hadronic events, OPAL found 78 candidate events and derived the partial widths, using $\Gamma_{\text{had}} = 1778 \pm$

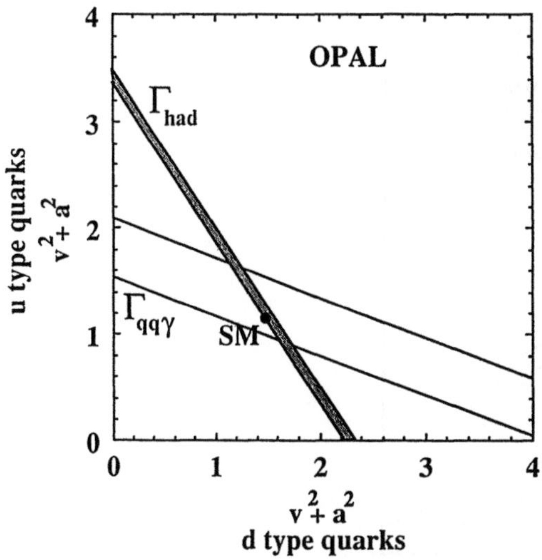

Figure 11: Correlation of the couplings for charge 1/3 and 2/3 quarks as obtained from the total hadronic width and from the analysis of the isolated photons. Also shown is the expectation from the standard model (black point).

26 MeV (OPAL [8]):

OPAL	$\Gamma_{d\bar{d}} = 369 \pm 67$ MeV
	$\Gamma_{u\bar{u}} = 330 \pm 99$ MeV

in agreement with the standard model predictions (see figure 11).

2.3 $b\bar{b}$ asymmetry and $B^0\overline{B^0}$ mixing

Forward-backward asymmetry of $Z^0 \rightarrow b\bar{b}$ decays has been studied by ALEPH [28], L3 [36] and OPAL [30]. From the charge of the high momentum, high p_T leptons, L3 and OPAL derived the following *observed* asymmetries (not corrected for $B^0\overline{B^0}$ mixing effect):

L3	$A_{FB}^{b\bar{b}} = 0.084 \pm 0.033$	
OPAL	$A_{FB}^{b\bar{b}} = 0.01 \pm 0.08$	($b \rightarrow \mu$ only)

Due to mixing in the $B^0\overline{B^0}$ system, some b quarks will turn into $\bar{b}$ antiquarks (and vice versa) before the decay. The observed asymmetry is therefore smaller than the actual asymmetry by a factor of $(1 - 2\chi)$, where χ is the probability that a hadron containing a b quark has oscillated into a hadron containing $\bar{b}$ at the time of its decay. By studying the numbers of like-sign and opposite-sign dilepton events, ALEPH [28] and L3 [36] determined the χ:

ALEPH	$\chi = 0.129 \, ^{+0.048}_{-0.044}$	
L3	$\chi = 0.11 \, ^{+0.08}_{-0.06}$	(from dimuon events only)

Only B_d and B_s mesons are subject to the mixing. The χ can be expressed as

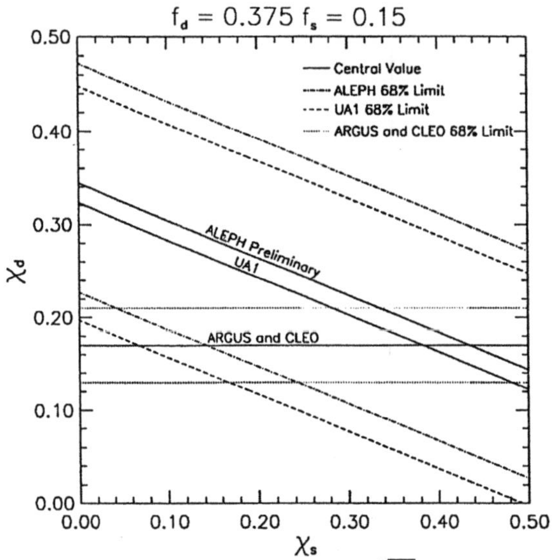

Figure 12: Comparison of ALEPH $B^0\overline{B^0}$ mixing result with those from UA1, ARGUS and CLEO.

$$\chi = f_d \chi_d + f_s \chi_s ,$$

where f_d and f_s are the fractions of B_d and B_s mesons produced, and χ_d and χ_s are the mixing parameters for the respective mesons. The result of ALEPH is shown in the fig. 12 assuming $f_d = 0.375$ and $f_s = 0.15$. Also shown in the figure are the results of UA1 [37], ARGUS [38] and CLEO [39]. The latter two are based on the study at $\Upsilon(4S)$ where no B_s are produced.

After the correction for $B^0\overline{B^0}$ mixing ALEPH and L3 quoted values for $A_{\rm FB}^{b\bar{b}}$:

ALEPH	$A_{\rm FB}^{b\bar{b}} = 0.181 \pm 0.073(\text{stat.}) \pm 0.059(\text{syst.})$
L3	$A_{\rm FB}^{b\bar{b}} = 0.109 \pm 0.044(\text{stat.})$

which agree with the standard model prediction of $\sim +10\%$.

2.4 τ polarization

The polarization of the τ lepton in $\tau^+\tau^-$ final states is expressed as

$$P_\tau = -\frac{2v_\tau a_\tau}{v_\tau^2 + a_\tau^2}$$

in the lowest order when averaged over $\cos\theta$. Compared to the forward-backward asymmetry $A_{\rm FB}^{ll} \propto \frac{v_l^2 a_l^2}{(v_l^2+a_l^2)^2}$, the P_τ has a particular advantage that it is linear in the small parameter v. The polarization is measured by using the momentum distributions of τ decay products. ALEPH measured [5] [40] all analyzable modes. The distribution of normalized pion energy from $\tau \to \pi\nu_\tau$ decay mode is shown in fig. 13. The following summarizes their results:

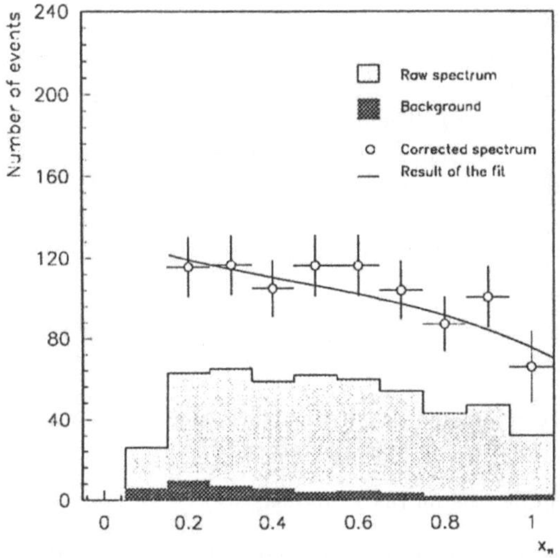

Figure 13: Spectrum of normalized pion energy E_π/E_τ from $\tau \to \pi\nu_\tau$ decays measured by ALEPH. After corrections for background and efficiency, the data is fitted to the predicted linear dependence plus radiative effects.

ALEPH				stat.		syst.
$e\nu\bar{\nu}$	:	-0.24	$\pm$	0.15	$\pm$	0.07
$\mu\nu\bar{\nu}$	:	-0.14	$\pm$	0.14	$\pm$	0.05
$\pi\nu$	:	-0.14	$\pm$	0.09	$\pm$	0.06
$\rho\nu$	:	-0.17	$\pm$	0.06	$\pm$	0.05
$a_1\nu$	:	-0.02	$\pm$	0.22	$\pm$	0.07

Combining all the results, they quote

$$P_\tau = -0.157 \pm 0.055,$$

where the correlation of the systematic errors between the various modes is not taken into account. From this they obtained:

$$\begin{aligned} v_\tau/a_\tau &= 0.079 \pm 0.028, \\ \sin^2\overline{\theta}_W &= 0.231 \pm 0.007. \end{aligned}$$

Note that the relative sign of v_τ and a_τ is thus determined, which is not possible from the lepton forward-backward asymmetries.

DELPHI gave me very preliminary results [41]:

DELPHI				stat.
$\mu\nu\bar{\nu}$	:	-0.165	$\pm$	0.170
$\pi\nu$	:	-0.222	$\pm$	0.185

where systematic errors are not yet evaluated.

3 Tests of QCD at LEP

The e^+e^- annihilation into hadrons has been providing a wonderful laboratory to test QCD at PETRA, PEP and TRISTAN energies. The initial state is quite simple compared to deep inelastic scattering or hadron-hadron collisions, where one has to deal with long distance properties of the theory like the structure functions.

The methodology of the QCD studies at LEP is similar to those at the lower energy e^+e^- machines. There are, however, particular advantages of studying QCD at LEP energies:

- We have large statistics from $Z^0 \to q\bar{q}$ decays with negligible background from other processes.
- At higher energies jets are more collimated and better reflect the kinematics of the primary partons, leading to smaller hadronization effects.
- Hard initial state radiation is suppressed on the Z^0 resonance.

Thus LEP is ideal for testing the quantities calculated with perturbative methods, that is, for determining the strong coupling constant α_s.

Only three topics are covered by my talk, namely,

- Determination of α_s from jet multiplicities
- Determination of α_s from energy-energy correlation (EEC)
- Test of triple gluon vertex

The last topic has been chosen because the test particularly benefits from the large statistics at LEP.

There are many other topics which are not covered by my talk. The following table summarizes the status of various QCD related results so far published or presented at various conferences. The result marked with 'prel.' was presented as preliminary at some conferences but is not yet committed to publication. There are still other on-going activities for topics not included in this table.

Topic	ALEPH	DELPHI	L3	OPAL
comparison of MC's, event shapes	√	√	√	√
α_s from jet multiplicities	√	√	√	√
α_s from EEC and/or AEEC	prel.	√		√
triple gluon vertex		√	√	√
gluon coherence effects				√
intermittency		√		
charged multiplicity distribution	√	√	√	√

3.1 Determination of α_s from Jet Multiplicities

At large energies, collimated jets of hadrons produced in e^+e^- annihilation reflect the underlying kinematics of the primary quarks and gluons (partons) and can be well described by QCD perturbation theory. The relative production rate of two, three or more hard partons is predicted by the perturbative QCD as a function of the parameter Λ

which determines the strong coupling constant α_s at a given energy scale μ^2. According to the concept of the asymptotic freedom of QCD, the α_s is expected to decrease logarithmically with increasing energy. Experimental studies of multijet production rates are well suited to test this necessary consequence of QCD and to determine the free parameter of the theory.

To compare to the predictions of perturbative QCD, observed particles in an event are combined into *jets* which are expected to represent the original parton structure of the event. For the jet finding algorithm the one introduced by JADE [42] is the most commonly used. The algorithm works as follows:

- Calculate the scaled invariant mass squared $y_{ij} = M_{ij}^2/E_{\rm vis}^2$, for all pairs of particles i and j of the event, where $M_{ij}^2 = 2E_iE_j(1-\cos\theta_{ij})$, $E_{\rm vis}$ is the visible energy of the event. Charged particles are usually assumed to be pions and neutrals to be photons.
- Find two particles i and j for which y_{ij} is the smallest.
- Combine them into a new pseudoparticle or 'cluster' of four-momentum $p_k = p_i + p_j$.
- Repeat the procedure until all y_{ij} exceed a certain cutoff $y_{\rm cut}$.
- Call the remaining 'clusters' *jets*.

Thus the number of resolvable jets is a function of the $y_{\rm cut}$.

Corrections are necessary for limited detector acceptance and resolution, initial-state photon radiation, and hadronization effect. The correction factors are obtained from two samples of Monte Carlo events: The first sample consists of the final partons before fragmentation process starts. The second sample includes hadronization, initial state photon radiation, simulation of the detector, and the same event reconstruction and selection as applied to the real data. Jet rates are calculated for each sample and the ratios of the two are taken to be the correction factors. A more sophisticated correction procedure is possible, e.g. [43] correction factors for detector effects and initial state photon, plus a matrix multiplication which takes into account the migration of events between different jet multiplicities during hadronization. For the Monte Carlo, QCD parton shower models (JETSET [44] and/or HERWIG [45]) are used which are known to describe well various event shape distributions [46] and also have been checked to reproduce the measured jet rates. However, they are based on a leading-log approximation and therefore cannot be used for extracting α_s.

The relative jet rates R_2, R_3 and R_4 are given by $O(\alpha_s^2)$ QCD calculations as follows:

$$\begin{aligned} R_2 &\equiv \frac{\sigma_2}{\sigma_{tot}} = 1 + C_{2,1}(y_{cut})\cdot\alpha_s(\mu^2) + C_{2,2}(y_{cut}, f)\cdot\alpha_s^2(\mu^2), \\ R_3 &\equiv \frac{\sigma_3}{\sigma_{tot}} = C_{3,1}(y_{cut})\cdot\alpha_s(\mu^2) + C_{3,2}(y_{cut}, f)\cdot\alpha_s^2(\mu^2), \\ R_4 &\equiv \frac{\sigma_4}{\sigma_{tot}} = C_{4,2}(y_{cut})\cdot\alpha_s^2(\mu^2), \end{aligned} \tag{1}$$

where σ_{tot} is the total hadronic cross section, σ_n are the cross sections for n-parton event production. Calculations of the coefficients $C_{n,k}$ in complete second order perturbation theory are available from Kramer and Lampe [47][48] and from Kunszt and Nason [49]; the latter are based on the original calculations of Ellis, Ross and Terrano [50]. The coefficients depend on the value of the jet resolution parameter $y_{\rm cut}$.

The coupling constant α_s can be written as a function of $\ln(\mu^2/\Lambda^2)$, where Λ is the QCD scale parameter which is to be determined by experiment and μ is the QCD renormalization scale. Here we use the functional form of α_s given in [51] to relate α_s and $\Lambda_{\overline{MS}}$ (the $\overline{\text{MS}}$ denotes the modified minimal-subtraction scheme):

$$\alpha_s(\mu^2) = \frac{12\pi}{(33-2N_f)\cdot\ln(\frac{\mu}{\Lambda_{\overline{MS}}})^2}\cdot\left(1-6\cdot\frac{153-19N_f}{(33-2N_f)^2}\cdot\frac{\ln(\ln(\frac{\mu}{\Lambda_{\overline{MS}}})^2)}{\ln(\frac{\mu}{\Lambda_{\overline{MS}}})^2}\right), \tag{2}$$

where the number of active quark flavours N_f is taken to be 5.

In e^+e^- annihilation, the renormalization scale μ was conventionally chosen to be the center-of-mass energy ($\mu^2 = s = E_{\text{cm}}{}^2$). Recent studies have shown, however, that a more general definition like $\mu^2 = f\cdot s$ with f =0.001 to 0.01 results in a better overall description of the observed jet rates [48][52]. The next-to-leading order coefficients like $C_{2,2}$ and $C_{3,2}$ exhibit an explicit dependence on the scale factor f [48]. With this additional occurrence of the renormalization scale, in addition to the term $\mu^2/\Lambda^2_{\overline{MS}}$ in (2), it is possible to optimize both μ^2 and the scale parameter $\Lambda_{\overline{MS}}$ at the same time. In leading order only, a change of μ^2 could always be absorbed by readjusting $\Lambda_{\overline{MS}}$. Note that the energy scale μ is not a physical parameter and theoretical results, if calculated to all orders in perturbation theory, should not depend on it. Finite order calculations, however, may depend on the choice of μ. The optimization of f can be regarded as a means to minimize the contributions from missing higher order terms. LEP experiments determine the value of $\Lambda_{\overline{MS}}$ first with the fixed scale factor $f = 1$ and derive the value of $\alpha_s(M_Z^2)$ using the formula (2). Next, f is allowed to vary in a fit or some small value of f is chosen to see how the resulting $\Lambda_{\overline{MS}}$ and thus $\alpha_s(M_Z^2)$ change. The change coming from this "(renormalization) scale uncertainty" can be regarded as a measure of ambiguities resulting from the missing higher order terms.

For combining unresolved partons into resolved ones, one uses algorithms based on an invariant mass cutoff as explained on the JADE jet algorithm. There is, however, a certain amount of freedom in the prescription for combining two unresolvable jets into a single jet. The calculation of the coefficients $C_{n,k}$ in (1) depends on the specific choice of the scheme and this is called "recombination scheme uncertainty". The ambiguity arises because the $O(\alpha_s^2)$ QCD calculations are performed for massless partons, whereas the effective mass of a jet formed by adding the four-momenta of two previously unresolved partons is non-zero. There are four most common schemes, called "E", "E0", "p", and "p0" schemes (see e.g. [49]). Their differences are:

	y_{ij}^2	p_k: 4-momentum of the combined jets
E	$(p_i+p_j)^2/E_{\text{cm}}^2$	p_i+p_j
E0	$(p_i+p_j)^2/E_{\text{cm}}^2$	E_i+E_j ; $\vec{p}_k = E_k\cdot(\vec{p}_i+\vec{p}_j)/\lvert\vec{p}_i+\vec{p}_j\rvert$
p	$(p_i+p_j)^2/E_{\text{cm}}^2$	$\vec{p}_k=\vec{p}_i+\vec{p}_j$; $E_k=\lvert\vec{p}_k\rvert$
p0	$(p_i+p_j)^2/E_{\text{vis}}^2$	$\vec{p}_k=\vec{p}_i+\vec{p}_j$; $E_k=\lvert\vec{p}_k\rvert$

where the E_{vis} in the p0 scheme is the actual total energy sum of the event recalculated after each recombination. The E0- and JADE-scheme are equivalent in $O(\alpha_s^2)$ calculations, where at maximum 4 partons exist. When applied to hadronic final states, the two schemes still yield almost identical results as far as the number of resolvable jets is concerned.

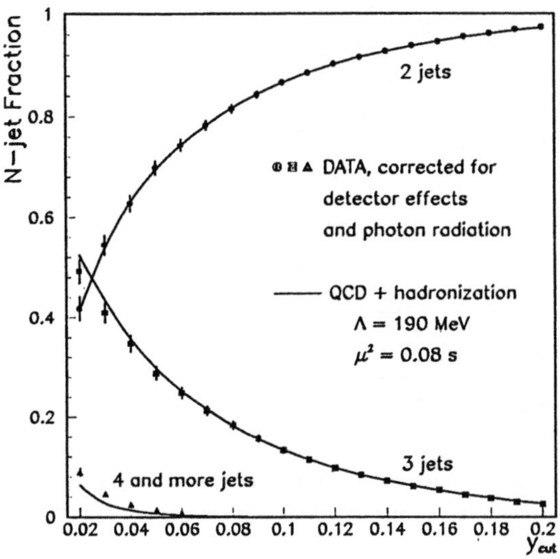

Figure 14: Comparison of corrected measured jet fractions and predicted jet rates in second order QCD for $\mu^2/s = 0.08$ and $\Lambda_{\overline{MS}} = 190$ MeV. The correction for hadronization is applied to the theoretical predictions, and not to the data.

ALEPH used the E0 algorithm and derived [53]:

$$\text{ALEPH} \quad \alpha_s(M_Z^2) = 0.121 \pm 0.002(\text{stat.}) \pm 0.003(\text{syst.}) \pm 0.007(\text{theor.})^{+0.007}_{-0.012}(\text{scale.}) \quad f = 0.25$$

where the theoretical error is due to perturbative higher orders and hadronization effects. The scale uncertainty was derived by changing μ between M_Z and the b-quark mass.

DELPHI's result [43] is:

$$\text{DELPHI} \quad \alpha_s(M_Z^2) = 0.114 \pm 0.003(\text{stat.}) \pm 0.004(\text{syst.}) \pm 0.012(\text{theor.}) \quad f = 1.$$

The theoretical error includes both the renormalization scale uncertainty which was derived from the α_s values for $f = 1$ and $f = 0.001$, and the renormalization scheme uncertainty from the different values of α_s when using E-, E0-, and p-schemes for the theoretical calculations.

Figure 14 shows the jet fractions measured by L3 Collaboration [54]. From the values at $y_{\text{cut}} = 0.08$ they determined:

$$\text{L3} \quad \alpha_s(M_Z^2) = 0.115 \pm 0.005(\text{exp.})^{+0.012}_{-0.010}(\text{theor.}) \quad f = 0.08,$$

and the theoretical expectations using this value are also shown in the figure. Their theoretical error includes the scale uncertainty which was derived from the α_s values for $f = 1$ and $f = 0.001$, and the renormalization scheme uncertainty from the different values of α_s when using E- and E0-schemes for the theoretical calculations. In the theoretical error they also include the uncertainty coming from fragmentation parameters.

OPAL systematically studied the recombination scheme dependence [55]. They used the four different schemes and applied them in a consistent manner for both the ex-

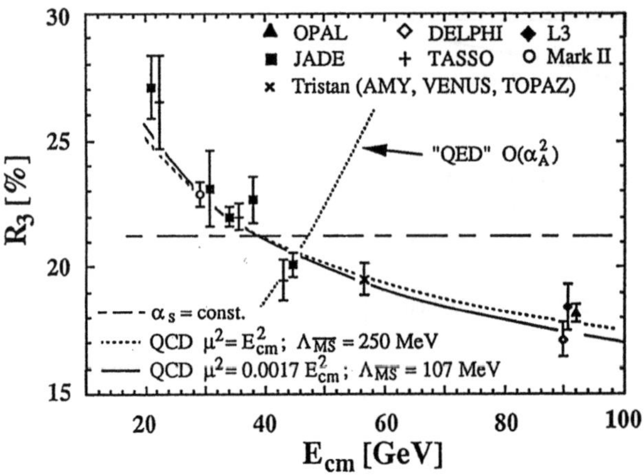

Figure 15: The energy dependence of the 3-jet event production rates at $y_{\rm cut} = 0.08$, compared with several assumptions about the energy dependence of α_s. The errors shown are statistical only.

perimental jet analysis and the corresponding analytic $O(\alpha_s^2)$ calculations. Within each scheme, hadronization corrections and their systematic uncertainties are evaluated. The corrections are found to be the smallest for the E0-scheme, and the largest for the E-scheme. Its systematic error is also the smallest for E0-scheme and for these reasons the E0-scheme is preferred in experimental studies. For each recombination scheme the $\Lambda_{\overline{MS}}$ is determined both for fixed renormalization scale $f = 1$ and for the case where the f is also treated as a free parameter in the fit. The values of fitted $\Lambda_{\overline{MS}}$ are much closer to each other with the optimized scales than with $f = 1$. The resulting optimized values of f are significantly different, ranging from the smallest $f = 0.00006 \pm 0.00001$ for E-scheme to the largest $f = 0.19^{+1.18}_{-0.19}$ for the p-scheme. As the measured value of $\alpha_s(M_Z^2)$, they took the arithmetic mean of the two $\alpha_s(M_Z^2)$ values for $f = 1$ and the optimized f. The half of the difference was taken to be the scale uncertainty. Their final results are the following:

OPAL E	$\alpha_s(M_Z^2) = 0.126 \pm 0.014$(total)
OPAL E0	$\alpha_s(M_Z^2) = 0.118 \pm 0.009$(total)
OPAL p	$\alpha_s(M_Z^2) = 0.118 \pm 0.008$(total)
OPAL p0	$\alpha_s(M_Z^2) = 0.118 \pm 0.008$(total).

The results are consistent with each other within their errors and they conclude that with a consistent treatment of both data and analytic calculations there remains no recombination scheme uncertainty.

The results on α_s from different collaborations agree well within their errors. The current size of errors is, however, mainly dominated by the theoretical errors coming from the renormalization scale uncertainty. Therefore higher order QCD calculations are necessary to further improve the precision of the determination of $\alpha_s(M_Z^2)$, or we should determine $\alpha_s(M_Z^2)$ by methods less sensitive to the missing higher order terms.

To see the energy dependence of α_s, the 3-jet fractions for $y_{\rm cut} = 0.08$ measured in e^+e^- annihilation [42][56][43][54][55] are plotted in fig. 15 [57] as a function of the center-of-mass energy. The possibility of an energy independent α_s can be ruled out, that is, α_s is really running as a manifestation of the non-abelian nature of QCD.

3.2 α_s from Energy-Energy Correlation

Energy-energy correlation (EEC) of hadronic events and its asymmetry (AEEC) were first introduced by Basham et al. [58] as experimental observables sensitive to the value of the strong coupling constant α_s. Experimentally the EEC is defined as the histogram of the angles between all combinations of pairs of particles in hadronic events, weighted with their energies, and averaged over all N events:

$$EEC(\chi) = \frac{2}{\Delta\chi \cdot N} \int_{\chi-\frac{\Delta\chi}{2}}^{\chi+\frac{\Delta\chi}{2}} \sum_{events}^{N} \sum_{i}^{N_{par}-1} \sum_{j>i}^{N_{par}} \frac{E_i E_j}{E_{vis}^2} \delta(\chi' - \chi_{ij}) d\chi',$$

where χ_{ij} is the angle between particles i and j, $\Delta\chi$ is the width of the histogram bin, E_i is the energy of particle i, N_{par} is the number of particles in the event, and the weights are normalized to the visible energy $E_{vis} = \sum_{i=1}^{N_{par}} E_i$.

Two-jet events yield a distribution with two peaks, one near $\chi = 0°$ corresponding to the angles between pairs of particles inside a jet, the other near $\chi = 180°$ corresponding to the angles between particles in opposite jets. On the other hand, events with hard gluon radiation contribute to the central region in an asymmetric fashion. This can be understood easily by considering a typical $q\bar{q}g$ event at the parton level, where two large angles and one small angle exist between the three partons. The shape of the EEC distribution is therefore correlated with the value of α_s.

The asymmetry in the energy-energy correlation (AEEC)

$$AEEC(\chi) = EEC(\pi - \chi) - EEC(\chi) \quad (0° < \chi \leq 90°)$$

removes contributions from the two-jet events and is particularly sensitive to α_s. It is insensitive to the tuning of the fragmentation parameters which only change the EEC distribution in a symmetric way. Also it has a smaller second order QCD correction than EEC.

Note that unlike the study of jet multiplicites, the EEC and the AEEC are not event-by-event analyses. Many experiments have studied the EEC and AEEC and extracted α_s at lower center-of-mass energies in e^+e^- annihilations [59]. It is expected that second order perturbative calculations yield more reliable results at Z^0 energies than at lower energies because of the smaller value of α_s. Also hadronization corrections are smaller at higher energies, resulting in less uncertainty in the comparison of theoretical calculations with data.

For comparison between data and theoretical predictions corrections are necessary for detector acceptance, detector resolution, initial-state photon radiation, and hadronization effect.

ALEPH uses a technique called CEEC (Cluster EEC) which does jet clustering before the EEC calculation in order to minimize the possible bias introduced by the hadronization correction. Their result [60] (preliminary) is:

$$\boxed{\text{ALEPH CEEC} \quad \alpha_s(M_Z^2) = 0.121 \pm 0.008\,^{+0.000}_{-0.015}\,(\text{scale}) \quad f = 1.}$$

The scale uncertainty is derived from the $\alpha_s(M_Z^2)$ value for $f = \mu^2/s = 0.002$.

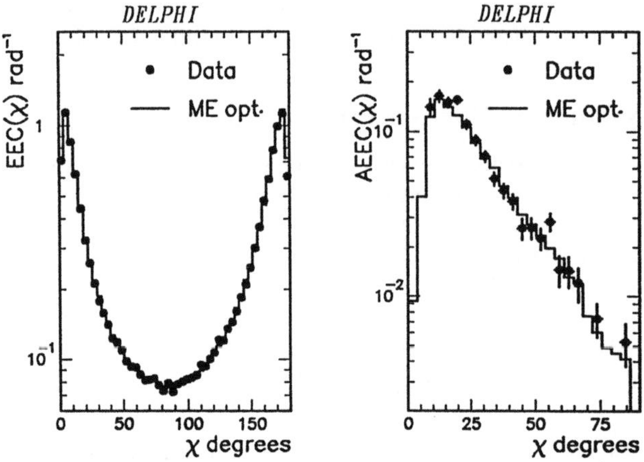

Figure 16: The corrected EEC and AEEC measured by DELPHI, compared with the expectations from the second order QCD matrix element option of the JETSET.

The distribution of EEC and AEEC measured by DELPHI is shown in fig.16. From a fit to the AEEC distribution for $28.8° < \chi < 90°$, they determined $\alpha_s(M_Z^2)$ [61]:

DELPHI AEEC	$\alpha_s(M_Z^2) = 0.106 \pm 0.003(\text{stat.}) \pm 0.003(\text{syst.})^{+0.003}_{-0.000}(\text{scale})$ $f = 0.002$.

OPAL used both EEC and AEEC to derive α_s [62]. The data were compared with four different $O(\alpha_s^2)$ analytic formulae [49] [63], and with the ERT [50] and GKS [64] $O(\alpha_s^2)$ matrix element formulae incorporated into the JETSET Monte Carlo:

OPAL EEC	$\alpha_s(M_Z^2) = 0.131 \pm 0.006\,(\text{exp.}) \pm 0.007\,(\text{theor.})$	$f = 1$
OPAL AEEC	$\alpha_s(M_Z^2) = 0.117\,^{+0.007}_{-0.009}\,(\text{exp.})\,^{+0.006}_{-0.002}\,(\text{theor.})$	$f = 1$

where the central value and experimental error are from the fits using the calculation of [49]. In the case of EEC, there are significant discrepancies between the different theoretical calculations. The discrepancies are not well understood. The theoretical error above is taken to be the largest of the differences between the result with [49] and those with other calculations [63]. The uncertainty is less in the case of AEEC due to the smaller second order correction.

Using the formula of [49] for which the second order term is expressed as a function of $f = \mu^2/s$, fits were done to EEC and AEEC with both $\Lambda_{\overline{MS}}$ and f as free parameters:

OPAL EEC	$\alpha_s(M_Z^2) = 0.117\,^{+0.006}_{-0.008}\,(\text{exp.})$	$f = 0.027 \pm 0.013$
OPAL AEEC	$\alpha_s(M_Z^2) = 0.117\,^{+0.007}_{-0.009}\,(\text{exp.})$.	

In the case of EEC the $\Lambda_{\overline{MS}}$ is quite sensitive to the choice of the scale, while the $\Lambda_{\overline{MS}}$ from AEEC is almost insensitive to the choice. An optimal value of f is therefore not quoted for AEEC. Note that the optimized scale μ^2 can be different for different physical observables. As mentioned before, by optimizing μ^2 one is trying to minimize the effect from missing higher order terms. The sizes of the effects may vary depending on the observables in question.

3.3 Test of Triple Gluon Vertex

One of the essential features of QCD is the self-coupling of gluons, a direct consequence of the non-abelian nature of this gauge theory. An expected phenomenon of "asymptotic freedom" has been confirmed by observing the decreasing α_s with increasing energy. Experimental study for the existence of the triple gluon vertex constitutes yet another important test of QCD. The triple-gluon vertex in e^+e^- annihilation enters in the second and higher orders of the α_s. Figure 17 (from ref.[65]) shows second order contributions yielding four-parton final states: (a) triple-gluon vertex, (b)(c) double bremsstrahlung, and (d) secondary $q\bar{q}$ production. In QCD the diagram (a) is predicted to be the dominant source of 4-jet events. Thus effect of the triple-gluon vertex can be studied in 4-jet events. Such tests become feasible with the large statistics of hadronic events at LEP. The first experimental study on the triple-gluon vertex in 4-jet final states was reported by AMY collaboration [66]. The study, however, suffered from low statistical significance.

One of the most promising signatures of the triple gluon vertex is the angular correlation of jet directions in the 4-jet final states. Several observables have been proposed which are sensitive to the specific helicity structure of the processes $g \to gg$ and $g \to q\bar{q}$ [67][68][69][70]. Bremsstrahlung gluons are polarized in the $q\bar{q}g$ plane. From helicity and angular momentum arguments, it follows that the $q\bar{q}$ directions from $g \to q\bar{q}$ tend to be perpendicular to the g polarization, while the gg from $g \to gg$ have a slight preference for being along the polarization vector.

- Bengtsson-Zerwas Angle:
  The angle $\chi_{\rm BZ}$, proposed by Bengtsson and Zerwas [70], is defined as the angle between the two planes spanned by the jet-momenta $\vec{p}_1$ and $\vec{p}_2$ and by $\vec{p}_3$ and $\vec{p}_4$, where the jets are ordered according to their energies. The most energetic jets 1 and 2 are likely to correspond to the primary quarks. The angle $\chi_{\rm BZ}$ is illustrated in Fig. 18 (a).

- Nachtmann-Reiter Angle:
  The angle $\theta^*_{\rm NR}$, originally proposed by Nachtmann and Reiter [68] and modified by Bengtsson [69] is defined as the angle between the two vectors $\vec{p}_1 - \vec{p}_2$ and $\vec{p}_3 - \vec{p}_4$, where the ordered jet momenta have the same convention as before. The angle $\theta^*_{\rm NR}$ is illustrated in Fig. 18 (b).

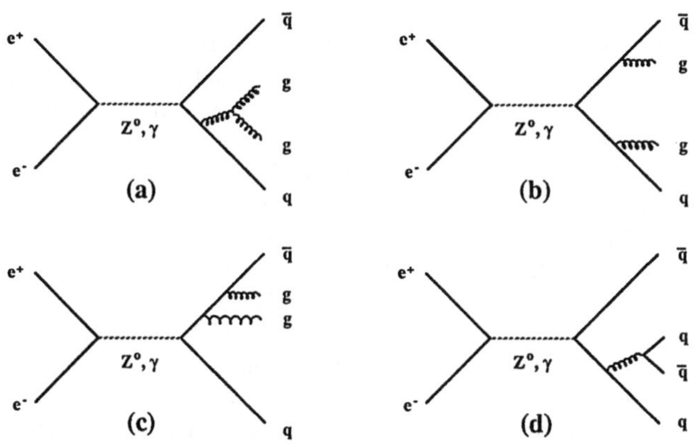

Figure 17: Basic Feynman diagrams leading to 4-parton final states.

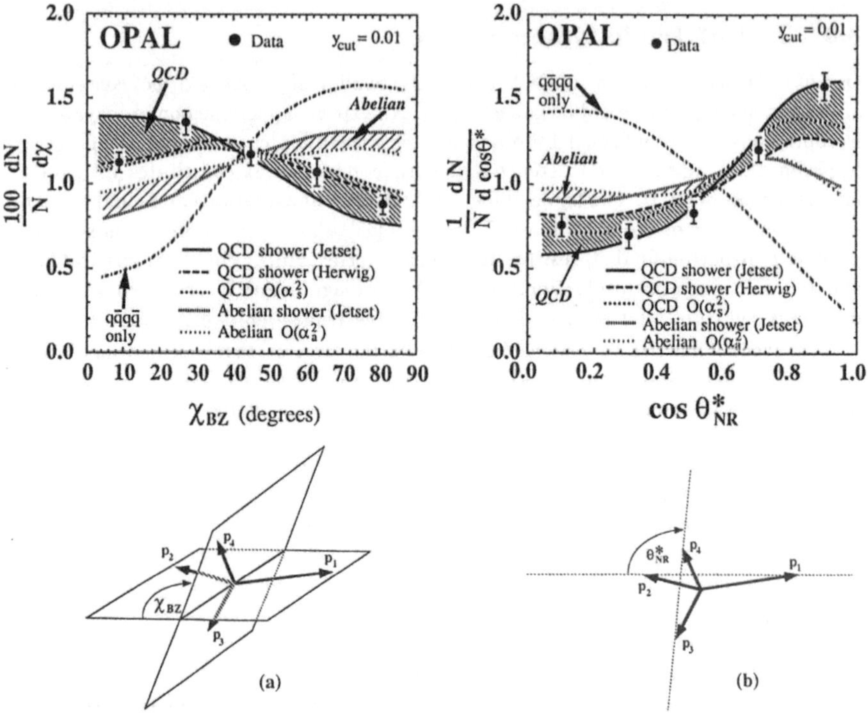

Figure 18: Distributions of (a) χ_{BZ} and (b) $\cos\theta^*_{NR}$ measured by OPAL for 4-jets events defined at $y_{cut} = 0.01$. The data are compared to the predictions of second order QCD and the abelian theory, and parton shower models. The shading shows the areas spanned by the different models.

- Körner-Schierholz-Willrodt angle:
  The ϕ_{KSW}, proposed by Körner, Schierholz and Wilrodt [67], is used for events for which there are two jets in both hemispheres defined by the thrust axis. The ϕ_{KSW} is the angle between the normals to the planes defined by the pair of jets in each hemisphere.

L3 [71] and OPAL [65] have compared the data with QCD predictions by using JETSET $O(\alpha_s^2)$ matrix element calculation, JETSET parton shower model, and HERWIG parton shower model (OPAL only). The data were also compared with the predictions of a fictitious abelian vector gluon theory in which the gluon self-coupling (diagram (a)) is absent. The model, implemented in the JETSET matrix element option, is constructed by simply replacing the group constants of QCD by those appropriate for the abelian case: $N_c = 3 \rightarrow 0$, $C_f = 4/3 \rightarrow 1$ and $T_r = N_f/2 \rightarrow 3N_f$, where N_c is the number of colours, C_f the fermionic Casimir operator, and N_f is the number of quark flavours (=5). The abelian model was also adapted to JETSET parton shower model, though the prescription is by no means unique. The alternative models are mainly used to demonstrate the sensitivity of the above observables to the characteristic features of QCD. The abelian theory is not a realistic theory. This has been already demonstrated in other studies such as the energy dependence of multi-jet production rates in e^+e^- annihilation.

The results of OPAL are shown in figure 18. Similar results were obtained by L3. L3 have also shown the measurement of ϕ_{KSW}, but this angle is not very sensitive to the difference between QCD and the abelian model unless quark and gluon jets are individually tagged [72]. In the figure, the different model predictions do not agree well in the detail, so the areas spanned by the various model predictions are hatched for better visibility. The data clearly favour the band predicted by QCD. The expectation from a hypothetical pure sample of $q\bar{q}q\bar{q}$ events is also shown in the figure 18. Further investigation [72] shows that it is mainly the contribution from this 4-quark final state (4.7 % in QCD and 31.4 % in the abelian theory at $y_{cut} = 0.01$) that causes the specific difference between QCD and the abelian theory in these observables, rather than the contribution from the triple gluon vertex. OPAL have therefore placed a limit of $< 9.1\%$ at 95% confidence level, on the relative production rate of $q\bar{q}q\bar{q}$ events.

DELPHI have taken a different approach [73]. Instead of comparing to specific non-QCD models, they make a fit to their data treating the group constants as free parameters. The relative weights of the diagrams of figure 17 depend on the group constants C_F, N_C and T_R and they took N_C/C_F and T_R/C_F as free variables. The relative weight of the triple gluon vertex diagram is proportional to the N_C/C_F. They fitted a two-dimensional distribution of two angles θ^*_{NR} and α_{34}. The α_{34} is defined as the angle between the two lowest energy jets 3 and 4 which preferentially correspond to the secondary partons. The idea is that the probability of the secondary jets going back-to-back is expected to be smaller in the diagram of triple gluon vertex than in double bremsstrahlung diagrams, and the α_{34} can be used for distinguishing between the two kinds. The following table shows their fitting result together with the expectation from the theories.

	N_C/C_F	T_R/C_F
Fit N_C/C_F and T_R/C_F	$2.55 \pm 0.55 \pm 0.4 \pm 0.2$	0.1 ± 2.4
QCD	2.25	1.875
Abelian model	0	15

In the fitted value of N_C/C_F, the first error is statistical, the second gives the systematic error for fragmentation and model dependence of the predictions, the third is an error in the correction of a bias in the fitting procedure. The fit indicates a significant contribution from triple-gluon vertex and this agrees with 2.25 ($N_C = 3$ and $C_F = 4/3$) expected for QCD.

4 Future LEP plans

Among the future LEP programs the first priority is given to the preparation for reaching center-of-mass energies beyond the threshold of W^+W^- pair creation, the so-called LEP II. Significant upgrade of the RF accelerating system is required because the energy lost by synchrotron radiation is proportional to $E^4_{\rm beam}$. In the table 3 the plan is shown of adding super-conducting RF cavities to presently available 128 normal-conducting copper cavities. The maximum attainable beam energy is also shown. From the past experience it is reasonable to assume that not all the cavities are operational. The values in parentheses are given on the assumption that the 1/8 of the cavities are unavailable. The W^+W^- threshold will be reached in 1994. After that there is still room for adding more super-conducting cavities. However the "nominal" LEP II scenario, which is

Table 3: Future LEP plans

Year		No. of Cu cavities	No. of S.C. cavities	Max. E_{beam}(GeV) (1/8 cavities off)	No. of bunches per beam
1990	at Z^0	128	4	~ 61 (~ 57)	4
1991	at Z^0	128	12	~ 63 (~ 61)	4
1992	at Z^0	128	32	~ 69 (~ 66)	8?
1993	at Z^0	128	64	~ 75 (~ 73)	8? or more (≤ 36)?
1994	LEP II	128	192	~ 92 (~ 90)	⋮
1995	LEP II	0 or > 0	256? 288?	$\sim 95 \sim 98$? ($\sim 91 \sim 96$?)	⋮
1996	LEP II	⋮	⋮	⋮	⋮
1997		Shutdown for LHC preparation			
1998 ...		LHC & LEP			

currently approved and funded, is 192 super-conducting cavities with 4 bunches per beam.

Until the LEP II becomes available, the machine will run at the Z^0 resonance aiming at collecting millions of Z^0 per experiment for more precise tests of the standard model and for discovering new phenomena.

The possibility of raising luminosity by increasing the number of bunches will be investigated. With the so-called Bretzel scheme for separating beams horizontally, the number of bunches could be increased, in principle, up to 36 per beam. The test of this scheme will start in 1991 and 8-bunch operation may be accomplished already in 1992. Doubling the number of bunches is particularly useful for LEP II, though the klystron system must be upgraded to provide the necessary beam power.

Based on the observation of the natural transverse beam polarization in 1990, precise calibration of the beam energy will be tried in 1991. Longitudinally polarized beams at Z^0 would provide the most precise measurements of the weak couplings. The program, however, will not be given a high priority until feasibility is demonstrated.

The 1997 is planned to be a long shutdown for preparation of Large Hadron Collider (LHC). One likely scenario from the 1998 onwards is that LEP and LHC run alternately to pursue their respective physics goals.

Acknowledgements

It is a pleasure to thank the organizers of the 5th Yukawa Memorial Symposium for their hospitality and for arranging such an enjoyable meeting. I would also like to thank the members of the LEP collaborations who provided the information and figures for this talk.

References

[1] ALEPH Collaboration, D. Decamp *et al.*: Phys. Lett. **B231**, 519 (1989);
ALEPH Collaboration, D. Decamp *et al.*: Phys. Lett. **B234**, 399 (1990);
ALEPH Collaboration, D. Decamp *et al.*: Phys. Lett. **B235**, 399 (1990).

[2] DELPHI Collaboration, P. Aarnio *et al.*: Phys. Lett. **B231**, 539 (1989);
DELPHI Collaboration, P. Aarnio *et al.*: Phys. Lett. **B241**, 425 (1990);
DELPHI Collaboration, P. Abreu *et al.*: Phys. Lett. **B241**, 435 (1990).

[3] L3 Collaboration, B. Adeva *et al.*: Phys. Lett. **B231**, 509 (1989);
L3 Collaboration, B. Adeva *et al.*: Phys. Lett. **B236**, 109 (1990);
L3 Collaboration, B. Adeva *et al.*: Phys. Lett. **B237**, 136 (1990);
L3 Collaboration, B. Adeva *et al.*: Phys. Lett. **B238**, 122 (1990).

[4] OPAL Collaboration, M.Z. Akrawy *et al.*: Phys. Lett. **B231**, 530 (1989);
OPAL Collaboration, M.Z. Akrawy *et al.*: Phys. Lett. **B235**, 379 (1990);
OPAL Collaboration, M.Z. Akrawy *et al.*: Phys. Lett. **B240**, 497 (1990);
OPAL Collaboration, M.Z. Akrawy *et al.*: Phys. Lett. **B247**, 458 (1990).

[5] ALEPH Collaboration, D. Decamp *et al.*: CERN-PPE/90-104, submitted to Z. Phys. C;
ALEPH Collaboration: Talks by J.R. Hansen and J.C. Brient at the 25th International Conference on High Energy Physics, Singapore, August 1990.

[6] DELPHI Collaboration: Talk by U. Amaldi at the 25th International Conference on High Energy Physics, Singapore, August 1990;
DELPHI Collaboration, P. Abreu *et al.*: Contributed paper to Singapore Conference (CERN-PPE/90-119).

[7] L3 Collaboration, B. Adeva *et al.*: L3 Preprints #8, #9 and #17, submitted to Phys. Lett. B;
L3 Collaboration: Talk by S.C.C. Ting at the 25th International Conference on High Energy Physics, Singapore, August 1990.

[8] OPAL Collaboration: Talk by T. Mori at the 25th International Conference on High Energy Physics, Singapore, August 1990.

[9] F. Dydak: *Results from LEP and the SLC* (summary talk) at the 25th International Conference on High Energy Physics, Singapore, August 1990.

[10] Talk by C.M. Hawkes at the 20th International Symposium on Multiparticle Dynamics, Dortmund, September 1990.

[11] *Z Physics at LEP 1*, edited by G. Altarelli, R. Kleiss and C. Verzegnassi, CERN 89-08 Vol. 1 (1989).

[12] ALIBABA program by W.J.P. Beenakker, F.A. Berends and S.C. van der Marck (Institut-Lorentz, University of Leiden).

[13] A. Borrelli *et al.*: Nucl. Phys. **B333**, 357 (1990).

[14] R.N. Cahn: Phys. Rev. **D36**, 2666 (1987).

[15] FORTRAN program ZAPP by G. Burgers.

[16] FORTRAN program MIZA by M. Martinez *et al.*
M. Martinez *et al.*: CERN-PPE/90-109, submitted to Z. Phys. C.

[17] D.C. Kennedy *et al.*: Nucl. Phys. **B321**, 83 (1989).

[18] M. Greco: Phys. Lett. **B177**, 97 (1986);
F. Aversa and M. Greco: Phys. Lett. **B228**, 134 (1989);
F. Aversa *et al.*: INFN-Frascati preprint LNF-90/049 (1990);
BHABHA program by M. Caffo, E. Remiddi and F. Semeria.

[19] The ZFITTER/ZBIZON program package of D. Bardin *et al.*: Berlin-Zeuthen preprint PHE-89-19 (1989), Z. Phys. **C44**, 493 (1989) and Comp. Phys. Comm. **59**, 303 (1990).

[20] Line shape program ZSHAPE, W.J.P. Beenakker, F.A. Berends and S.C. van der Marck (Institut-Lorentz, University of Leiden).

[21] Talk by D. Schaile at the DESY Theory Workshop, Hamburg, October 1990.

[22] OPAL Collaboration, T. Mori: Private communication.

[23] CDF Collaboration, P. Shalbach: Proceedings of the APS conference, Washington DC, April 1990.

[24] UA2 Collaboration, J. Alitti *et al.*: Phys. Lett. **B241**, 150 (1990).

[25] CDHS Collaboration, H. Abramowicz *et al.*: Phys. Rev. Lett. **57**, 298 (1986);
A. Blondel *et al.*: Z. Phys. **C45**, 361 (1990).

[26] CHARM Collaboration, J.V. Allaby *et al.*: Phys. Lett. **B177**, 446 (1986) and Z. Phys. **C36**, 611 (1987).

[27] M. Martinez: presented by F. Dydak in his talk at the 25th International Conference on High Energy Physics, Singapore, August 1990.

[28] ALEPH Collaboration, D. Decamp *et al.*: Phys. Lett. **B244**, 551 (1990);
ALEPH Collaboration: Talk by R.P. Johnson at the 25th International Conference on High Energy Physics, Singapore, August 1990.

[29] L3 Collaboration: Talk by V. Innocente at the 25th International Conference on High Energy Physics, Singapore, August 1990.

[30] OPAL Collaboration: Talk by A. Jawahery at the 25th International Conference on High Energy Physics, Singapore, August 1990.

[31] DELPHI Collaboration: Talk by W. Adam at the 25th International Conference on High Energy Physics, Singapore, August 1990;
DELPHI Collaboration, P. Abreu *et al.*: Contributed paper to Singapore Conference (CERN-PPE/90-118).

[32] TASSO Collaboration, W. Braunschweig *et al.*: DESY 88-159 (1988).

[33] DELPHI Collaboration, P. Abreu *et al.*: CERN-PPE/90-123, submitted to Phys. Lett. B.

[34] CLEO Collaboration, D. Bortoletto *et al.*: Phys. Rev. **D37**, 1719 (1988) and Phys. Rev. **D39**, 1471 (1989).

[35] OPAL Collaboration, M.Z. Akrawy *et al.*: Phys. Lett. **B246**, 285 (1990);
OPAL Collaboration, Talk by A. Jawahery at the 25th International Conference on High Energy Physics, Singapore, August 1990.

[36] L3 Collaboration, Talk by J.G. Branson at the 25th International Conference on High Energy Physics, Singapore, August 1990.

[37] UA1 Collaboration, C. Albajar *et al.*: Phys. Lett. **B186**, 247 (1987).

[38] ARGUS Collaboration, H. Albrecht *et al.*: Phys. Lett. **B192**, 245 (1987).

[39] CLEO Collaboration, M. Artuso *et al.*: Phys. Rev. Lett. **62**, 2233 (1989).

[40] ALEPH Collaboration: Talks by A. Stahl, S. Orteu and F. Zomer at Workshop on Tau Lepton Physics, Orsay, September 1990.

[41] DELPHI Collaboration, P. Vaz: Private communication.

[42] JADE Collaboration, W. Bartel *et al.*: Z. Phys. **C33**, 23 (1986);
JADE Collaboration, S. Bethke *et al.*: Phys. Lett. **B213**, 235 (1988).

[43] DELPHI Collaboration, P. Abreu *et al.*: CERN-EP/90-89, submitted to Phys. Lett. B.

[44] T. Sjöstrand: Comp. Phys. Comm. **39**, 347 (1986);
T. Sjöstrand and M. Bengtsson: Comp. Phys. Comm. **43**, 367 (1987);
M. Bengtsson and T. Sjöstrand: Nucl. Phys. **B289**, 810 (1987).

[45] G. Marchesini and B.R. Webber: Nucl. Phys. **B310**, 461 (1988);
G. Marchesini and B.R. Webber: Cavendish-HEP-88/7.

[46] ALEPH Collaboration, D. Decamp *et al.*: Phys. Lett. **B234**, 209 (1990);
ALEPH Collaboration: Talk by M. Schmelling at Rencontres de Moriond, March 1990;
DELPHI Collaboration, P. Aarnio *et al.*: Phys. Lett. **B240**, 271 (1990);
OPAL Collaboration, M.Z. Akrawy *et al.*: Z. Phys. **C47**, 505 (1990).

[47] G. Kramer and B. Lampe: J. Math. Phys. **28**, 945 (1987) and preprints DESY 86-103, DESY 86-119 (1986).

[48] G. Kramer and B. Lampe: Z. Phys. **C39**, 101 (1988).

[49] Z. Kunszt and P. Nason [conv.]: in *Z Physics at LEP 1*, edited by G. Altarelli, R. Kleiss and C. Verzegnassi, CERN 89-08 Vol. 1 (1989).

[50] R.K. Ellis, D.A. Ross and A.E. Terrano: Nucl. Phys. **B178**, 421 (1981).

[51] Particle Data Group: Phys. Lett. **B239**, III.50 (1990).

[52] S. Bethke: Z. Phys. **C43**, 331 (1989).

[53] ALEPH Collaboration, D. Decamp *et al.*: CERN-PPE 90-176, submitted to Phys. Lett. B.

[54] L3 Collaboration, B. Adeva *et al.*: L3 Preprint #011, submitted to Phys. Lett. B.

[55] OPAL Collaboration, M.Z. Akrawy *et al.*: CERN-PPE/90-143 and an erratum to it, submitted to Z. Phys. C.

[56] TASSO Collaboration, W. Braunschweig *et al.*, Phys. Lett. **B214**, 286 (1988);
MARK-II Collaboration, S. Bethke *et al.*: Z. Phys. **C43**, 325 (1989);
AMY Collaboration, I.H. Park *et al.*: Phys. Rev. Lett. **62**, 1713 (1989);
S. Iwata: Talk at the 9th International Conference on Physics in Collision, Jerusalem, June 1989;
VENUS Collaboration, K. Abe *et al.*: Phys. Lett. **B240**, 232 (1990).

[57] S. Bethke: Private communication.

[58] C.L. Basham, L.S. Brown, S.D. Ellis, S.T. Love, Phys. Rev. Lett. **41**, 1585 (1978), Phys. Rev. **D17**, 2298 (1978) and Phys. Rev. **D19**, 2018 (1979).

[59] MARK-II Collaboration, D. Schlatter *et al.*: Phys. Rev. Lett. **49**, 521 (1982);
PLUTO Collaboration, Ch. Berger *et al.*: Z. Phys. **C12**, 297 (1982);
CELLO Collaboration, H-J. Behrend *et al.*: Z. Phys. **C14**, 95 (1982);
MARK-J Collaboration, B. Adeva *et al.*: Phys. Rev. Lett. **50**, 2051 (1983);
JADE Collaboration, W. Bartel *et al.*: Z. Phys. **C25**, 231 (1984);
CELLO Collaboration, H.J. Behrend *et al.*: Phys. Lett. **B138**, 311 (1984);
TASSO Collaboration, M. Althoff *et al.*: Z. Phys. **C26**, 157 (1984);
MAC Collaboration, E. Fernandez *et al.*: Phys. Rev. **D31**, 2724 (1985);
MARK-J Collaboration, B. Adeva *et al.*: Phys. Rev. Lett. **54**, 1750 (1985);
TASSO Collaboration, W. Braunschweig *et al.*: Z. Phys. **C36**, 349 (1987);
MARK-II Collaboration, D.R. Wood *et al.*: Phys. Rev. **D37**, 3091 (1988);
TOPAZ Collaboration, I. Adachi *et al.*: Phys. Lett. **B227**, 495 (1989).

[60] ALEPH Collaboration: Talk by H. Hu at the 20th International Symposium on Multiparticle Dynamics, Dortmund, September 1990.

[61] DELPHI Collaboration, P. Abreu *et al.*: CERN-PPE/90-122, submitted to Phys. Lett. B.

[62] OPAL Collaboration, M.Z. Akrawy *et al.*: CERN-PPE/90-121, submitted to Phys. Lett. B.

[63] A. Ali and F. Barreiro: Phys. Lett. **B118**, 155 (1982) and Nucl.Phys. **B236**, 269 (1984);
D.G. Richards, W.J. Stirling and S.D. Ellis: Phys. Lett. **B119**, 193 (1982) and Nucl. Phys. **B229**, 317 (1983):
N.K. Falck and G. Kramer: Z. Phys. **C42**, 459 (1989).

[64] F. Gutbrod, G. Kramer and G. Schierholz: Z. Phys. **C21**, 235 (1984).

[65] OPAL Collaboration, M.Z. Akrawy *et al.*: CERN-PPE/90-97, submitted to Z. Phys. C.

[66] AMY Collaboration, I.H. Park *et al.*: Phys. Rev. Lett. **62**, 1713 (1989).

[67] J.G. Körner, G. Schierholz and J. Willrodt: Nucl. Phys. **B185**, 365 (1981).

[68] O. Nachtmann and A. Reiter: Z. Phys. **C16**, 45 (1982).

[69] M. Bengtsson: Z. Phys. **C42**, 75 (1989).

[70] M. Bengtsson and P.M. Zerwas: Phys. Lett. **B208**, 306 (1988).

[71] L3 Collaboration, B Adeva *et al.*: Phys. Lett. **B248**, 227 (1990).

[72] S. Bethke, A. Ricker and P.M. Zerwas: Archen preprint PITHA 90/14 (1990).

[73] DELPHI Collaboration, P. Abreu *et al.*: CERN-PPE/90-174, submitted to Phys. Lett. B.

Role of Radiative Corrections in the Electroweak Theory

A. Sirlin

Department of Physics, Columbia University, New York, NY 10027, USA and
Department of Physics, New York University, New York, NY 10003, USA

Abstract. The role of radiative corrections in electroweak physics is examined. Special emphasis is given to their significance in neutral currents, m_W, m_Z and Z^0 physics, as well as at low energies. The on-shell and $\overline{MS}$ renormalization frameworks and the basic radiative corrections Δr, $\Delta\hat{r}$ and $\Delta\hat{r}_W$ are outlined. The comparison between theory and experiment, upper bounds on m_t, the effect of a fourth generation with Dirac or Majorana neutrinos and novel parametrizations of new physics are discussed.

1 Introduction

It seems fair to say that the experimental developments over the last few years concerning Z^0 and W physics mark the onset of an era of great precision at high energies.

In the past, high precision in electroweak physics has also been achieved at low energies. For example, the Michel parameter, derived from the muon decay spectrum, is

$$\rho_{\text{Michel}} = 0.7518 \pm 0.0026, \tag{1}$$

and the test of unitarity of the Kobayashi-Maskawa matrix (modern version of the universality of the weak interactions) presently reads

$$| V_{ud} |^2 + | V_{us} |^2 + | V_{ub} |^2 = 0.9970 \pm 0.0016. \tag{2}$$

The theoretical error in eq.(2) has decreased this year because of recent work by Jaus and Rasche [1]. The remaining discrepancy may be partly due to uncertainties in the nuclear mismatch correction. In fact, the error quoted in eq.(2) is more conservative than in recent publications [1,2] because we have assigned a larger uncertainty to this effect. It is also interesting to point out that Wilkinson has recently proposed a novel and largely empirical analysis of the nuclear mismatch problem that leads to [3]

$$| V_{ud} |^2 + | V_{us} |^2 + | V_{ub} |^2 = 0.9989 \pm 0.0012.$$

Springer Proceedings in Physics, Vol. 65 Present and Future of High-Energy Physics
Editors: K.-I. Aoki and M. Kobayashi

It is important to note that, if the radiative corrections were not included in the analysis, the r.h.s. of eq.(2) would be $\simeq 1.037$ which would require a drastic modification of the Standard Model (SM); furthermore, one would obtain $\rho_{\text{Michel}} \simeq 0.71$ which differs by about 13σ from the predictions of the two-component theory of the neutrino! Thus, radiative corrections play a significant role in low energy electroweak physics. Accurate comparisons as the unitarity test can be used to derive constraints on new physics. For example, in some models with additional U(1) factors the extra Z's give rise to box diagram contributions that distinguish μ and β decays, decreasing $| V_{ud} |^2 + | V_{us} |^2 + | V_{ub} |^2$ [4]. In order not to spoil the unitarity test they must be sufficiently massive. In particular, calling $Z(\beta) = Z_\psi \sin\beta + Z_\chi \cos\beta$ the lighter additional Z in the $SU(3)_C \times SU(2)_L \times U(1) \times U(1)_\chi \times U(1)_\psi$ theory originating from the breaking of E_6, assuming that the second extra Z is much more massive and employing Wilkinson's value for $| V_{ud} |^2 + | V_{us} |^2 + | V_{ub} |^2$, one obtains

$$\cos^2\beta[1 + (5/27)^{1/2}\tan\beta]\ln x/(x-1) < 0.096 \text{ (90\% C.L.)},$$

where $x = m_{Z(\beta)}^2/m_W^2$. For $Z(0) = Z_\chi$ (the additional Z occurring in $SO(10) \to SU(5) \times U(1)_\chi$) this implies the very sharp constraint $m_{Z_\chi} >$ 501GeV. Similar bounds are obtained for $-23^o \lesssim \beta \lesssim 47^0$. On the other hand, for $Z(\pi/2) = Z_\psi$ (the additional Z in $E_6 \to SO(10) \times U(1)_\psi$), we have universal couplings to ordinary quarks and leptons and, as a consequence, no bounds are obtained. The unitarity test can also be used to obtain constraints on compositeness. For example, Veltman has pointed out that the accurate agreement with unitarity poses a severe problem for composite models of W's[5].

2 Neutral Currents, m_W, m_Z, Z^0 Physics

As the level of precision has greatly improved over the last few years, the role of the radiative corrections in the comparison between theory and experiment has become increasingly important. As we want to compare many different phenomena over a huge range of energy scales, from $q^2 \simeq 0$ in atomic parity violation to $q^2 = m_Z^2$ and beyond in $Z^0$and W physics, it is very useful to have simple, efficient renormalization frameworks. In this talk we will concentrate on two formulations: *i*) on-shell, which may be defined as a class of schemes in which e (the conventional charge of the positron), m_W and m_Z (the physical masses) are taken as the basic renormalized parameters of the theory and *ii*) the $\overline{MS}$ scheme. Within class *i*) there are several approaches [6]-[10]. Perhaps the epitome of this formulation is to be found in the "Simple Renormalization Framework" of ref.[6]. Assuming that the Higgs bosons transform as $SU(2)_L$ doublets or singlets, the following relations are valid

$$\cos\theta_W = m_W/m_Z, \tag{3}$$

$$e = gs, \tag{4}$$

$$\frac{G_\mu}{\sqrt{2}} = \frac{g^2}{8m_W^2(1-\Delta r)}, \tag{5}$$

where $s^2 \equiv 1 - c^2$ is an abbreviation for $\sin^2\theta_W$, g is the $SU(2)_L$ coupling, $G_\mu = 1.16639(2)\cdot 10^{-5}\text{GeV}^{-2}$ is the μ-decay coupling constant and Δr is the radiative correction to μ-decay after subtracting the photonic corrections of the local V-A theory. We note that in this framework eqs. (3) and (4), which are valid at the tree-level, are maintained to higher orders; thus, (3) may be regarded as the definition of $\cos\theta_W$ and (4) as that of g. On the other hand, (5) is modified by Δr, a basic correction of the electroweak theory. It can be written as

$$\Delta r = \frac{Re A_{WW}(m_W^2) - A_{WW}(0)}{m_W^2} - \frac{2\delta g}{g} + \delta C_b + \delta C_v, \tag{6}$$

$$\frac{2\delta g}{g} = \frac{2\delta e}{e} - \frac{c^2}{s^2}X, \tag{7}$$

$$X = Re\left[\frac{A_{ZZ}(m_Z^2)}{m_Z^2} - \frac{A_{WW}(m_W^2)}{m_W^2}\right], \tag{8}$$

where δC_b and δC_v stand for box and vertex corrections, respectively, δe is the charge renormalization counterterm and the A's represent unrenormalized self-energies evaluated at the indicated momenta. Combining eqs.(3-5) and solving for s^2, we obtain

$$s^2 = \frac{1}{2}\left\{1 - \left[1 - \frac{4A^2}{m_Z^2(1-\Delta r)}\right]^{1/2}\right\}, \tag{9}$$

where

$$A = \left(\pi\alpha/\sqrt{2}G_\mu\right)^{1/2} = 37.2803\text{GeV}.$$

For $m_t < 90$GeV, in the SM $\Delta r \gtrsim 0.06$, but it is sensitive to short distance physics. How large is Δr experimentally? Solving eqs. (3-5) for Δr, we have

$$\Delta r = 1 - \frac{A^2}{m_Z^2 c^2 s^2} = 1 - \frac{A^2}{m_W^2(1 - m_W^2/m_Z^2)}. \tag{10}$$

Using $m_W/m_Z = 0.8775 \pm 0.0051$ (CDF)[11], 0.8831 ± 0.0055(UA2)[12], one obtains $m_W/m_Z = 0.8801 \pm 0.0037$(AVE). Combination with $m_Z = 91.172 \pm 0.031$GeV [13] and eq.(10) leads to $m_W = 80.24 \pm 0.34$GeV and a first determination $(\Delta r)_I = 0.042 \pm 0.020$. The CDHS and CHARM measurements of $R_\nu = \sigma(\nu_\mu + N \to \nu_\mu + \cdots)/\sigma(\nu_\mu + N \to \mu^- + \cdots)$ provide an independent determination of $\sin^2\theta_W$ which depends weakly on the top quark mass m_t. Using this information, the CDF bound $m_t > 89$GeV[14] and the m_Z value one obtains approximately $m_W = (80.03 \pm 0.32)$GeV. Combination with the previous value and eq.(10) leads to $m_W = (80.13 \pm 0.23)$GeV and a more precise determination $(\Delta r)_{II} = 0.049 \pm 0.013$.

Theoretically, for $m_t < 100$GeV, Δr is dominated by $-\mathrm{Re}\Pi^{(\mathrm{r})}_{\gamma\gamma}(m_Z^2) \simeq 0.06$. For large m_t, there are important negative contributions. Asymptotically

$$(\Delta r)_{\rm top} \simeq -\frac{3\alpha}{16\pi s^4}\frac{m_t^2}{m_Z^2} + \cdots \quad (m_t^2/m_W^2 \gg 1), \tag{11}$$

$$(\Delta r)_H \simeq \frac{11\alpha}{48\pi s^2}\ln\left(\frac{m_H^2}{m_Z^2}\right) + \cdots \quad (m_H^2/m_Z^2 \gg 1), \tag{12}$$

where H refers to the Higgs boson. Recently there have been a number of accurate calculations of Δr , as a function of m_t and m_H , using m_Z, G_μ and α as inputs [15-17]. An interesting theoretical feature of these calculations is that, following the work of refs. [18,19], they incorporate the terms of $O\big(\alpha^2(m_t^2/m_W^2)^2\big)$, as well as the leading higher order logarithms. The approach of Degrassi et al. [16] incorporates also the subleading logarithms of $O\big(\alpha^2 \ln m_Z/m_f\big)$. Table 1, based on ref.[16], illustrates Δr and the predicted value of m_W, as functions of m_t, for $m_H = 100$GeV and $m_Z = 91.17$GeV.

Table 1: Δr and m_W, extracted from (G_μ, m_Z, α) as a function of m_t, for $m_H = 100$GeV and $m_Z = 91.17$GeV (from ref.[16]).

m_t (GeV)	$10^2\Delta r$	m_W (GeV)
90	6.08	79.91
120	5.13	80.08
150	4.07	80.27
180	2.83	80.48
210	1.35	80.71
240	−0.40	80.98

Comparing Table 1 with $(\Delta r)_I$ and $(\Delta r)_{II}$ and recalling the constraint $m_t > 89$GeV, we can obtain m_t upper bounds for $m_H = 100$GeV. Table 2 illustrates these upper bounds at 90% CL, as well as the corresponding quantities for $m_H = 500, 1000$ GeV.

Table 2: 90% C.L. m_t upper bounds derived from $(\Delta r)_I$ and $(\Delta r)_{II}$ for several values of m_H. All masses are in GeV.

m_H	${m_t}^{\max}$ (from $(\Delta r)_I$)	${m_t}^{\max}$ (from $(\Delta r)_{II}$)
100	209	172
500	219	185
1000	225	191

Langacker has recently carried out a detailed global analysis of m_Z, m_W, neutral currents, the total Z^0 width Γ_Z, the average leptonic width $\Gamma_{l^+l^-}$ and $R = \Gamma_h/\Gamma_{l^+l^-}$, the ratio of hadronic and leptonic widths [20]. A simultaneous fit to m_t and $\sin^2\theta_W$ leads to the upper bounds listed in Table 3.

Table 3: 90% and 95% C.L. m_t upper bounds for several m_H values (frcm the global analysis of ref [20]). All masses are in GeV.

m_H	${m_t}^{\max}$ (90% C.L.)	${m_t}^{\max}$ (95% C.L.)
40	168	178
250	178	189
1000	192	202

We note that the 90% C.L. for $m_H = 1000\text{GeV}$ is very close to that derived from $(\Delta r)_{II}$ (Table 2). The reason is that the most stringent m_t bounds arise from m_Z, m_W, R_ν. Assuming $40\text{GeV} < m_H < 1\text{TeV}$ and leaving m_t arbitrary, the simultaneous fit of all the data to $\sin^2\theta_W$ and m_t leads to [20]

$$
\begin{aligned}
m_t &= (139^{+33}_{-39} \pm 16)\text{GeV}, && (13)\\
\sin^2\theta_W &= 0.2272 \pm 0.0040, && (14)\\
\sin^2\hat{\theta}_W(m_Z) &= 0.2328 \pm 0.0010, && (15)\\
\Delta r &= 0.049 \pm 0.011, && (16)
\end{aligned}
$$

where $\sin^2\hat{\theta}_W(m_Z)$ is the $\overline{MS}$ parameter to be discussed later and the central values are evaluated for $m_H = 250\text{GeV}$. The $\pm 16\text{GeV}$ error in (13) reflects our ignorance of m_H within the assumed range, while the errors in eqs. (14-16) include the m_t, $m_H$uncertainties. We note that (16) is quite close to the approximate estimate $(\Delta r)_{II}$ given before.

As recently discussed by Bardeen et al. [21], Miransky et al. [22], Marciano [23] and at this Symposium [24], dynamical symmetry breaking scenarios involving $< \bar{t}t >$ condensates lead to $m_t \simeq (220 - 230)\text{GeV}$ for $\Lambda \simeq (10^{19} - 10^{15})\text{GeV}$ (Λ is the scale of "new physics" responsible for the formation of the condensate). These values of m_t are somewhat larger than the current upper bounds, but one should keep an open mind about this interesting theoretical possibility. Marciano has also emphasized that with a fourth generation, maximal $t'-t$ mixing and the same range of Λ, the predicted values become $m_t \simeq m_{t'} \simeq 140\text{GeV}$, $m_{b'} \simeq 160\text{GeV}$ [23], which are in good agreement with current constraints.

One way to illustrate the significant role of the quantum effects is to compare the values of m_W and m_Z predicted using R_ν as input with those derived in the absence of the radiative corrections. Table 4 shows that the latter procedure is in disagreement with experiment, while inclusion of the corrections leads to consistent predictions at the current level of precision.

Table 4: Predicted values of m_W and m_Z , using R_ν as input. Columns 2 and 3 give the radiatively corrected predictions for $m_t = 100$ and 200 GeV, respectively, and m_H= 250 GeV[20].

	$m_t = 100$ (GeV)	$m_t = 200$ (GeV)	no rad. corr	Exp. Value
m_W(GeV)	79.7 ± 0.9	78.7 ± 0.9	75.9 ± 0.9	80.2 ± 0.3
m_Z(GeV)	91.0 ± 0.8	89.7 ± 0.8	87.1 ± 0.7	91.17 ± 0.03

At present it is better to use the accurately measured m_Z as input and predict other observables as a function of m_t. Table 5 compares predicted and experimental values of several observables at the Z^0 peak. It is apparent that there is good agreement in the range $100 \leq m_t \leq 200\text{GeV}$.

Table 5: Comparison of several observables at the Z^0 peak (not all independent) with SM predictions for $m_t = 100, 200\text{GeV}$ and $m_H = 250\text{GeV}$ [20](σ_h is the cross-section into final hadronic states). The predictions use the values of $\sin^2\theta_W$ determined from eq.(9).

m_t (GeV)	Γ_Z (GeV)	$\Gamma_{l^+l^-}$ (MeV)	Γ_h (MeV)	Γ_{inv} (MeV)	σ_h (nb)
100	2.480 ± 0.003	83.4 ± 0.1	1732 ± 2	498.6 ± 0.5	41.45 ± 0.03
200	2.506	84.1	1751	503.2	41.40
Exp.	2.498 ± 0.020	84.0 ± 0.9	1755 ± 21	496 ± 18	41.63 ± 0.60

If there are exotic Higgs representations such as triplets, at the tree level $\rho_0 = m_W^2/m_Z^2 \cos^2\theta_W \neq 1$ and the gauge sector involves one additional parameter. As the theoretical expressions involve mainly the combination

$$\rho_{\text{eff}} \equiv \rho_0(1 + 3G_\mu m_t^2/8\sqrt{2}\pi^2),$$

fits that leave ρ_0, $\sin^2\theta_W$ and m_t arbitrary lead to much weaker constraints on m_t because a large m_t effect can be compensated with $\rho_0 < 1$ (generated, for example, by a triplet $(\phi^{++}\phi^{+}\phi^{0})^T$). Even if one allows the range $\rho_0 < 1$, the logarithmic dependence in Δr implies $m_t < 410$GeV. The partial width $\Gamma_{b\bar{b}}$ is useful in this connection because it involves vertex corrections that have a significant m_t dependence. As a consequence, if one includes in the analysis the recent data on the Z^0 widths one finds $m_t < 344$GeV (90% CL) for arbitrary ρ_0[20]. The fit also leads to $\rho_{\text{eff}} = 1.006 \pm 0.003$.

For the future, if $\delta m_W \simeq \pm 50$MeV is achievable, it would imply $\delta\Delta r \simeq \pm 0.0035$. For comparison, the Δr variation in the range 40GeV $\leq m_H \leq$ 1TeV is $\simeq 12\%$. Perhaps we can get useful information about m_H after the top quark is discovered! As an example, Table 6 illustrates predicted values of m_W and the on-resonance left-right asymmetry $A_{LR}^{res}(\mu^+\mu^-)$ for $m_t = 150$GeV[16,25]. It is useful to recall that the expected accuracy is $\delta A_{LR} \simeq \pm 0.003$, perhaps better.

Table 6: Predicted values of m_W and $A_{LR}^{\text{res}}(\mu^+\mu^-)$ for $m_t = 150$GeV (from refs.[16,25]).

m_H (GeV)	$A_{LR}^{\text{res}}(\mu^+\mu^-)$	m_W (GeV)
40	0.142	80.31
100	0.139	80.27
250	0.135	80.21
500	0.132	80.16
1000	0.129	80.10

3 Effect of a Fourth Generation

It has been known for a long time that over most of the $m_{t'}$ and $m_{b'}$ parameter space the contribution to Δr from an additional fermion generation (t' and b' refer to the corresponding $I_3 = +1/2$ and $I_3 = -1/2$ quarks) is ≤ 0 [26]. This in turn is related to the well-known behavior of the loop corrections to the ρ-parameter for large isodoublet mass splittings [27]. However, even when one takes into account the current experimental lower bounds on $m_{t'}$ and $m_{b'}$, there remains a narrow strip along the $m_{t'} = m_{b'}$ diagonal in the $m_{t'}$, $m_{b'}$ plane where the fourth generation leads to a rather small positive contribution to Δr[26]. Such effect is of opposite sign to that arising from m_t and leads to a relaxation of the upper bound $(m_t)_{\text{max}}$. For Dirac neutrinos, the leptonic contribution of the fourth generation follows a similar pattern. A recent update gives for the maximum possible value of the overall quark and lepton contribution of the fourth generation [28]

$$(\Delta r^{(4)})_{\max} = 3.4 \cdot 10^{-3} \;\; (\text{ Dirac } \nu \text{ }). \tag{17}$$

It turns out that eq.(17) still holds if the 4th generation mixes arbitrarily with the third, provided its mixing with the first two is neglible! Comparison with the tables of ref. [16] shows that, for $(m_t)_{\max} \simeq 200\text{GeV}$, eq.(17) allows a compensating increase in $(m_t)_{\max}$ of at most $\simeq 7\text{GeV}$.

Suppose now that the right handed neutrino ν_R of the fourth generation is endowed with a Majorana mass entry of $O(G_\mu{}^{-1/2})$ while the Dirac mass entries $m_D^{(\nu)}$ and $m_D^{(l)}$ (l is the charged lepton) are of the same order of magnitude. This scenario has been recently proposed as a "natural" framework for justifying the existence of "heavy" additional neutrinos, not contributing to Γ_Z[29]. Diagonalizing the ν mass matrix one has two Majorana neutrinos of masses m_1 and $m_2 = (m_D^{(\nu)})^2/m_1$. The analysis shows that in this scenario we encounter a qualitatively different situation: for arbitrary $m_D^{(\nu)}$, the leptonic contributions to $\Delta r(\Delta\rho)$ are not mathematically bounded above (below) [28]. Restricting $m_D^{(\nu)} \leq 300\text{GeV}$ on triviality grounds [29] leads to

$$(\Delta r^{(4)})_{\max} = 1.7 \cdot 10^{-2} \;\; (\text{ Majorana } \nu\text{'s }) \tag{18}$$

and a compensating increase in $(m_t)_{\max}$ of at most 31 GeV[28].

Another way of relaxing $(m_t)_{\max}$ has been recently discussed by Denner et al. [30]. These authors point out that in two-Higgs doublet models there is a region of parameter space (see also ref. [31]) in which the scalars can give large positive contributions to Δr. However, this region is not allowed in the interesting supersymmetric scenario [31].

4 The $\overline{MS}$ Parameter $\sin^2\hat{\theta}_W(m_Z)$

The idea has been proposed by many people of introducing effective parameters that "absorb" to a considerable extent the m_t and m_H dependences of the radiative corrections in Z^0 amplitudes. Recently it has been emphasized that there is an important parameter of the theory very well suited for this purpose, to wit $\hat{s}^2 \equiv \sin^2\hat{\theta}_W(m_Z)$ ($\overline{MS}$ parameter, defined by modified minimal subtraction, evaluated at the m_Z scale) [32]. It has several desirable properties: i) it occurs naturally in GUTs predictions ii) the electroweak form factor $\hat{\kappa}^2(q^2)$ multiplying $\hat{s}^2$ in Z^0 amplitudes is independent of m_H and depends weakly (logarithmically) on m_t [33] iii) it is gauge-independent iv) it can be employed in any renormalizable extension of the SM. In order to introduce $\hat{s}^2$ we recall that the $O(\alpha)$ contributions proportional to $s^2 = 1 - m_W^2/m_Z^2$ in the Z^0 amplitude are of the form [34]

$$s^2 \left\{ 1 - (\frac{c}{s}) \hat{A}_{\gamma Z}(q^2)/q^2 - V(q^2) \right.$$
$$\left. - Re \left[\frac{c^2}{s^2} X + (\frac{c}{s})(A_{\gamma Z}(m_Z^2) + A_{\gamma Z}(0))/m_Z^2 \right] \right\},$$

where X is defined in (8), $V(q^2)$ denotes vertex corrections, $A_{\gamma Z}$ is the unrenormalized γZ mixing self energy and $\hat{A}_{\gamma Z}$ is its renormalized counterpart. According to the prescriptions of ref. [34] both $A_{\gamma Z}$ and $\hat{A}_{\gamma Z}$ depend weakly (logarithmically) on m_t and $\mathrm{Re}\hat{A}_{\gamma Z}(m_Z^2) = 0$. One recognizes that $(c^2/s^2)X$ is the counterterm for s^2 and therefore

$$s^2[1 - (c^2/s^2)X] = s_0^2,$$

the unrenormalized parameter. Next we take two steps: a) we remove $(c^2/s^2)X$ by expressing the amplitude in terms of s_0^2; b) we choose the counterterms present in s_0^2 to perform the $\overline{MS}$ renormalization, i.e. we subtract the terms involving $\delta = (n-4)^{-1} + [\gamma - \ln(4\pi)]/2$ in the divergent parts and set the 't Hooft mass scale $\mu = m_Z$. In terms of $\hat{s}^2$, the above term becomes

$$\hat{s}^2 \left\{ 1 - (c/s)\hat{A}_{\gamma Z}(q^2)/q^2 - V(q^2) \right.$$
$$\left. -(c/s)Re(A_{\gamma Z}(m_Z^2) + A_{\gamma Z}(0))_{\overline{MS}}/m_Z^2 \right\},$$

where $\overline{MS}$ means both the $\overline{MS}$ subtraction and the choice $\mu = m_Z$. Comparing the two expressions we see that the term $(c^2/s^2)X$ that contains the m_H and dominant m_t dependence has been absorbed in the $\overline{MS}$ redefinition of s^2! At the same time we note that, to one loop order,

$$\hat{s}^2 = [1 - (c^2/s^2)X_{\overline{MS}}]s^2. \qquad (19)$$

To link directly $\hat{s}^2$ to m_Z we first note that eq.(10) can be written as

$$m_Z^2 = \frac{A^2}{c^2 s^2 (1 - \Delta r)}.$$

This suggests the introduction of a new radiative correction $\Delta\hat{r}$ such that

$$\hat{c}^2 \hat{s}^2 (1 - \Delta\hat{r}) = c^2 s^2 (1 - \Delta r). \qquad (20)$$

Combining (19) and (20) we obtain, to one-loop order,

$$\Delta\hat{r} = [\Delta r - (c^2/s^2 - 1)X]_{\overline{MS}}. \qquad (21)$$

Comparing eq.(21) with eqs. (6-8) we see that we obtain $\Delta\hat{r}$ from Δr by erasing the large enhancement factor $c^2/s^2 \simeq 3.4$ in the term involving

X and performing the $\overline{MS}$ renormalization in the resulting expression. This leads to

$$\Delta\hat{r} = \left\{ Re\left(\frac{A_{ZZ}(m_Z^2)}{m_Z^2} - \frac{A_{WW}(0)}{m_W^2}\right) - \frac{2\delta e}{e} + \frac{\alpha}{4\pi\hat{s}^2}\left[(\frac{7}{2} - 6\hat{s}^2)\frac{\ln\hat{c}^2}{\hat{s}^2} + 6\right]\right\}_{\overline{MS}}. \quad (22)$$

In evaluating $\Delta\hat{r}$ it is natural to use α and $\alpha/\hat{s}^2$ as expansion parameters and replace $m_W^2 \to m_Z^2\hat{c}^2$ (this is particularly important in the leading term proportional to m_t^2 and it is justified theoretically in ref. [16]).

It is also useful to determine $\hat{s}^2$ from m_W[35]. The fact that eq.(10) can be written as

$$m_W^2 = A^2/s^2(1 - \Delta r)$$

suggests the introduction of a new radiative correction $\Delta\hat{r}_W$ such that

$$\hat{s}^2(1 - \Delta\hat{r}_W) = s^2(1 - \Delta r).$$

Recalling eqs. (6-8) and eq.(19) one obtains

$$\Delta\hat{r}_W = \left[\frac{ReA_{WW}(m_W^2) - A_{WW}(0)}{m_W^2} - \frac{2\delta e}{e}\right]_{\overline{MS}} + \frac{\alpha}{4\pi\hat{s}^2}\left[(\frac{7}{2} - 6\hat{s}^2)\frac{\ln\hat{c}^2}{\hat{s}^2} + 6\right]. \quad (23)$$

Defining $\tau \equiv m_t^2/m_Z^2$, $\xi \equiv m_H^2/m_Z^2$, the leading asymptotic behaviors are

$$\Delta\hat{r} = \begin{cases} -3\alpha\tau/16\pi\hat{s}^2\hat{c}^2 & (\tau \gg 1), \\ (\alpha\ln\xi/2\pi\hat{s}^2)[(5/12\hat{c}^2) - 3/8] & (\xi \gg 1), \end{cases} \quad (24)$$

$$\Delta\hat{r}_W = \begin{cases} (\alpha\ln\tau/4\pi\hat{s}^2)(1 - 16\hat{s}^2/9) & (\tau \gg 1), \\ (\alpha\ln\xi/2\pi\hat{s}^2)(1/24) & (\xi \gg 1). \end{cases} \quad (25)$$

We see that the large τ behavior of $\Delta\hat{r}$ is given by a power law while that of $\Delta\hat{r}_W$ is logarithmic. As a consequence, the dependence of $\Delta\hat{r}_W$ on m_t (and to a lesser extent on m_H) is considerably weaker than that of $\Delta\hat{r}$. In turn, because of the absence of the c^2/s^2 enhancement factor in the term proportional to τ , the latter's dependence on m_t is much slower than that in Δr. As a consequence, $\hat{s}^2$ is presently known with considerably better accuracy than s^2[Cf. eqs. (14,15)].

Recently it has been shown that by including the irreducible two-loop corrections to the ρ parameter [19] in $\Delta\hat{r}$, the relation

$$\hat{s}^2\hat{c}^2 = A^2/m_Z^2(1-\Delta\hat{r})$$

automatically contains the corrections of order $\alpha^2 \ln^2(m_Z/m_f)$, $\alpha^2 \ln(m_Z/m_f)$ and $\alpha^2(m_t^2/m_Z^2)^2$, where m_f represents a generic fermion mass [16]. An equivalent statement holds for the relation

$$\hat{s}^2 = A^2/m_W^2(1-\Delta\hat{r}_W).$$

This control of the leading and some of the subleading corrections of $O(\alpha)$ makes the basic corrections $\Delta\hat{r}$ and $\Delta\hat{r}_W$ specially useful for precision studies of radiative corrections at large m_t. In particular, one can derive simple relations between Δr, s^2, $\hat{s}^2$, $\Delta\hat{r}$ and $\Delta\hat{r}_W$ which bridge the $\overline{MS}$ and on-shell schemes and lead to an accurate analysis of the m_W-m_Z interdependence [16].

As mentioned before, the form factor $\hat{\kappa}$ multiplying $\hat{s}^2$ in Z^0-mediated amplitudes depends mildly on m_t and is independent of m_H. For this reason $\hat{s}^2$ is also a very convenient parameter for precision studies at the Z^0 peak and to correlate different observables [32]. A comparative analysis of $\overline{MS}$ and on-shell formulations of the radiative corrections to $e^- + e^+ \to f + \bar{f}$ is given in ref.[25].

It is also interesting to note that the current determination of $\sin^2\hat{\theta}_W(m_Z)$ (eq.(15)) is consistent with the prediction of the simplest supersymmetric extension of SU(5) for reasonable values of the SUSY-breaking scale [20].

Finally, it is worthwhile to call attention to recently proposed parameterizations of "new physics" [36-38]. Because in general heavy particles do not decouple in spontaneously broken theories, "new physics" characterized by mass scales $\gg m_Z$ may give rise to significant contributions in the radiative corrections to ordinary processes. In many cases it is expected that such contributions appear mainly in the self-energies. Peskin and Takeuchi [36] have proposed to express this effect in terms of two parameters, S and T, that represent the contributions of the new physics to "weak-isospin conserving" and "violating" combinations of self-energies, respectively. The radiative corrections to the various observables can then be written as the SM contributions plus various linear combinations of S and T. In particular, it has been emphasized that one generation technicolor theories give large and fairly model independent contributions to the "weak-isospin conserving" S [36,39]. Marciano and Rosner [37] have carried the analysis in the $\overline{MS}$ framework and pointed out that present accurate experiments on atomic parity violation are particularly sensitive to this parameter. Kennedy and Langacker [38] have used a more general parametrization involving S, T and a third quantity and by performing a global fit to all the available data have attempted to determine them.

It is apparent that the present experimental information already severely constrains the one generation technicolor models.

5 Acknowledgements

This work was supported in part by National Science Foundation Grant PHY-8715995 and Department of Energy Grant DE-AC02-76 ER02271.

References

[1] W. Jaus and G. Rasche, Phys. Rev. D 41, 166 (1990).

[2] G. Rasche and W.S. Woolcock, "The Determination of V_{ud} and a Test of the Unitarity of the Quark Mixing Matrix",Institut für Theoretische Physik Universität Zurich report (1990).

[3] D.H. Wilkinson, "The Vector Coupling Constant and the Kobayashi-Maskawa Matrix", University of Sussex report (1989); talk delivered at the PANIC Conference, Boston, Massachusetts (1990) and private communication.

[4] W.J. Marciano and A. Sirlin, Phys. Rev. D35, 1672 (1987).

[5] M. Veltman, private communication.

[6] A. Sirlin, Phys. Dev. D22, 971 (1980).

[7] K. Aoki, Z. Hioki, R. Kawabe, M. Konuma and T. Muta, Supplement of the Progress of Theoretical Physics, No. 73, 1 (1982).

[8] M. Böhm, W. Hollik and H. Spiesberger, Fortschr. Phys. 34, 687 (1986); W. Hollik, Fortschr. Phys. 38, 165 (1990).

[9] D.C. Kennedy and B.W. Lynn, Nucl. Phys. B321, 83 (1989); B322, 1 (1989).

[10] A.A. Akhundov, D.Y. Bardin and T. Riemann, Nucl. Phys. B276, 1 (1986).

[11] CDF, preliminary result presented at Neutrino'90, CERN, Geneva, June 1990.

[12] UA2: J. Alitti et al., CERN report CERN-EP/90-22.

[13] E. Fernandez, talk delivered at Neutrino'90, CERN, Geneva, June 1990.

[14] K. Sliwa, in Z^0 Physics, Proceedings of the XXVth RENCONTRE DE MORIOND, ed. J. Tran Thanh Van, Editions Frontieres, Gif-sur-Yvette Cedex-France, 1990, p. 459.

[15] G. Burgers and F. Jegerlehner, in Z^0 Physics at LEP1, eds. G. Altarelli, R. Kleiss and C. Verzegnassi, CERN 89-08, Vol. 1, p. 55 (1989).

[16] G. Degrassi, S. Fanchiotti and A. Sirlin, "Relations Between the On-Shell and $\overline{MS}$ Frameworks and the m_W-m_Z Interdependence", NYU report (May 1990), to be published in Nuclear Physics B.

[17] Z. Hioki, University of Tokushima report TOKUSHIMA 90-03.

[18] M. Consoli, W. Hollik and F. Jegerlehner, Phys. Lett. B227, 167 (1989).

[19] J.J. Van der Bij and F. Hoogeveen, Nucl. Phys. B283, 477 (1987).

[20] P. Langacker, University of Pennsylvania report UPR-0435T (1990) and private communication.

[21] W.A. Bardeen, C.T. Hill and M. Lindner, Fermilab report FERMI-PUB-89/127-T (1989).

[22] V. Miransky, M. Tanabashi and K. Yamawaki, Mod. Phys. Lett. A4, 1043 (1989); Phys. Lett. B221, 177 (1989).

[23] W.J. Marciano, Brookhaven Nat. Lab. report BNL-43230 (1989).

[24] W.A. Bardeen, contribution to this Symposium.

[25] G. Degrassi and A. Sirlin, Max-Planck-Institut report MPI-PAE/PTh 48/90 (1990).

[26] S. Bertolini and A. Sirlin, Nucl. Phys. B248, 589 (1984) and references cited therein.

[27] M. Veltman, Nucl. Phys. B123, 89 (1977).

[28] S. Bertolini and A. Sirlin, unpublished (1990).

[29] C.T. Hill and E.A. Paschos, Phys. Lett. B241, 96 (1990).

[30] A. Denner, R.J. Guth and J.H. Kühn, Max-Planck-Institut report MPI-PAE/PTh 77/89 (1989).

[31] S. Bertolini, Nucl. Phys. B272, 77 (1986).

[32] A. Sirlin, Phys. Lett. B232, 123 (1989) and in Z^0 Physics, Proceedings of the XXVth RENCONTRE DE MORIOND, ed. J. Tran Thanh Van, Editions Frontieres, Gif-sur-Yvette Cedex-France, 1990, p. 103.

[33] W.J. Marciano and A. Sirlin, Phys. Rev. Lett. 46, 163 (1981); S. Sarantakos, W.J. Marciano and A. Sirlin, Nucl. Phys. B217, 84 (1983).

[34] A. Sirlin, Nucl. Phys. B332, 20 (1990).

[35] S. Fanchiotti and A. Sirlin, Phys. Rev. D41, 319 (1990).

[36] M.E. Peskin and T. Takeuchi, Phys. Rev. Letters 65, 964 (1990).

[37] W.J. Marciano and J.L. Rosner, Enrico Fermi Institute report EFI-90-55.

[38] D.C. Kennedy and P. Langacker, University of Pennsylvania report UPR-0436T.

[39] M. Golden and L. Randall, Fermilab report FERMILAB-PUB-90/83-T (1990); B. Holdom and J. Terning, ITP-Santa Barbara report NSF-ITP-90-108 (1990); A. Dobado, D. Espriu and M.L. Herrero, CERN report CERN-TH.5785/90; C. Roiesnel and Tran N. Truong, "Technicolor Corrections to Electroweak Parameters", Ecole Polytechnique Report (1990).

Electroweak Symmetry Breaking: Top Quark Condensates

W. A. Bardeen

Fermi National Accelerator Laboratory, P.O. Box 500, Batavia, IL 60510, USA

Abstract. The fundamental mechanisms for the dynamical breaking of the electroweak gauge symmetries remain a mystery. This paper examines the possible role of heavy fermions, particularly the top quark, in generating the observed electroweak symmetry breaking, the masses of the W and Z bosons and the masses of all observed quarks and leptons.

1. Introduction.

Identifying the physical mechanism for the spontaneous breaking of the electroweak gauge symmetries has been a central challenge of elementary particle physics. During the present decade we should begin to establish the experimental evidence for the specific realization chosen by nature. Theoretical speculations have covered a wide range. The Standard Model presumes the existence of elementary scalar fields which condense to produce the observed symmetry breaking and result in a single physical Higgs particle whose mass is not determined. Alternative proposals have involved more complex Higgs sectors with multiple physical Higgs scalars, supersymmetric models with many additional physical states, or dynamical symmetry breaking models, such as technicolor, which rely on condensates of new technifermions and an entirely new sector of strong dynamics. The specific models are motivated by a variety of physical issues including renormalization, strong CP, natural gauge hierarchies, supersymmetry and superstrings. These models generally require the introduction of many new fundamental particles and their interactions.

This paper will focus on a different alternative which relies only on the presently observed particles and a heavy top quark to generate the electroweak symmetry breaking. Our work is motivated by Y. Nambu [1] who suggested that short range, attractive interactions between the fermions could generate the symmetry breaking, in analogy to the BCS theory of superconductivity. We will explore the physical consequences of these ideas. Fermion condensates of a heavy top quark will play a central role, and the expected masses for the top quark and physical Higgs particle, a top quark - antitop quark bound state, are principal predictions of the model. We will emphasize the

Springer Proceedings in Physics, Vol. 65 **Present and Future of High-Energy Physics**
Editors: K.-I. Aoki and M. Kobayashi

reliability of these predictions for the minimal model and examine possible extensions to more complex situations.

The presentation of this paper follows the results of Bardeen, Hill and Lindner (BHL) [2]. The role of top quark condensates in electroweak symmetry breaking was also the central theme of the top mode model of Miransky, Tanabashi and Yamawaki (MTY) [3]. The more general theme of composite gauge vector bosons and composite Higgs bosons has been a recurrent theme in the theoretical literature [4].

After a brief discussion of the elements of the original BCS mechanism and its relativistic NJL generalization, the dynamical basis of the top condensate model is analyzed by comparing three different approaches to understanding the essential elements of the minimal version of the theory. In the following section, various critical elements of the theory are discussed along with possible generalizations of the minimal model.

2. The BCS and NJL Models.

The electroweak interactions are generated by gauge interactions where the electroweak symmetries are spontaneously broken by the vacuum structure. All of the masses of the observed elementary particles, from gauge bosons to quarks and leptons, are generated by this symmetry breaking. The Standard Model relies on a fundamental Higgs field to provide the symmetry breaking. Dynamical symmetry breaking replaces the elementary Higgs field by condensates of the more fundamental degrees of freedom. Dynamical symmetry breaking forms the basis of the BCS theory of superconductivity [5].

2.1. The BCS Theory.

In the BCS theory of superconductivity, the complex interactions between the electrons result in a residual local, attractive interaction between the electrons. This attractive interaction can cause the electrons to bind into Cooper pairs. Dynamical symmetry breaking occurs when the energy of a Cooper pair becomes negative, and the normal vacuum becomes unstable to spontaneous creation of electron pairs. The vacuum structure is modified, and a new BCS ground state forms with a condensate of electron pairs, $\langle e_{\uparrow} e_{\downarrow} \rangle \neq 0$. A gap forms at the fermi surface, the electron becomes "massive", and the Meisner effect excluding the magnetic field from a superconductor reflects the dynamical mass generated for the magnetic part of the photon interactions. It is clear that the basic elements of the BCS theory could provide the framework for dynamical symmetry breaking in the electroweak interactions as suggested by Nambu [1].

2.2. The NJL Model.

The attractive, local interaction which induced the dynamical symmetry breaking was given a relativistic generalization through the Nambu - Jona-Lasinio (NJL) model [6]. This model considers the effects of a local, chiral invariant interaction between the fermions of the theory. The model is described by the following Lagrangian,

$$L = \overline{\Psi}\,\{i\partial\}\,\Psi + G\,(\overline{\Psi}_L\Psi_R)\cdot(\overline{\Psi}_R\Psi_L)\,, \qquad (1)$$

where a cutoff, Λ, must be introduced to define the quantum theory. The interaction is attractive for $G > 0$.

If the interactions are not sufficiently attractive, $G < G_c$, the vacuum structure will choose the symmetric phase for the chiral symmetries. The fermions will remain massless, $m_f = 0$, and the fermions will not form chiral condensates, $\langle\overline{\Psi}_L\Psi_R\rangle = 0$.

For sufficiently attractive couplings, $G > G_c$, the four fermion interactions will induce a dynamical symmetry breaking. In the broken vacuum, the fermions will develop masses and chiral condensates which imply a gap equation the fermion masses, $m_f = -G\langle\overline{\Psi}_L\Psi_R\rangle \neq 0$. The symmetry breaking implies the existence of Nambu-Goldstone bosons. A scalar boundstate of the massive fermions is formed with the usual NJL relation, $m_s = 2\,m_f$.

The NJL model is usually "solved" by using a bubble approximation for the dynamics. Fermions can develop dynamical masses, but all vertex corrections are surpressed. This approximation corresponds to the use of the BCS wavefunction in superconductivity. The gap equation follows from the mass relation, $m_f = -G\langle\overline{\Psi}_L\Psi_R\rangle \neq 0$, where the condensate is computed using an internal free fermion loop with the dynamically generated mass, m_f, and the cutoff, Λ. In bubble approximation, the fermion - anti-fermion scattering amplitude is computed as a sum of diagrams where the fermion bubbles are iterated in the direct channel. Because of the sensitivity to the cutoff, the bubble contributions to the condensate must be computed consistently with those in the scattering amplitudes, otherwise the cutoff will introduce an explicit breaking of chiral symmetries. A direct calculation using the bubble approximation confirms the existence of the appropriate Goldstone bosons and the NJL prediction of the scalar meson boundstate mass. The NJL model has been previously invoked for generating composite structures [4] and provides the fundamental basis for models involving condensates of the top quark.

3. Top Quark Condensate Models.

In the Standard Model, the electroweak symmetries are spontaneously broken by condensates of an elementary scalar Higgs field. As the allowed mass for the top quark has systematically increased, it is natural to speculate on a possible connection between a large top quark mass and source of electroweak symmetry breaking. Top quark condensate models carry this idea to the extreme. The Higgs sector of the Standard Model is totally eliminated in favor of local, attractive interactions between the fermions of the theory which will induce the electroweak symmetry breaking as in the NJL model. Because of its large mass, the top quark plays the central dynamical role. In the minimal model, electroweak symmetry breaking follows from top quark condensate alone.

The analysis of this section follows that of Bardeen, Hill and Lindner (BHL) [2] which was motivated directly by the suggestions of Nambu [1]. The idea of top quark condensates as the mechansim of electroweak symmetry breaking was also advocated by Miransky, Tanabashi and Yamawaki (MTY) [3]. MTY use a somewhat different dynamical basis for their analysis of the four-fermion interactions than that presented in the BHL paper and reach somewhat different results.

As mentioned above, the Higgs sector of the Standard Model is replaced by local, attractive interactions of the fundamental fermion fields. In the minimal model, the top quark plays an essential role in generating the electroweak symmetry breaking and the masses for the physical W and Z gauge bosons. The minimal model is described by the Lagrangian,

$$L_{fermion} = L_{kinetic} + G\,(\overline{\Psi}_{La}{}^{A}\, t_R{}^{a})\cdot(\overline{t}_{Rb}\, \Psi_{LA}{}^{b}), \tag{2}$$

where the composite operators are defined using a cutoff but preserving the electroweak gauge symmetries. $L_{kinetic}$ contains the kinetic terms for the fermions with the usual gauge couplings of the Standard Model. The four-fermion interactions in Eq.(2) represent the residual attractive interactions generated by a more fundamental dynamics existing above the cutoff scale. The four-fermion theory is not renormalizable, and physical quantities will be expected to depend on the cutoff scale even after renormalization of the independent coupling constants. Additional four-fermion interactions with weaker couplings could be added to generate masses for the lighter quarks and leptons but will have little effect on the dynmical symmetry breaking.

If these interactions are sufficiently attractive, $G > G_c$, then the electroweak symmetries will be spontaneously broken generating a mass for the top quark, $m_{top} > 0$, and a nontrivial top quark condensate, $\langle \overline{t}\, t \rangle \neq 0$. This symmetry breaking will also induce masses for the electroweak gauge bosons, $m_W \neq 0$ and $m_Z \neq 0$. We will also find that the physical Higgs particle of the Standard Model will be formed as a top quark - antitop quark bound state.

3.1. Bubble Approximation (NJL).

The standard method for analyzing the Nambu - Jona-Lasinio model (NJL) makes use of the bubble approximation. This method can be used as a first appoximation to the top quark condensate model as it contains the basic features of the composite structure produced by the dynamical structure of the theory. Phenomenological predictions will require the more complete analysis given in subsequent sections. The bubble approximation can be viewed as the large N_c limit of the theory where N_c is the number of colors, and $G*N_c$ is held fixed but all gauge couplings are neglected. The bubble theory has an exact solution in leading N_c approximation and $1/N_c$ corrections can be systematically computed.

The top quark mass is determined by the appropriate solution of the gap equation,

$$\begin{aligned} m_t &= -(1/2)\, G \langle \bar{t}\, t\rangle \\ &= G\,(N_c/8\pi^2)\,\{\, \Lambda^2 - m_t^2 \log(\Lambda^2/m_t^2)\,\}\, m_t \end{aligned} \tag{3}$$

with solutions,

$$m_t = 0$$

or (4)

$$m_t^2 \log(\Lambda^2/m_t^2) = \Lambda^2 - 8\pi^2/(N_c*G)$$

with the massive solution being the preferred vacuum solution. If the top quark mass is to be much below the cutoff, $m_{top} << \Lambda$, then a fine tuning of the four-fermion coupling, G, is required to cancel the quadratic cutoff dependence. Dynamical symmetry breaking can only occur if $G > G_c = (8\pi^2/Nc)\,(1/\Lambda^2)$.

In the broken symmetry phase, the vector bosons become massive through the effects of the vacuum polarization diagrams including the contributions of the fermion bubbles which are needed to preserve the transversality consistent with the underlying electroweak gauge symmetry. The inverse W-boson propagator is given by

$$(1/g_2^2) D_W{}^{-1\mu\nu}(k) = (k^\mu k^\nu/k^2 - g^{\mu\nu})\,\{\,(1/g_2^2)\,k^2 - f_W^2(k)\,\}, \tag{5}$$

where the effective decay constant, f_W, is

$$f_W^2(k) = N_c\,(1/16\pi^2) \int_0^1 dx\; x\, m_t^2 \log(\Lambda^2/(x m_t^2 - x(1-x)k^2)) \tag{6}$$

and $m_W^2 = g_2^2 f_W^2(m_W)$. The Z boson mass can also be computed in terms of its effective decay constant.

The computed values of the effective decay constants can be determined by the observed Fermi constant

$$W: \; f_W^2 = (N_c/32\pi^2)\, m_t^2 \{ \log(\Lambda^2/m_t^2) + 1/2\} = 1/4\sqrt{2}\, G_F,$$

$$Z: \; f_Z^2 = (N_c/32\pi^2)\, m_t^2 \{ \log(\Lambda^2/m_t^2) \} \approx f_W^2. \qquad (7)$$

For a large cutoff, $\Lambda \approx 10^{15}$ GeV, the bubble theory predicts a value for the top quark mass, m_{top} = 163 GeV, while m_{top} = 1 TeV for a smaller cutoff, $\Lambda \approx 10$ TeV. The bubble theory also predicts the mass of the physical Higgs scalar boson. It is given by $m_{higgs} = 2\, m_{top}$ which is the result usually quoted in the pure NJL theory.

We have seen that the elimination of the elementary Higgs sector has resulted in predictions for masses of both the top quark and the physical Higgs particle. Although qualitatively correct, the above predictions are strongly modified by the full electroweak dynamics.

3.2. Effective Field Theory.

From the bubble theory, we can infer that the effective low energy theory is the full Standard Model with composite Higgs fields. When viewed as the Standard Model, the coupling constants run with momentum scale at low energy, and the Higgs becomes static at high energy.

The effective field theory can be defined through the introduction of a static Higgs field, $H_A(x)$,

$$L_{fermion} = L_{kinetic} + G\, (\overline{\Psi}_{La}{}^{A}\, t_R{}^{a})\cdot(\overline{t}_{Rb}\, \Psi_{LA}{}^{b})$$

$$= L_{kinetic} - (\overline{\Psi}_{La}{}^{A}\, t_R{}^{a}) H_A - H^{+A}(\overline{t}_{Rb}\, \Psi_{LA}{}^{b}) - (1/G)\, H^{+}H. \qquad (8)$$

Instead of integrating out the static Higgs field to produce the four-fermion interaction, we can instead integrate out the short distance physics, replacing the cutoff scale Λ with a lower normalization scale μ. The short distance physics will generate contributions to the effective action defined at scale μ,

$$L_{fermion} = L_{kinetic} - (\overline{\Psi}_{La}{}^{A}\, t_R{}^{a}) H_A - H^{+A}(\overline{t}_{Rb}\, \Psi_{LA}{}^{b})$$

$$+ Z_H\, (D_\mu H)^2 - (1/2)\, \lambda_0\, (H^{+}H)^2 - (1/G + \Delta M^2)\, (H^{+}H) + O(1/\Lambda^2). \qquad (9)$$

In the bubble theory, the induced couplings are given by

$$Z_H = (N_c/16\pi^2)\log(\Lambda^2/\mu^2),$$
$$\lambda_0 = (2N_c/16\pi^2)\log(\Lambda^2/\mu^2), \tag{10}$$

and ΔM^2 has a quadratic dependence on the cutoff. From these results we can infer compositeness conditions on the running coupling constants,

$$Z_H = 1/g_t^2 \to 0, \quad \mu^2 \to \Lambda^2, \tag{11a}$$

$$\lambda_0 = \lambda/g_t^4 \to 0, \quad \mu^2 \to \Lambda^2. \tag{11b}$$

These conditions are exact in the bubble theory and are abstracted to the full theory where they should refect the approximate behavior of the effective running couplings. If Z_H becomes sufficiently large, then the effective top quark Yukawa coupling, g_t, is small, and the effective field theory below scale μ is the weakly coupled Standard Model with a dynamical Higgs field.

3.3. Renormalization Group.

The long distance behavior of the four-fermion theory can be described by a weakly coupled gauge theory with a composite Higgs field. Renormalization group methods are an efficient way to sum infinite sets of diagrams. The leading terms are just the leading log contributions which are expected to dominate if there is a large hierarchy of scales, ie. $m_{top} << \Lambda$. The renormalization group can be used to evolve the running couplings to high scales where they must be matched to the appropriate boundary conditions of the composite theory.

We can compare various treatments of the coupling constant evolution:

1] Bubble (NJL) theory includes only the fermion loop contributions. This theory generates a composite Higgs but suppresses gauge and Higgs loop contributions.

2] The usual large Nc limit of QCD requires $N_c \to \infty$, with $G*N_c$ and $\alpha_3 * N_c$ fixed, neglecting all other gauge couplings in loops. This theory includes all planar QCD corrections, and the low energy behavior is affected by infrared fixed points and is ultimately a theory of hadrons, not quarks and gluons.

3] The full Standard Model includes the effects of virtual Higgs contributions as well as the full gauge boson corrections. The theory is dominated by infrared fixed points of the renormalization group.

Figure 1 compares the three treatments for the running of $Z_H = 1/g_t^2$, from the Z boson mass scale to the cutoff scale, $\Lambda = 10^{15}$ GeV. This

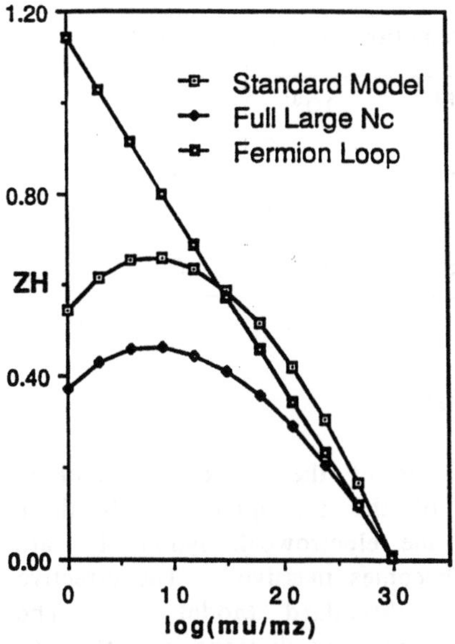

Figure 1. Higgs wavefunction/ top Yukawa coupling evolution.

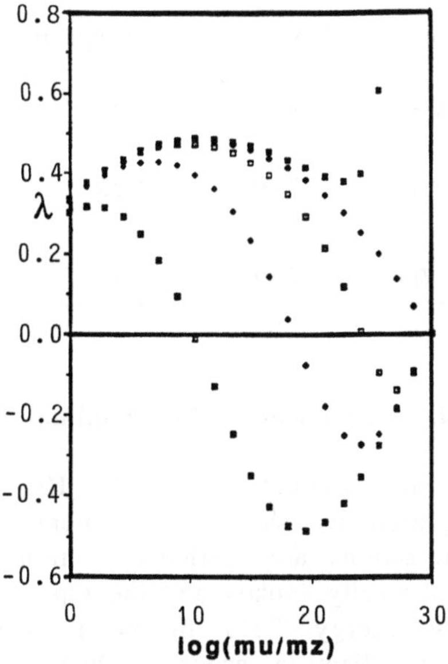

Figure 2. Higgs self-coupling evolution.

coupling directly determines the top quark mass as $m_{top} = g_t(m_{top})*v$. The turnover observed at low scales reflects the infrared fixed point behavior. The naive composite boundary condition was used at high scales although the perturbative renormalization group methods will break down as g_t becomes large. For a cutoff of 10^{15} GeV, the top quark mass predictons are

$$m_t(N_c \to \infty) \approx 270\text{GeV} > m_t(SM) \approx 230\text{GeV} > m_t(NJL) \approx 165\text{GeV}.$$

Figure 2 compares various renormalization group trajectories for the coupling constant of the Higgs self-interaction which determines the physical Higgs particle mass. The infrared fixed point structure sharply focusses the running at low energy and provides a precise prediction of the mass. The trajectories which flow to negative couplings at high energy are ruled out by the expected vacuum instability of these solutions at short distance.

Using the naive compositeness conditions but the full Standard Model evolution, BHL [2] obtained the following predictions for the top and physical Higgs particle masses for various composite scales, Λ. From a study of the compositeness conditions and the Standard Model evolution, it is expected that the theoretical ambiguities in the top quark mass predictions are only a few GeV [2].

Table 1. Top quark and Higgs mass predictions in minimal model.

Λ(GeV)	10^{19}	10^{15}	10^{13}	10^{9}	10^{4}
m_{top}	218	229	237	264	455
m_{higgs}	239	259	268	310	605

3.4. Conclusions for Minimal Model.

In the minimal model, the Higgs sector of the Standard Model is replaced by short range interactions of the top quark. If these interactions are sufficiently attractive, the electroweak symmetries are dynamically broken and the top quark becomes massive. The effective low energy field theory is the full Standard Model. The renormalization group methods give precise predictions for the top quark and the physical Higgs particle. The top quark must be quite heavy, $m_{top} > 220$ GeV. A high composite scale is favored, 10^{15} GeV → 10^{19} GeV, which might be identified with GUT or string model physics. The physical Higgs particle is expected to be only slightly heavier than the top quark, $m_{higgs} \approx 1.1\ m_{top}$, in contrast to the NJL (bubble) prediction of $m_{higgs} \approx 2\ m_{top}$. The infrared fixed point structure of the renormalization group equations stabilizes the predictions of the minimal model.

The minimal model predictions can be compared with constraints on the top quark mass coming from the various precision tests of the Standard Model. CDF provides a direct lower limit for the top quark mass, $m_{top} > 91$ GeV [7]. The strongest upper limits on the top quark mass come from deep inelastic neutrino scattering experiments and the collider measurements of the W boson mass. Langacker has reported the results of global fits to the present electroweak data [8]. He obtains the following upper limits,

$$m_{top} < 180 \text{ GeV (90\%CL)}, < 190 \text{ GeV (95\%CL)}, 210 \text{ GeV (99\%CL)},$$

where a Higgs particle mass of 250 GeV is assumed. This analysis depends on the careful understanding of the systematic errors for both theory and experiment for a wide range of processes. It is clear that the present analysis favors a lighter top quark than the expectations of the minimal top quark condensate model. However, we must await the discovery of the top quark as the determination of its mass will have crucial implications for the minimal top quark condensate model and perhaps the structure of Standard Model radiative corrections.

4. Comments and Extensions.

The minimal model of electroweak symmetry breaking discussed in the previous section can be related to a number of other approaches. It is important to examine the theoretical structure that makes it possible to have rather precise predictions of the top quark and Higgs particle masses. There are also many alternatives to the minimal model which still rely on the basic idea of short range, attractive interactions to generate the composite Higgs structure. In this section we will make a number of comments on the theoretical foundations of the model and discuss some of the most obvious extensions of the theory including a fourth generation and supersymmetry.

4.1. Infrared Fixed Points and Triviality.

The low energy behavior of the minimal model is governed by the full dynamics of the usual Standard Model. The low energy predictions are stabilized by the infrared fixed points (or more precisely pseudo-fixed points) of Standard Model renormalization group equations.

Fixed points were originally analyzed by Pendleton and Ross [9] and shown to provide a relation between the top quark Yukawa coupling constant and the running of the gauge coupling constants. If the evolution is to match the compositeness condition at high energy, then a different, but similar, relation between the couplings is achieved and the pseudo-fixed point discovered by Hill [10] dominates the low energy behavior. Figure 3 show the running top quark mass, or Yukawa

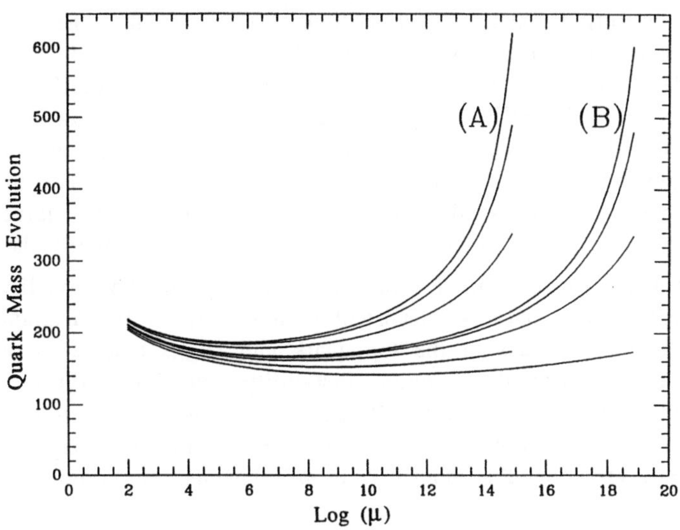

Figure 3. The quark mass evolution showing the pseudo-fixed point behavior for composite scales: 10^{15} (A) and 10^{19} (B) GeV.

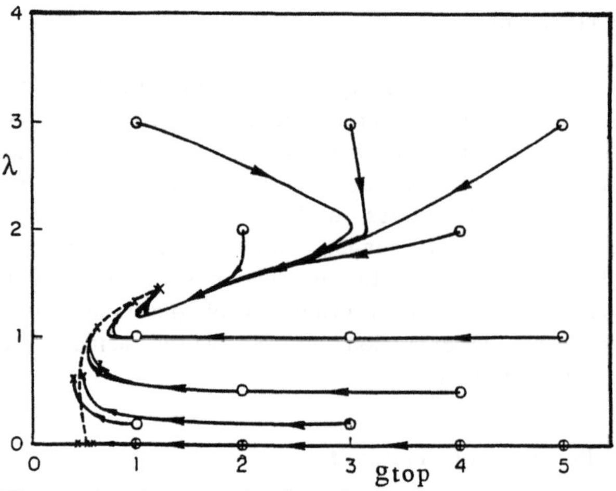

Figure 4. Renormalization flow to the pseudo-fixed point [11].

coupling constant, as a function of normalization scale using a variety of couplings at a high energy cutoff scale, (A): $\Lambda = 10^{15}$ GeV or (B): $\Lambda = 10^{19}$ GeV. The renormalization flow of the top quark and Higgs coupling constants to the pseudo-fixed point was analyzed by Hill, Leung and Rao [11] and is shown in Figure 4 for a variety of intial conditions. The pseudo-fixed point structure makes the low energy predictions very insensitive to high energy boundary conditions and the precise value of the cutoff.

The Standard Model is said to be a trivial quantum field theory [12] as the running coupling constants grow at high energy and that the low energy couplings would vanish in a theory without cutoff. In this sense the Standard Model is not fully renomalizable as the cutoff can not be removed. Triviality diagrams, as in Figure 5, show the limitations on the low energy parameters, m_{top} and m_{higgs}, which follow from requiring that the effective theory remain perturbative up to the cutoff scale. Since the minimal model requires the naive compositeness condition, $Z_H \to 0$ or $g_t^2 \to \infty$ as $\mu \to \Lambda$, be satisfied, the top quark condensate model will lie on the boundary of the triviality diagram. In fact, it will only be consistent with the tip of the diagram with the largest allowed values of the top quark and Higgs masses for a given cutoff scale because the vacuum instability noted in Figure 2 eliminates possible solutions for lighter Higgs particle masses. For a wider class of models, the boundaries of the triviality domains may be interesting to analyze for the possible interpretations of composite structure.

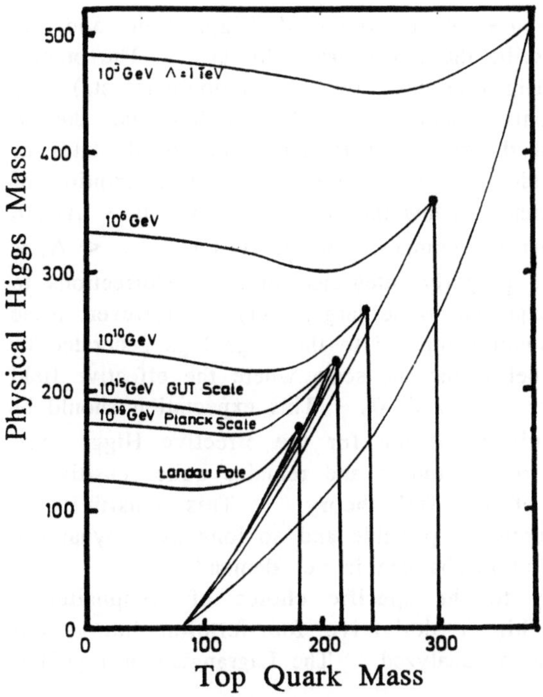

Figure 5. The triviality diagram for the Standard Model [12]. For each cutoff, the physical values of the top quark and Higgs particle masses must lie within the triviality domain. The compositeness condition is shown for each cutoff by the vertical line .

4.2. Compositeness Conditions.

In the bubble (NJL) approximation, an exact connection was made between the fundamental four-fermion dynamics at short distance and the effective Standard Model theory relevant to the long distance dynamics. We have abstracted an ultraviolet boundary condition on the running of the Standard Model coupling constants to reflect the composite structure. This connection is made in a domain where both the effective Standard Model couplings, $g_t \to \infty$, and the four-fermion couplings, $G \to G_c$, are becoming nonperturbative. Physics near the composite scale, Λ, is expected be very sensitive to renormalization effects, strong operator mixing, etc.

The basic physical structure of the theory will be preserved so long as the critical coupling, $G \to G_c$, remains a second order phase transition. The fine tuning required to produce an electroweak scale much below the composite scale can always be achieved. The second order transition implies the existence of a dynamical Higgs field and the effective field theory to describe the low energy physics. The precise bound state structure (one Higgs doublet, two Higgs doublets, etc) may depend on the nonperturbative aspects of the fundamental theory. However, the effective field theory which includes the bound states as independent degrees of freedom, should provide a good description of the dynamics at scales sufficiently below the composite scale, $\mu < \Lambda/10$, $\Lambda/100$. The physics near the composite scale, $\Lambda/10 < \mu < \Lambda$, is nonperturbative and must be properly integrated out. Corrections to the bubble theory can be expected to be large, O(1). However, these effects are expected to be small compared to the large logs generated by integrating out the physics below the the scale where the effective field theory becomes perturbative, eg. $\Lambda/10$. This expectation should be valid for the running couplings but not for the effective Higgs mass parameter which is subject to fine tuning and remains quite sensitive to even small modifications of the full theory. This sensitivity is irrelevant so long as fine tuning is possible and so long as a dynamical explanation of the fine tuning mechanism is not demanded.

To test the sensitivity to the specific choice of compositeness conditions, a model with higher derivative four-fermion interactions suggested by Suzuki [13] can be analyzed. The Lagrangian of Eq.(2) is replaced by

$$\begin{aligned} L &= L_0 + G_0\cdot\{ (\overline{\Psi}_L t_R+(\chi/\Lambda^2)\cdot D\overline{\Psi}_L D t_R)(\overline{t}_R \Psi_L+(\chi/\Lambda^2)\cdot D\overline{t}_R D\Psi_L \} \\ &= L_0 - (1/G_0)\cdot\{H^+H\} - (\overline{\Psi}_L t_R+(\chi/\Lambda^2)\cdot D\overline{\Psi}_L D t_R)H \\ &\qquad - H^+(\overline{t}_R\Psi_L+(\chi/\Lambda^2)\cdot D\overline{t}_R D\Psi_L)\,. \end{aligned} \tag{12}$$

By integrating out the high momentum components of the fermion loops, the effective action becomes

$$\begin{aligned} L \to L_0 &- (\overline{\Psi}_L t_R+(\chi/\Lambda^2)\cdot D\overline{\Psi}_L D t_R)H - H^+(\overline{t}_R\Psi_L+(\chi/\Lambda^2)\cdot D\overline{t}_R D\Psi_L) \\ &+ Z_H\cdot\{DH^+DH\} - m^2\{H^+H\} - (1/2)\cdot\lambda_0\cdot\{(H^+H)^2\}\,, \end{aligned} \tag{13}$$

where the running couplings are given by

$$\begin{aligned} Z_H &= (N_c/(4\pi)^2)\cdot\{ \ln(\Lambda^2/\mu^2) - 2\cdot\chi + \chi^2/4 \}\,, \\ \lambda_0 &= 2\cdot(N_c/(4\pi)^2)\cdot\{ \ln(\Lambda^2/\mu^2) - 4\cdot\chi + 3\cdot\chi^2 - (4/3)\cdot\chi^3 + \chi^4/4 \}\,, \\ m^2 &= 1/G_0 + \cdots \quad \text{(fine tuned)} \end{aligned} \tag{14}$$

using bubble approximation for the explicit calculations. For scales sufficiently below the composite scale, the higher derivative Yukawa

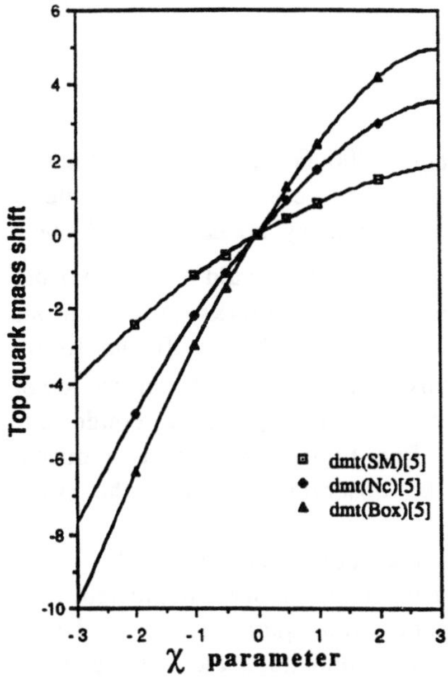

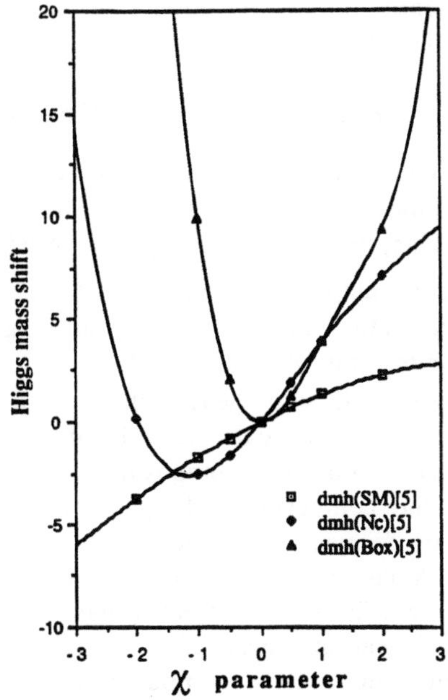

Figure 6. Top quark mass shift. Figure 7. Higgs mass shift. Shift due to modified UV boundary conditions for various treatments of the coupling constant evolution (SM, large Nc, box (NJL)) where χ is the coupling constant of the higher derivative interactions.

couplings may be neglected, $D_\mu/\Lambda << 1$, and the theory evolves as the normal Standard Model as in the case of minimal model. However, the presence of the higher derivative interactions has modified the compositeness boundary conditions.

Using Eq.(14) for the evolution between scales Λ and $\Lambda/5$, the low energy effective theory can be computed using various approximations for the evolution (NJL(bubble), large Nc, Standard Model) below the scale, $\Lambda/5$. Figures 6 and 7 show the effects of the higher derivative interactions on the predictions of the top quark mass and the Higgs mass. For reasonable variations of the higher derivative coupling strength, $0 < \chi < O(1)$, the predictions of the Standard Model evolution are very stable with at most a few GeV shift in the masses. It is the fixed point structure of the full Standard Model evolution that provides this stability. This example is used only to indicate the possible effects of the physical evolution of the effective theory near the composite scale. As mentioned earlier, the initial evolution from the composite scale is likely to require nonperturbative analysis. This initial evolution modifies the boundary conditions for the subsequent

Standard Model evoluton but has a limited effect on the ultimate predictions.

4.3. Fourth Generation Models.

If the top quark is found to be light, $m_{top} < 200$ GeV, then the top quark dynamics can not produce all the observed electroweak symmetry breaking, for $\Lambda < m_{planck}$. Additional symmetry breaking could come from a number of sources. An obvious extension would be to consider condensates involving a fourth generation of quarks and leptons assuming the masses satisfy the ρ parameter bounds. If the fourth generation is very heavy, then the composite scale could be much lower than the GUT scale considered for the minimal top quark condensate model. For low composite scales, the fine tuning problem is reduced, and the composite scale physics could be observable through the study of rare decays, FCNC, etc.

A degenerate fourth generation was considered by BHL [2]. The top contribution to electroweak symmetry breaking was neglected, and the degenerate mass for the fourth generation quark and the Higgs particle mass were computed for different composite scales, Λ, and are shown in Table 2.

Table 2. Mass predictions for the fourth generation model.

Λ (GeV)	10^{19}	10^{15}	10^{13}	10^{6}	10^{4}
m_{quark}	199	206	212	277	388
m_{higgs}	235	248	258	365	553

A fourth generation model with maximal mixing of the fourth generation quarks with the top quark was considered by Marciano [14]. He found that the top quark and the fourth generation up quark were nearly degenerate with a mass of 140 GeV. The fourth generation bottom quark was somewhat heavier at 160 GeV. Clearly the precise nature of the weak mixing will have an important impact on the predictions for a fourth generation model, and the compositeness conditions will only partly constrain these mixings.

The recent bounds from LEP on the number of light neutrinos implies that the fourth generation neutrino, if it exists, must be rather heavy, $m_{\nu 4} > 45$ GeV. Heavy neutrinos could result from mixing structure in the neutrino mass matrix [15]. With the addition of right-handed neutrinos, it might be natural to expect that the dirac masses of the neutrinos are comparable to the charged lepton masses. A large

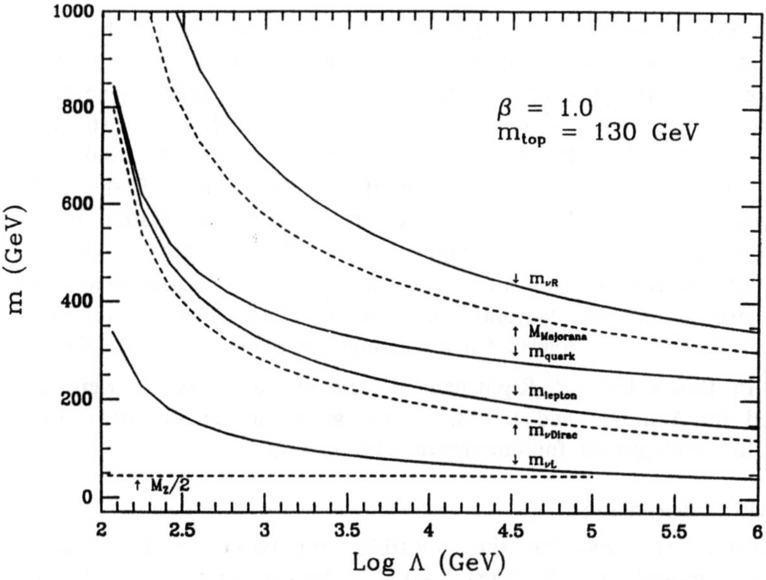

Figure 8. Quark and lepton masses of the four generation model with neutrino and quark condensates as functions of the cutoff.

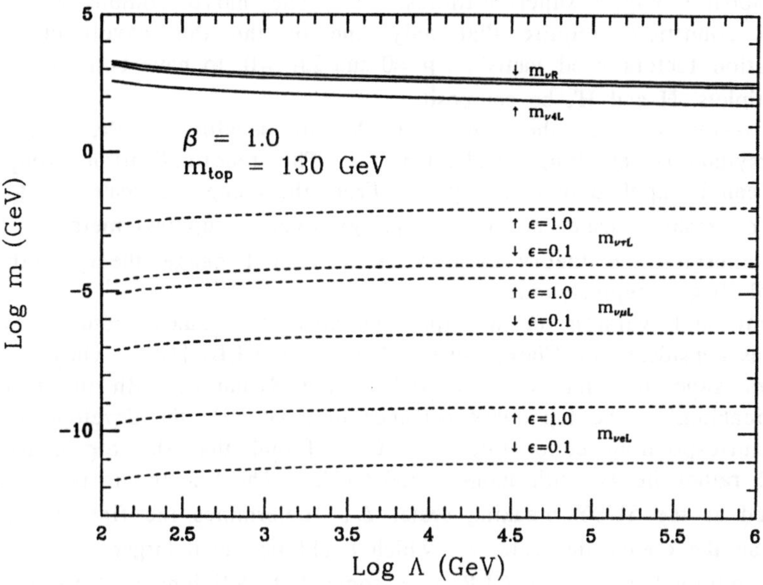

Figure 9. Neutrino mass spectrum of the four generation model assuming comparable Dirac masses for charged and neutral leptons.

Majorana mass for the neutrinos which does not break the usual electroweak symmetries could then be invoked which would suppress the masses of the observed neutrinos of the first three generations leaving the fourth generation neutrino heavy. Hill et al [16] have suggested a class of models where the Majorana mass results from neutrino condensates produced by attractive interactions of the right-handed neutrinos. Their predictions for the fourth generation masses are given as a function of the composite scale, Λ, in Figure 8 while the spectrum of neutrino masses are given in Figure 9. These results are for the case where the Majorana and normal Higgs VEV's are taken equal, $\beta = V_m/V_h = 1$; note that a low composite scale, $\Lambda < 10^6$ GeV, is required in this scheme. Right-handed neutrino condensates were also considered by Achiman and Davidson [17] as a mechanism for neutrino mixing with implications for composite DFS axions.

4.4. SUSY Extensions.

It is natural to consider the possible extension of the fermion condensate model to theories with supersymmetry. At the fundamental level, a supersymmetric extension of the local four-fermion interaction replaces the fundamental Higgs fields. This minimal supersymmetric extension of the BHL model would generate two composite Higgs supermultiplets. The naive compositeness boundary conditions require that only one of the the wavefunction normalization factors need vanish, $Z_H = 0$ and $Z_{H'} \neq 0$, to have both Higgs supermultiplets, H and H', be composite.

The results of even the minimal model are sensitive to the nature of supersymmetry breaking mechanisms. The renormalization group methods can be applied in two stages. From the composite scale, Λ, to the SUSY breaking scale, Δ, the couplings evolve supersymmetrically. At low energies the theory evolves as a normal gauge theory with additional Higgs representations.

A minimal version of a supersymmetric top quark condensate model was considered by Clark, Love and Bardeen (CLB) [18]. Only the top quark supermultiplet was involved in the dynamics with minimal SUSY breaking. The top quark becomes massive with the development of the corresponding condensate. CLB found that the top quark remained rather heavy with $m_{top} \approx 200$ GeV. The fine tuning problem is reduced as the SUSY breaking scale, Δ, determines the fine tuning rather than the composite scale, Λ, which could be much larger.

The minimal model of CLB was somewhat deficient as only the effective Higgs coupled to the top quark received a vacuum expectation value. The bottom quark would remain massless even if the effective Yukawa coupling were nonzero. Additional SUSY breaking must be added, as in the usual SUSY models, to the original interactions to produce vacuum expectation values for both effective Higgs multiplets.

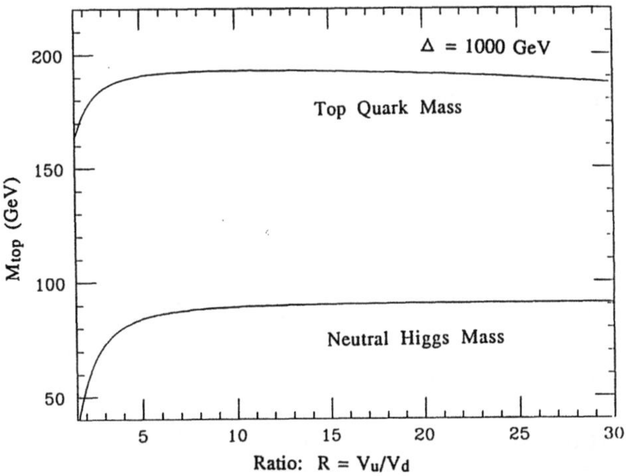

Figure 10. Top quark and physical Higgs particle masses for the minimal SUSY condensate model as a function of the ratio of the vacuum expectation values of the effective Higgs fields.

The prediction of the top quark mass is reduced by the ratio of the VEV seen by the top quark, V_u, to the VEV seen by the gauge bosons, $V = \sqrt{V_u^2+V_d^2}$. The top quark mass depends on both the composite scale and SUSY breaking scale. The predicted values are given in Table 3 where the mass value must still be multiplied by the VEV ratio, m_{top} = [Table 3] * (V_u/V).

With this standard SUSY breaking scheme, the top quark mass predictions [19] are shown in Figure 10 as a function of the Higgs VEV ratio, V_u/V_d. Also plotted are the mass predictions for the lightest neutral Higgs particle. These results can be related to the boundaries of standard renormalization group studies of supersymmetric models [20]. In the minimal models considered above, the Higgs particle mass must be less than the Z boson mass, 91 GeV, and more than the recent LEP lower bounds, 40 GeV [21]. These results imply that the top quark should be heavier than about 170 GeV in the minimal composite SUSY model with the standard SUSY breaking scheme.

Supersymmetry is usually invoked to help explain the gauge hierarchy problem. The composite Higgs models considered in this section do not directly address this problem although the fine tuning scale is set by the SUSY breaking scale instead of the higher compositeness scale. However this scale dependence also implies that the effective couplings of the fundamental, higher dimension operators may have greatly enhanced couplings if they are to generate the composite Higgs structure; this may cause problems with the perturbative unitarity of the theory. Although these models focus on

Table 3. SUSY model predictions of the top quark mass (GeV) as functions of the composite scale Λ and SUSY breaking scale, Δ.

$\Lambda \backslash \Delta$	10^2	10^4	10^6	10^8	10^{10}
10^6	259	290			
10^9	222	241	254	263	
10^{13}	203	215	224	231	236
10^{15}	198	208	216	222	227
10^{19}	191	200	206	211	214

composite Higgs structure, other supermultiplets could be considered for compositeness on the same basis.

4.5. Multiple Composite Higgs.

Another natural extension of the minimal top quark condensate model is the possible formation of additional composite Higgs bound states. This extension can be achieved by considering additional local, four-fermion interactions between the fundamental fermion fields. These additional interactions can produce attractive interactions in more than one channel and result in new bound states. Normally one would expect that additional fine tuning would be required for the new states to generate physics at a sufficiently low scale.

The simplest model involves interactions of both the top and bottom quarks as given in the following Lagrangian,

$$L_{fermion} = G_u\,(\overline{\Psi}_{La}{}^{A}\, t_R{}^{a})\cdot(\overline{t}_{Rb}\, \Psi_{LA}{}^{b}) + G_d\,(\overline{\Psi}_{La}{}^{A}\, b_R{}^{a})\cdot(\overline{b}_{Rb}\, \Psi_{LA}{}^{b})$$

$$+ G_{ud}\,(\overline{\Psi}_{La}{}^{A}\, t_R{}^{a})\cdot(\overline{\Psi}_{Lb}{}^{B}\, b_R{}^{b})\varepsilon_{AB} + \text{h.c.}\ , \qquad (15)$$

where the last term is needed to provide mixing and break the chiral symmetries which would result in electroweak axions that are presently ruled out by experiment. With sufficient fine tuning, this theory generates two Higgs doublets, $H_{uA} = (\overline{t}_{Rb}\ \Psi_{LA}{}^{b})$ and $H_{bA} = (\overline{\Psi}_{Lb}{}^{B}\ b_R{}^{b})\varepsilon_{AB}$. Whether the degree of fine tuning required to keep both Higgs multiplets light is possible to achieve may require a nonperturbative study of the phase transition structure of the fundamental four-fermion theories. Studies using bubble approximation and modified renormalization group methods seem to indicate a consistent picture with additional light composite Higgs states. Two doublet models were considered by a number of authors [22] with consequences for the mass predictions of the top and bottom quarks as well as the spectrum of observable Higgs particles. The compositeness conditions place interesting constraints on the effective potential of the composite Higgs fields with implications for their low energy dynamics.

As mentioned before, four generation models require a mechanism for the producing a massive fourth neutrino. This may be accomplished [15,16] by introducing right-handed neutrinos. Possible condensates of these right handed neutrinos would produce electroweak singlet composite Higgs fields [16,17]. Hill, Luty and Paschos [16] have studied a unified picture of singlet and nonsinglet condensates which could play a role in realistic four generation models. Achiman and Davidson [17] have emphasized the possible role of right-handed neutrino condensates in producing composite DFS axions and their role in model building.

4.6. UV Fixed Points and Reduction.

We have emphasized the role of infrared (pseudo-) fixed points in generating stable predictions for the top quark condensate models. The pseudo-fixed points [10] control the renormalization group evolution at low energy where the gauge coupling constants control the running of the top quark Yukawa coupling constant. If the composite scale is large, then the top quark mass predictions are insensitive to the composite boundary conditions and are dictated by the infrared pseudo-fixed points.

A more ambitious analysis of the relation between the various electroweak coupling constants is made in the reduction approach advocated by Zimmerman, et al. [23] and Marciano[24]. In this approach, it is assumed that the Standard Model couplings are not all independent but have a functional interdependence. For example, the top quark Yukawa coupling constant is determined as a function of the gauge coupling constants. The renormalization group is used to determine the possible relationships between the various Standard Model couplings. For the top quark, the results are similar to those obtained in the orginial fixed point analysis of Pendleton and Ross [9].

Focussing on the top quark mass, the renormalization group provides a relation between the running top quark Yukawa coupling, K_t, and the color coupling constant, α_3. The renormalization group equations can be integrated to give

$$K_t = 2\,\alpha_3^{8/7} \,/\, (C + 9\,\alpha_3^{1/7}) + \text{electroweak corrections}\,, \qquad (16)$$

where C is an integration constant with special values, 0 and ∞. $C = \infty$

might correspond to the situation of the light quarks. For C = 0, the coupling relation becomes analytic, and C > 0 is required if the Eq.(15) is to be nonsingular. If we identify C = 0 with the top quark situation, then the top quark mass is predicted to be 90-95 GeV [23, 24] which is just at the present lower limit of the direct search by CDF [7]. The same analysis would predict the Higgs particle mass of about 64 GeV which is still somewhat above the recent LEP lower bounds [21]. Refinements are not expected to change these predictions by large amounts.

The reduction approach chooses the special value, C = 0, to preserve the analytic structure of Eq.(16). From the renormalization group point of view, this constraint comes from requiring a nonsingular behavior up to infinite energies where α_3 vanishes. This contrasts with the top quark condensate models where the coupling constant becomes large at the composite scale which is taken to be less than the Planck scale. Hence, the solutions with C < 0 are required for top condensate models. The reduction method requires knowledge of the gauge coupling constant, α_3, for physical values corresponding to energies far beyond the Planck scale. Hence we can only view the constraints imposed by the reduction method as mathematical conditions imposed on the theory (perhaps as a result of hidden symmetries) and not as conditions on the physical running couplings. We will soon be able to determine the validity of the top quark mass predicition as the limits from CDF are improved or the top quark is discovered.

4.7. Schwinger-Dyson Equation Approach.

We have emphasized the use of renormalization group methods to give reliable predictions for the low energy parameters of the top condensate theory. The renormalization group is used to sum the leading contributions from an infinite set of Feynman diagrams describing the short distance physics. An alternate approach involves the direct solution of the Schwinger-Dyson equations for the behavior of the top quark self-energy function and use it to predict the top quark mass and the related electroweak symmetry breaking.

The Schwinger-Dyson equations have been studied by Barrios and Mahanta [25] and by King and Mannan [26]. Both groups study the effects of local four-fermion interactions on the solutions to the Schwinger-Dyson equations with the gauge interactions included in ladder approximation. The four-fermion coupling must be fine-tuned to generate a light top quark, $m_{top} << \Lambda$. The color gauge interactions increase the predicted value of the top quark mass from that obtained from the pure four-fermion theory in bubble approximation. The results are given in Table 4 for various values of the cutoff scale, Λ. Barrios and Mahanta have compared their Schwinger-Dyson results with the standard renormalization group procedure, as used by BHL [2],

Table 4. Top mass predictions from the Schwinger-Dyson approach compared with the renormalization group results.

Λ (GeV)	$m_t(\alpha_3=0)$	$m_t(\alpha_3)$	$m_t(RG)$
10^5	379.5	442.9	438.2
10^9	229.2	306.2	310.6
10^{15}	165.2	257.3	261.6
10^{19}	143.9	242.0	246.1

and achieve good agreement between the two methods so long as the same physics is considered.

The Schwinger-Dyson equation method as implemented by both groups [25,26] includes only the effects of the local four-fermion interactions and the color gauge interactions, both in ladder approximation. These results must be compared to the equivalent renormalization group calculations which are similar to the "large Nc" results of BHL. As emphasized by BHL, the full low energy dynamics must be included to make meaningful physical predictions. The virtual Higgs contributions and the full electroweak gauge interactions were essential to obtain reliable predictions of the top quark and Higgs particle masses. These physical contributions must be included in the Schwinger-Dyson approach before its results can be compared directly to data. The Schwinger-Dyson approach must also confront possible nonperturbative aspects of the short distance physics associated with the four-fermion interactions which may also affect the intial evolution of the renormalization group method. The low energy predictions are expected to be somewhat insensitive to the short distance structure because of the infrared (pseudo-) fixed point behavior at long distance, but it may be difficult to separate these effects in the more direct analysis of Schwinger-Dyson equations.

4.8. Long Distance Contributions.

The renormalization group method can be used to systematically study the contributions of physics below the composite scale. The BHL analysis uses the one loop anomalous dimensions to compute the effective action to use at low energies, $\mu \approx m_Z$. The low energy radiative corrections must be used for accurate comparison of the theory with experiment. For example, top quark self-energy will receive a low energy contribution from its QCD interactions which result in a mass shift,

$$\delta m_{top} = m_0\,(\alpha_3/3\pi)\,(\,4 + 3\,\ln(\mu^2/m_0^2)\,)\,, \qquad (17)$$

where μ is the low energy normalization scale, α_3 is the QCD coupling constant and m_0 is the lowest order top quark mass. In the BHL calculation, the ln contributions were absorbed by evolving the top quark Yukawa coupling constant to the top quark mass scale instead of the m_Z normalization scale. A remaining $O(\alpha_3)$ contribution could affect the top quark mass predictions as emphasized by Kugo [27]. However, these QCD corrections may be largely cancelled by similar electroweak corrections as they are for the log corrections which are largely cancelled because of the infrared fixed point structure. A consistent calculation must incorporate the two loop contributions to the renormalization group evolution as well as the finite part of the one loop effects at the low energy scale. Since the two loop anomalous dimensions are known, the top quark mass predictions can be systematically improved.

5. Conclusions.

We have shown that the BCS mechanism can be used to connect the electroweak scale (m_W, m_Z) with the masses of the top quark and the physical Higgs particle. Electroweak symmetry breaking is produced by condensates of the top quark which are triggered by attractive, local interactions of the top quark. Stable predictions for the top quark and Higgs masses are achieved using the full Standard Model evolution which reflects the presence of infrared (pseudo-) fixed points in the renormalization group equations.

The minimal model predicts a heavy top quark, $m_{top} > 200$ GeV. For large composite scales, the physical Higgs particle is predicted to be only slightly heavier than the top quark, $m_{higgs} \approx 1.1 * m_{top}$ which contrasts with the NJL prediction of $m_{higgs}/m_{top} = 2$. The renormalization group analysis results in a rather precise prediction of the top quark mass, $m_{top} = 229 \pm 5$ GeV ($\Lambda \approx 10^{15}$ GeV) where the error reflects an estimate of the theoretical error coming from the composite boundary conditions. To achieve a relatively light top quark in the minimal model, the composite scale must be taken to be quite large and could be associated with the GUT scale, 10^{15} GeV, or the Planck scale, 10^{19} GeV. Even with this choice of composite scale, the minimal top quark model predicts masses which are somewhat larger than the range estimated by recent fits to the Standard Model radiative corrections, $m_{top} = 137 \pm 40$ GeV [8]. While the minimal top condensate model seems to be somewhat disfavored by the present data, we should wait until the top quark is discovered before reaching final conclusions about the validity of the minimal model or the normal radiative corrections.

There are many extensions to the minimal model which preserve the basic idea of electroweak symmetry breaking being generated by new short distance dynamics rather than additional fundamental degrees of freedom. The four generation version of the theory relaxes the constraint on a heavy top quark but requires a mechanism for producing a heavy neutrino for the fourth generation. Right-handed neutrinos could have an important dynamical role in generating mixing and producing a spectrum of neutrino masses. Condensates of right-handed neutrinos could have important phenomenological consequences and could be responsible for a new mechanism for axions.

Fermion condensates may also generate a more complex Higgs structure. Models with two Higgs doublets have been studied in some detail. The compositeness conditions place interesting constraints on the effective potentials of the dynamical Higgs fields. As stated above, neutrino condensates may also play a role in the low energy dynamics.

We have briefly discussed the relation of the condensate models with alternative approaches including coupling constant reduction which has much different predictions for the top quark and Higgs masses and the direct Schwinger-Dyson equation method which yields the same results as the renormalization group methods so long as the same physical input is achieved. We have not discussed early work [4] which focussed on the possible role of four-fermion interactions in generating composite vector mesons using methods associated with the NJL models. The role of electroweak gauge symmetry plays a crucial role in these models, and the dynamical structure of these models remains to be understood.

There are many remaining questions for the implementation of condensate models. The models require fine-tuning to produce an electroweak scale that is much below the composite scale as was the case in the normal Standard Model. It is interesting to speculate about possible mechanisms which could generate this fine-tuning dynamically. The physics at the composite scale is not renormalizable and is presumably generated by physics beyond the composite scale such as the fragments of a GUT model or superstring theory. These theories should generate many higher dimension interactions which are mostly irrelevant to the low energy physics. If the coupling constants of these interactions are dynamically determined, it is possible to imagine feed-back mechanisms could produce the apparent fine-tuning needed to produce the desired infrared structure of the full theory. The top quark condensate model makes definite predictions for the top quark and Higgs particle masses the flavor mixing structure observed in the quark sector can only be accommodated. The understanding of flavor mixing whether it is in the explored domain of the quarks or the unexplored domain of the neutrinos remains an outstanding problem for any fundamental theory.

6. Acknowledgements.

I would like to thank Nishinomiya City for its hospitality in sponsoring the 1990 Nishinomiya Yukawa Memorial Symposium. I would also like to thank Dr. Ken-Ichi Aoki and the other organizers of the scientific programs for a stimulating and productive experience in Japan. This work would not have been possible without the continuing collaboration with Chris Hill and my other colleagues on the many different aspects of this research.

7. References.

1. Y. Nambu, Proceedings of the 1988 Kazimierz Workshop, Z. Ajduk et al., eds. (World Scientific, 1989); Proceedings of the 1988 Nagoya Workshop, M. Bando et al., eds. (World Scientific, 1989); EFI Preprint 89-08 (1989) (unpublished).

2. W. Bardeen, C. Hill and M. Lindner, Phys. Rev. D41, 1647(1990).

3. V. Miransky, M. Tanabashi and K. Yamawaki, Mod. Phys. Lett. A4,1043(1989); Phys. Lett. B221, 177(1989).

4. H. Terazawa, Phys. Rev. D22,2921(1980); H. Terazawa, Y. Chikashige and K. Akama, Phys. Rev. D15, 480(1977), F. Cooper, G. Guralnik, and N. Snyderman, Phys. Rev. Lett. 40, 1620(1978); T. Eguchi, Phys. Rev. D14, 2755(1976); K. Kikkawa, Prog. Theor. Phys. 56, 947(1976); T. Kugo, Prog. Theor. Phys. 55, 2032(1976); J.D. Bjorken, Ann. Phys. 24, 174(1963).

5. J. Bardeen, L. Cooper and J. Schrieffer, Phys. Rev. 108, 1175(1957).

6. Y. Nambu and G. Jona-Lasinio, Phys. Rev. 122, 345(1961); W. Bardeen, C. Leung, and S. Love, Phys. Rev. Lett. 56, 1230 (1986).

7. CDF Collaboration (F. Abe, et al.), Phys. Rev. Lett. 64, 142(1990); FERMILAB-PUB-90-137-E (1990) (to be published in Phys. Rev. D).

8. D. Kennedy and P. Langacker, "Precision Electroweak Experiments and Heavy Physics: A Global Analysis", U. Pennsylvania Preprint UPR-0436T (1990); P. Langacker, "Precision Tests of the Standard Model", U. Pennsylvania Preprint UPR-0435T (1990).

9. B. Pendleton and G. Ross, Phys. Lett. 98B, 291(1981).

10. C. Hill, Phys. Rev. D24, 691(1981).

11. C. Hill, C. Leung and S. Rao, Nucl. Phys. B262, 517(1985).

12. L. Maiani, G. Parisi and R. Petronzio, Nucl. Phys. B136, 115(1978); N. Cabbibo et al, Nucl. Phys. B158,295(1979); M. Lindner, Z. Phys. C31, 295(1986).

13. M. Suzuki, Mod. Phys. Lett. A5,1205(1990); 1990 Nagoya Workshop (1990).

14. W. Marciano, "Dynamical Symmetry Breaking and the Top Quark Mass", Brookhaven Preprint, September, 1990.

15. C. Hill and E. Paschos, Phys. Lett. B241, 96(1990).

16. C. Hill, M. Luty and E. Paschos, "Electroweak Symmetry Breaking by Fourth Generation Condensates and the Neutrino Spectrum", FERMILAB-PUB-90-212-T (1990).

17. Y. Achiman and A. Davidson, "Dynamical Axion of Dynamical Electro-Weak Symmetry Breaking", Wuppertal Preprint WU B90-14 (1990).

18. T. Clark, S. Love and W. Bardeen, Phys. Lett. B237, 235(1990).

19. K. Sasaki, M. Carena, C. Wagner, T. Clark and W. Bardeen, (in prep.)

20 H.P. Nilles, Phys. Rep. 110, 1(1984); J. Bagger and S. Dimopoulos, Phys. Rev. Lett. 55 920(1985); H. Okada and K. Sasaki, Phys. Rev. D40, 3743(1989).

21. OPAL Collaboration (M. Akrawy, et al.), "Search for the Minimal Standard Model Higgs Boson in e+e- Collisions at Lep", CERN-PPE-90-150 (1990).

22. M. Luty, Phys. Rev. D41, 2893(1990); M. Suzuki, Phys. Rev. D41, 3457(1990); C. Froggatt, I. Knowles and R. Moorhouse, Phys.Lett. B249, 273(1990) ; J. Fröhlich and L. Lavoura, "Bootstrap Dynamical Symmetry Breaking with Top and Bottom Quarks", Dortmund Preprint DO-TH-90/11 (1990); C. Hill, M. Luty and E. Paschos, "Electroweak Symmetry Breaking by Fourth Generation Condensates and the Neutrino Spectrum", FERMILAB-PUB-90-212-T (1990).

23. J. Kubo, K. Sibold and W. Zimmerman, Phys. Lett. B220, 85(1989).

24. W. Marciano, Phys. Rev. Lett. 62 (1989).

25. F. Barrios and U. Mahanta, ITP Santa Barbara Preprint, UCTP-104/90 (1990).

26. S. King and S. Mannan, Phys.Lett. B241, 249(1990).

27. T. Kugo, contribution to Nishinomiya/Kyoto Symposium (1990).

Searches at LEP

D. Treille

CERN-PPE Division, CH-1211 Geneva 23, Switzerland

Introduction

After one year of physics at LEP and $\sim$ 700k Z^0 decays registered by its four experiments, no new particle nor new phenomenon has been unveiled. At the per cent level the Z^0 is standard. Several potentially interesting channels have been explored to the 10^{-4} level or less without success.

To conclude that there is nothing to discover at LEP would however be premature and rather short minded. Interesting modes can still appear with much smaller branching ratios: among many other possibilities a classical Higgs of $\sim$ 55 GeV or a manifestation of Z compositeness through its decay into 3 γ are not expected to show up before one reaches the 10^{-6} level or so.

Moreover if a theory like the Minimal Supersymmetric Standard Model is a reality one has a good chance to discover something at LEP I + LEP 200 since such a theory absolutely needs a scalar particle, at least, below or not far above the Z mass.

1 Indirect Searches

As it is well known and abundantly illustrated in this school, another kind of systematic exploration is made at LEP through a set of accurate measurements.

If new particles occur at energies which are not accessible directly at LEP, one has still the possibility to observe them indirectly through their manifestation as virtual intermediate states appearing in loop diagrams. Indeed the structure of electroweak interactions is such that, at the difference of what happens in pure electromagnetic ones, heavy particles do not decouple from the lower energy world.

If the top is not accessible directly at LEP, which is now very likely, this particle will then be the most conspicuous example of such a state not produced but influencing considerably several observations. The effect of virtual top exchange on the boson masses or the width of the Z into $b\bar{b}$ (Fig. 1) are well known. Besides the top, which will hopefully be discovered in hadronic machines during LEP lifetime, many other possible virtual effects can show up as deviations of observables from their standard value. A complete strategy to study such deviations and identify their physical origin is now existing: it rests on quite sophisticated linear combinations of several observables (asymmetries,

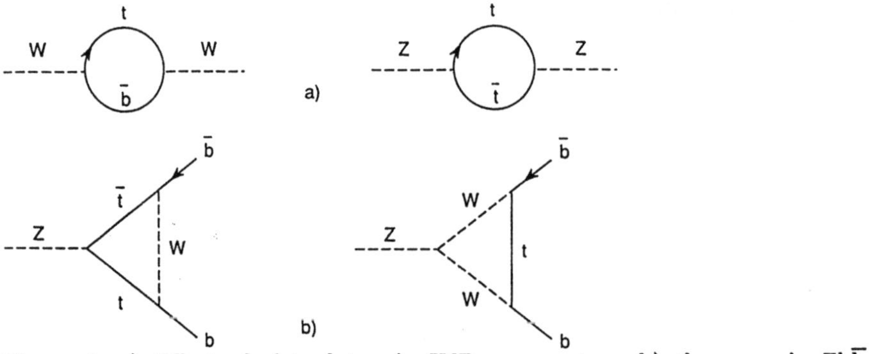

Figure–1: a) Effect of virtual top in IVB propagators, b) the same in $Zb\bar{b}$ vertex.

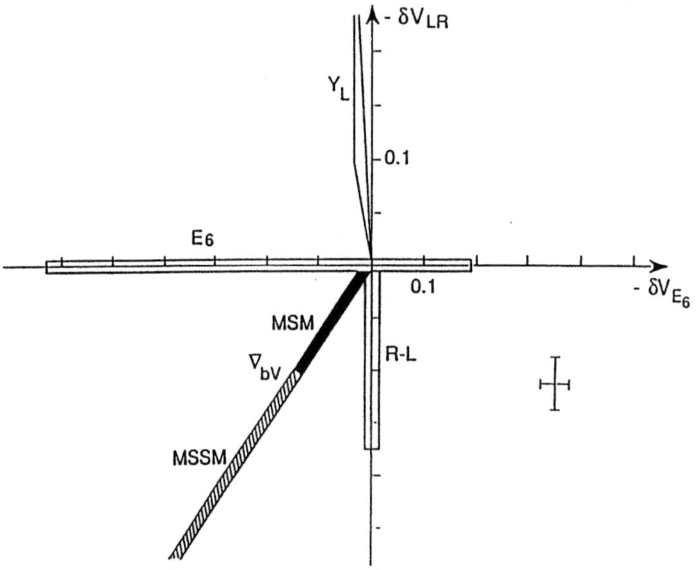

Figure–2: The strategy to identify new physics. The δV_i are the deviations, from their Standard Model value, of quantities built up from appropriate linear combinations of LEP observables: $\Gamma_{b\bar{b}}$, $A_{b\bar{b}}$, M_Z, M_W, etc. The cross shows the experimental error expected from High Luminosity LEP.

partial widths, masses of IVB) and its power depends crucially on the accuracies with which these quantities will be measured. Figure 2 [1] illustrates this strategy on which we come back in the Appendix 3 summarizing the future options. So, whatever be the issue of direct searches, one is certain with LEP to perform meanwhile, if the luminosity–and in some cases the available energy–are sufficient, a set of very interesting measurements.

2 Direct Searches : Generalities

Let us now come back to this program of direct searches, which are obviously a must for every new machine, as long as possibilities are still open experimentally.

The occurence of non standard final states or effects can be looked for by an inspection of global quantities describing the Z resonance, like the total width, the invisible width and the total hadronic cross section.

This last quantity can be expressed as:

$$\sigma_h \propto \frac{\Gamma_e \Gamma_h}{\Gamma_{tot}^2} .$$

Its sensitivity to the presence of a new channel will depend on the type of final state involved. If the channel is of the hadronic type, it will affect both the denominator and numerator and

$$\frac{\Delta\sigma_h}{\sigma_h} \sim \frac{\Delta\Gamma}{\Gamma} .$$

If on the contrary it is an "invisible" one the sensitivity will be twice as large

$$\frac{\Delta\Gamma}{\Gamma} = \frac{1}{2}\frac{\Delta\sigma_h}{\sigma_h}$$

Presently the experimental error $\Delta\sigma_h/\sigma_h$ is 1.3% and the theoretical one: 0.5%.

Similarly one can feel the presence of an invisible channel through its contribution to Γ_{inv}. With the present uncertainty the room left for novelties in Γ_{inv} is $\lesssim$ 30 MeV at the 95% CL (per experiment).

The impact of such global inspections is limited but nevertheless quite useful in specific cases.

The bulk of searches is, however, obtained from the study of specific final states with characteristic topological features.

It would be tedious to list all the cuts used to isolate such final states and a synthetic description is quite preferable. In fact, most of the searches belong to one (or more) of three generic categories.

A) Acoplanar two prongs leptonic states: such final states occur when "escaping" particles (ν, $\tilde{\nu}$, lightest supersymmetric particle, LSP) are emitted in leptonic decays. This can concern heavy leptons, supersymmetric states, light Higgses, etc.

The Standard Model background for such a configuration in a perfect detector is negligible. If the detector has "holes" or "cracks" for photon detection the radiative leptonic modes of the Z are a potential background. Initial state radiation on the Z is limited by the resonant situation and gives mostly acolinearity. Final state radiation on the contrary is dangerous: for a heavy radiator, like a τ, the radiation is no more peaked along the lepton trajectory and its angular distribution relative to the emittor can be expressed as:

$$\frac{dn}{d\cos\theta} \alpha \frac{\sin^2\theta}{\left(\sin^2\theta + \frac{\cos^2\theta}{\gamma^2}\right)^2},$$

where $\gamma \equiv E/m$ is the Lorentz factor of the emittor. However with such a simple final state and in the case of single γ emission one can protect oneself by using constraints: the missing momentum should not point to a notoriously weak region of the detector.

B) If in the final state under study leptonic or semileptonic decays are supposed to take place, the signature expected is the presence of isolated particles, sticking out of the hadronic jets. The isolation variable usually considered is

$$\rho_i = \min_{\substack{jet\ j \\ j=1,n}} \sqrt{2E_i(1-\cos\theta_{ij})},$$

where i denotes the candidate particle, E_i its energy, and θ_{ij} its angle with the axis of jet j. The inclusion of particle i in or its exclusion from the nearest jet is a matter of debate.
The standard model background is here at the level of $\sim 10^{-3}$. The isolated tracks of background events are not always leptons and one could improve the situation by requiring a leptonic identification. However many hypothetical states could decay into a τ and such a leptonic requirement would suppress the abundant $\tau \rightarrow h^{\pm} + ...$ monotrack source: it is therefore generally not applied.

C) A purely hadronic decay of a pair of particles into four jets is generally the most abundant expected final state. The standard model background is high since 10% of hadronic Z final states will appear as four jets.

One has to use the equality of the masses of the two hypothetical particles (if appropriate). Jet-jet masses must be reconstructed; constraints to the beam energy are exploited by a rescaling of these effective masses, and peaks are looked for in two dimensional plots. The typical jet-jet mass resolution is a few GeV/c^2. In the absence of reference physical signal, one has, however, to trust the result of Monte Carlo simulations for such an estimate.

In the remaining of the description of searches, I shall for each one simply refer to the type of topological configuration it involves (as A, B, C).

3 What are we looking for?

The most obvious candidates are the standard missing objects: the tau neutrino and the top quark.

For the former LEP has not much to say; it has simply confirmed that there seems to be three and only three light neutrino species. It is likely that a dedicated and clever fixed target experiment will be necessary to bring a direct evidence for the ν_τ. For the later we will review the situation in Ch. 4 and see what LEP can bring.

Then comes the standard scalar sector. The standard Higgs search will be described in Section 5 and future prospects discussed in Appendix 3.

A straightforward extension of the SM would be to add further families (with the constraint of the number of light neutrinos). Several kinds of heavy neutral leptons can be envisaged, per se, or to provide solutions to various problems (generation of neutrino mass,...). This will be explored in Section 6.

Finally, as often repeated, we know that the SM suffers from serious defects which have to be cured by new ideas.

The most promising is supersymmetry about which LEP has to tell us a lot (Ch7). Another option, less clearly formulated theoretically, is compositeness: possible hints from LEP in this respect will be described in Section 9.

4 The Top

The top must exist. The studies of the b quark, at various e^+e^- machines including LEP, have demonstrated that it is the lower member of a doublet, waiting for a partner. The present limit on $B \to \ell^+\ell^- X$ cannot be accomodated for in the absence of a GIM mechanism involving the top. Other arguments (cancellation of anomalies, ...) could be used as well.

That the top has not been seen up to now can be the result of its high mass or of a non standard decay pattern.

For a standard top the present mass limit is 89 GeV (95% CL) from CDF. It is expected that a mass limit of 120 GeV can be reached by the end of 1992 if an integrated luminosity of 10 pb^{-1} (10 times more than up to now) is registered. The chance to discover it at the Tevatron are high. LEP–even LEP 200–cannot compete.

If the top decay is non standard the situation is different. A non standard decay means $t \not\to bW^*$ and can represent many possibilities. One is $t \to bH^\pm$ through a charged Higgs: for $m_t < m_W + m_b$ the standard decay can be negligible, while for $m_t > m_W + m_b$ it is probably not small (2-body phase space). We can note (see Section 7) that for $m_H < m_W$, such a scalar cannot be the Minimal SUSY charged Higgs which is bound to be heavier than the W.

From the hadronic colliders the mass limit for a non standard top, obtained by measuring the width of the W into $l\nu$, is 45 GeV.

LEP has already reached a similar limit. One way is through the exclusion of a $H^\pm$ up to M, which excludes a top up to $M + m_B$. Another method consists in excluding directly a top decaying with such a topology. The search for $H^\pm$ at LEP [2] has used topologies A, B and C, depending on the unknown branching ratio $H \to \tau\nu$. The cross section for H^+H^- pair production is given by formula A4 of Appendix 1, a particular case of the general formula A1. Figure 3 shows

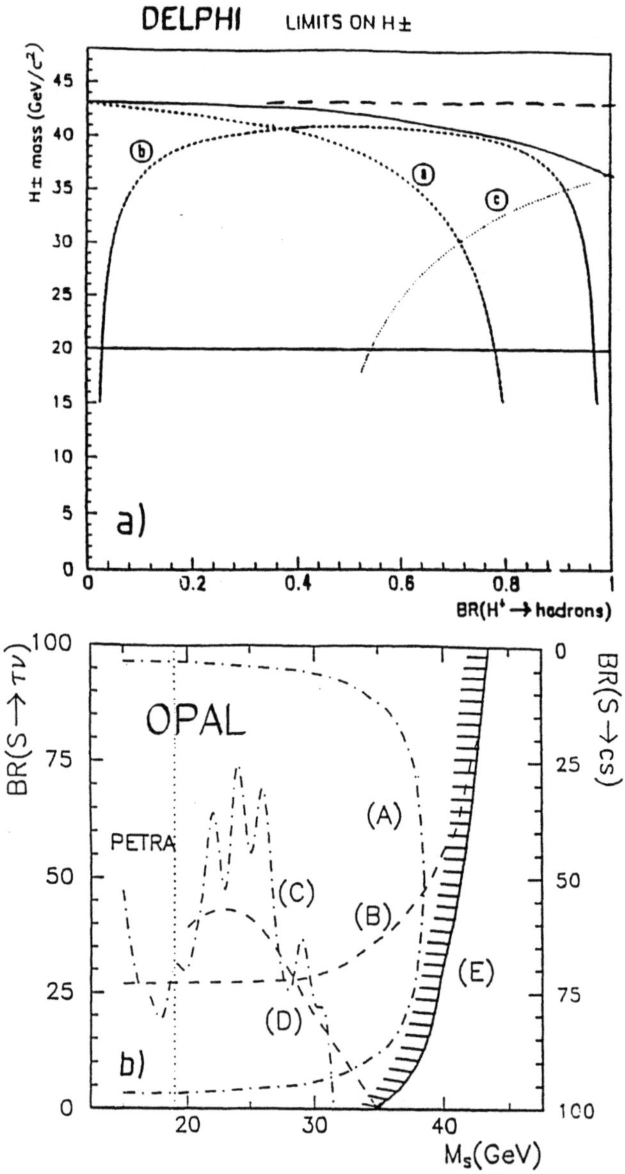

Figure–3: Charged Higgs mass limit: a) from DELPHI, b) from OPAL.

the results of DELPHI and OPAL, very similar to the ones of ALEPH and L3. Both show the interplay of the three topologies: charged Higgs are excluded up to 43 GeV when $\tau\nu$ dominates. One can plot this limit on Fig. 4 [3] which represents the most recent exclusion contour of the collider in the $M_t - M_H$ plane. The direct exclusion of a top decaying into $bH^{\pm}$ can be illustrated by

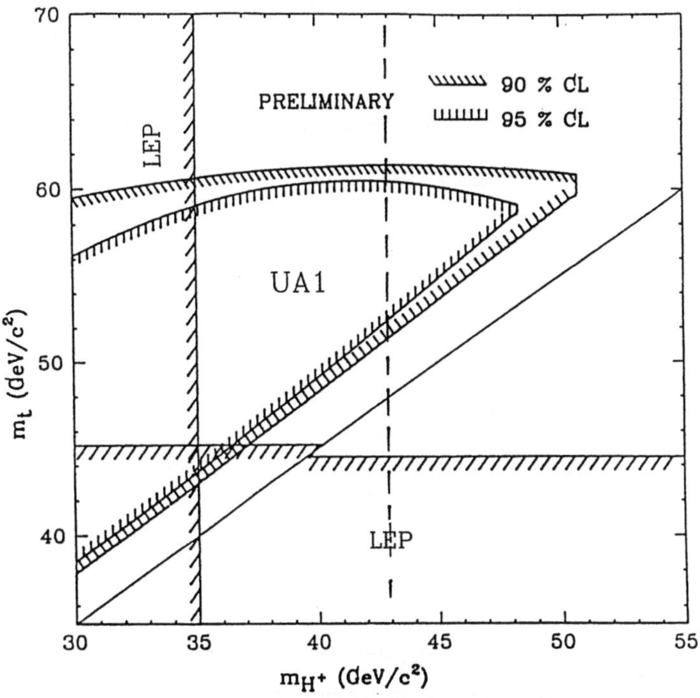

Figure–4: Limits on $M_{H}\pm - M_{top}$ from UA1.

the work of OPAL [4], which is a systematic search for heavier quarks. Figure 5 shows how the acoplanarity of the hadronic Z decays should be modified by the existence of a heavy quark (top or b′, the lower member of the doublet of a fourth family). Acoplanarity is defined here as

$$A \equiv 4\mathrm{Min}(\Sigma\,|p_{i\perp}|\,/\Sigma\,|p_i|)^2\ .$$

The Monte Carlo curve of Fig. 5 is obtained using ZHADRO for the amount of heavy quark pairs present in Z decay (most important is the QCD correction near threshold) and JETSET + HERWIG for the fragmentation. The experimental results clearly exclude such a production and lead to the mass limits shown in Fig. 6. The exclusion of t → bH has been obtained assuming m_H = 23 GeV and 100% of H → c$\bar{s}$ but the authors have shown that the result is quite independent of these assumptions. No top, even non standard, can exist below 45 GeV.

Hopefully LEP 200 will allow to push that limit to close to 90 GeV [5].

Meanwhile what LEP will bring is a more and more accurate determination of the top mass through its virtual contribution to loop diagrams. We can note from Fig. 6 that a b′ quark is excluded as well even if its decay is dominantly a flavour changing neutral current (FCNC) process.

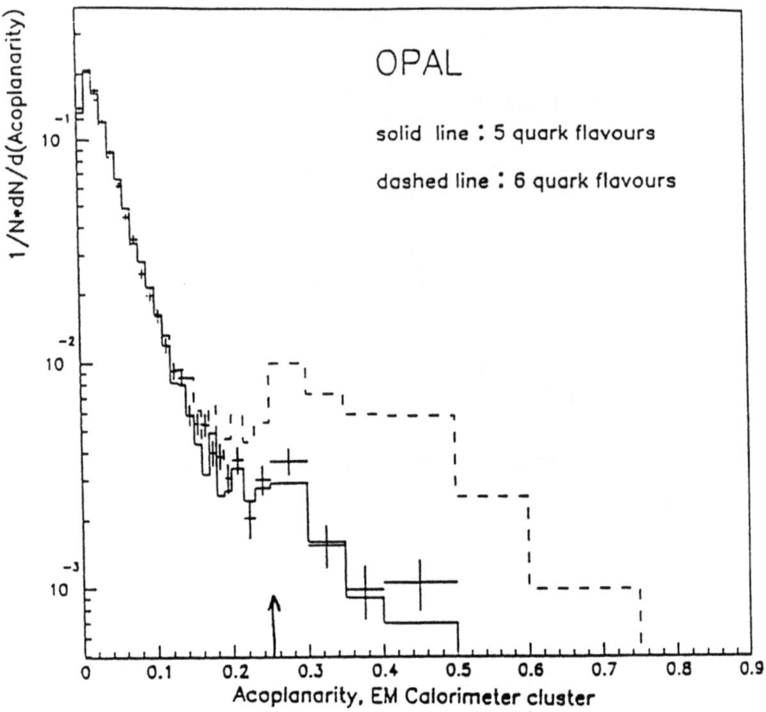

Figure–5: Acoplanarity of hadronic events with and without the 6th flavour.

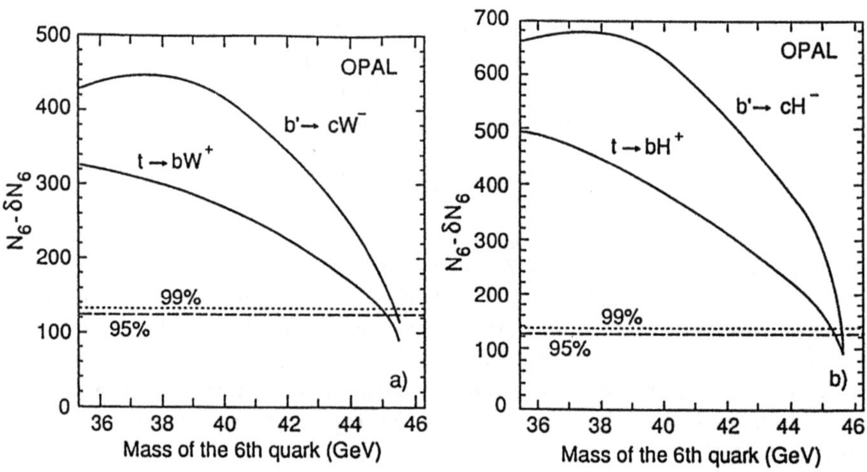

Figure–6: OPAL limit on t and b′.

5 The standard Higgs

We admit here that the Higgs mechanism is the process which generate masses in the Standard Model; we also assume that the degree of freedom left over after spontaneous breaking of the symmetry and mass generation for the IVB corresponds indeed to an elementary neutral scalar boson.

The production and decay of this boson are quite familiar. The Bjorken mechanism (Fig. 7) leads to a double bump cross section depending on which Z is real (Fig. 8). Up to ~ 50 GeV the Higgs production is larger when one sits on the Z peak. For masses ≥ 50 GeV it is more profitable (at equal luminosity) to sit above the Z, at an energy roughly given by

$$\sqrt{s} \simeq M_H + M_Z + (10 - 15GeV) \ .$$

Radiative corrections dominated by initial state radiation (see formula of App. 1) are shown in (Fig. 9) and their shape can be intuitively understood quite readily.

The Higgs boson will decay into the most massive fermion–antifermion pair kinematically available. For clarity of the final state the Z (or $Z^\star$) produced in association is required to decay into a lepton pair ($\ell^+\ell^-$ *or* $\nu\overline{\nu}$).

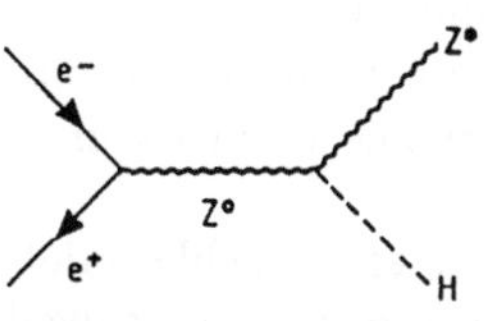

Figure–7: The Higgs production mechanism.

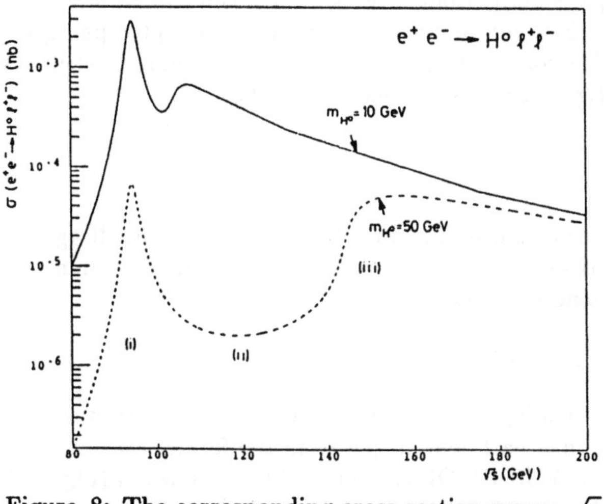

Figure–8: The corresponding cross-section versus $\sqrt{s}$.

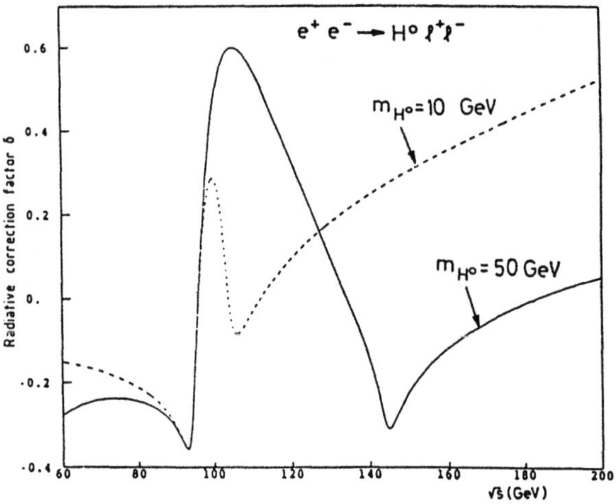

Figure–9: Radiative correction to the Higgs cross-section.

Light Higgs

The Higgs mass is an unknown quantity which can range from zero to ~ 1 TeV or so. We will come back later to the upper mass accessible at LEP.

Below $2m_\mu$ the Higgs boson can only decay, depending on its mass, into $\gamma\gamma$ and ee. Its lifetime and therefore its flight path increases when its mass decreases.

There is a gradual change, with increasing mass, from an invisible objet, escaping detection and only identified as missing momentum, to a long lived one, decaying into an e^+e^- pair in the detector. In the former case a search for topology A gives the exclusion curve labeled A in Fig. 10a. In the latter case the search for e^+e^- pairs born "from nothing" in the tracking region of the detectors and either alone ($Z \to \nu\overline{\nu}$) or associated to a charged lepton pair gives the exclusion curve B. The combination of both methods definitely excludes the existence of a light Higgs boson, as shown in Fig. 10a-d [6].

Heavier Higgs

As the mass increases other channels ($\mu\mu, \pi\pi, \tau\tau, c\overline{c}$) open up. Such Higgs decays have been excluded as well. Above 10 GeV the $b\overline{b}$ decay mode dominates. The final state one is looking for is either

$$\mathrm{jet} - \mathrm{jet} - \ell^+\ell^- \ (\ell = \mathrm{e}, \mu)$$

or jet-jet acoplanar with missing $\nu\overline{\nu}$, the latter case having a cross section $2 \times 3 = 6$ times bigger than each lepton species in the former case.

The exclusion curves of DELPHI, OPAL and L3 [7] are shown in (Fig. 11). Clearly the most efficient channel is: jet-jet-$\nu\overline{\nu}$, as expected. The horizontal

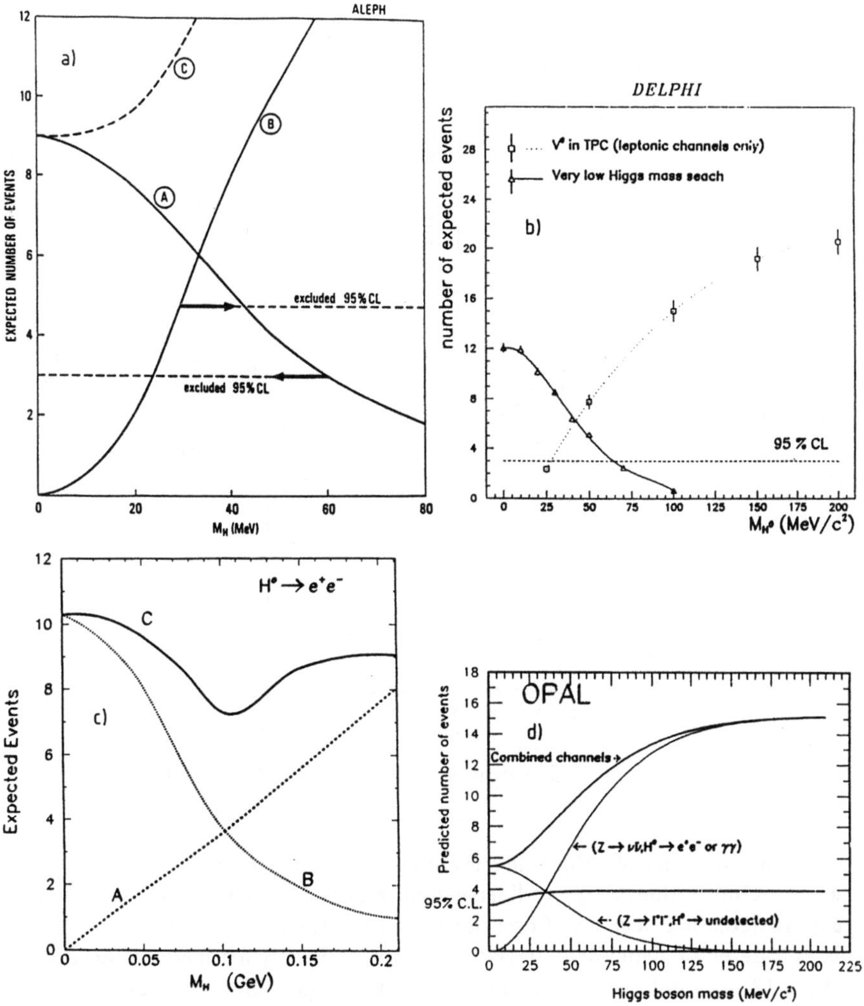

Figure–10: Limit on the light classical Higgs boson: a) ALEPH, b) DELPHI, c) L3, d) OPAL.

line at $\sim$ 3 is the 95% CL curve corresponding to a Poisson law and the case where no candidate is observed and no background expected.

In Figs. 12 and 13 ALEPH results [8] are described. The curve of visible energy in the ALEPH detector (Fig. 12a), after some harmless cuts against 2γ background, is totally free of any tail on the low side (Fig. 12b), the region where a Higgs of 40 GeV (associated to $Z \rightarrow \nu\overline{\nu}$) would show up (Fig. 12c). This is a good demonstration of the hermeticity of the ALEPH detector, which allows to set as a mass limit on the Higgs scalar (Fig. 13)

$$m_H > 42 \text{ GeV} .$$

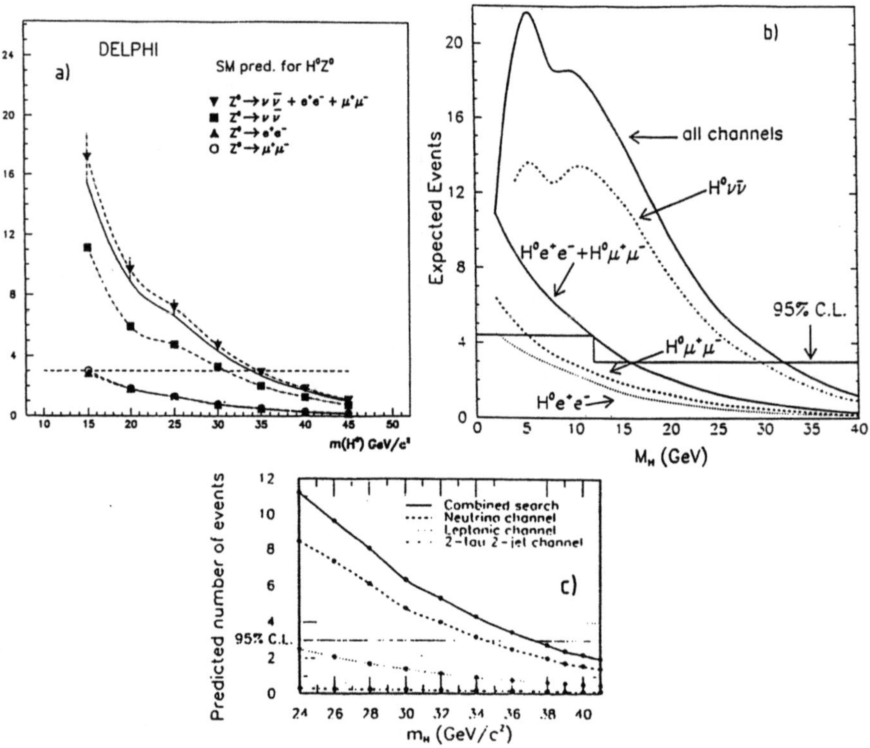

Figure–11: Heavy Higgs limit from: a) DELPHI, b) L3, c) OPAL.

Beyond this kind of value one should pay attention to two varieties of background. One is due to $q\bar{q}$ radiative final states (Fig. 14) with the photon escaping detection because of non hermeticity of the electromagnetic coverage: this is a detector dependent background, and ALEPH for instance according to Fig. 12 does not seem to be in immediate danger. Here it is unlikely that kinematical constraints could save the situation if the photon is lost. We shall see that such a final state, which is only a potential danger for the H $Z^\star$ final state, is a disaster for $Z \rightarrow H\gamma$ and totally forbids any hope to observe the standard value of this mode which would be the dominant one on the Z for high Higgs masses.

Another background for $HZ^\star$ ($Z^\star \rightarrow \ell^+\ell^-$) is direct 4 fermion production in Z decay [9]. Due to the actual process (Fig. 15) the kinematical features are fortunately quite peculiar: the mass of fermion pairs (for instance $q\bar{q}$) peaks either at low values or near the maximum available (Fig. 16). If one superimposes the signal expected for a Higgs (the effective mass resolution is assumed here to be ~ 4 GeV for $q\bar{q}$), it would be drowned in the background at ~ 50 GeV: but a cut on the dilepton mass, in case of $HZ^\star$ ($Z^\star \rightarrow \ell^+\ell^-$), can save the situation by reducing the background without doing much harm to the signal.

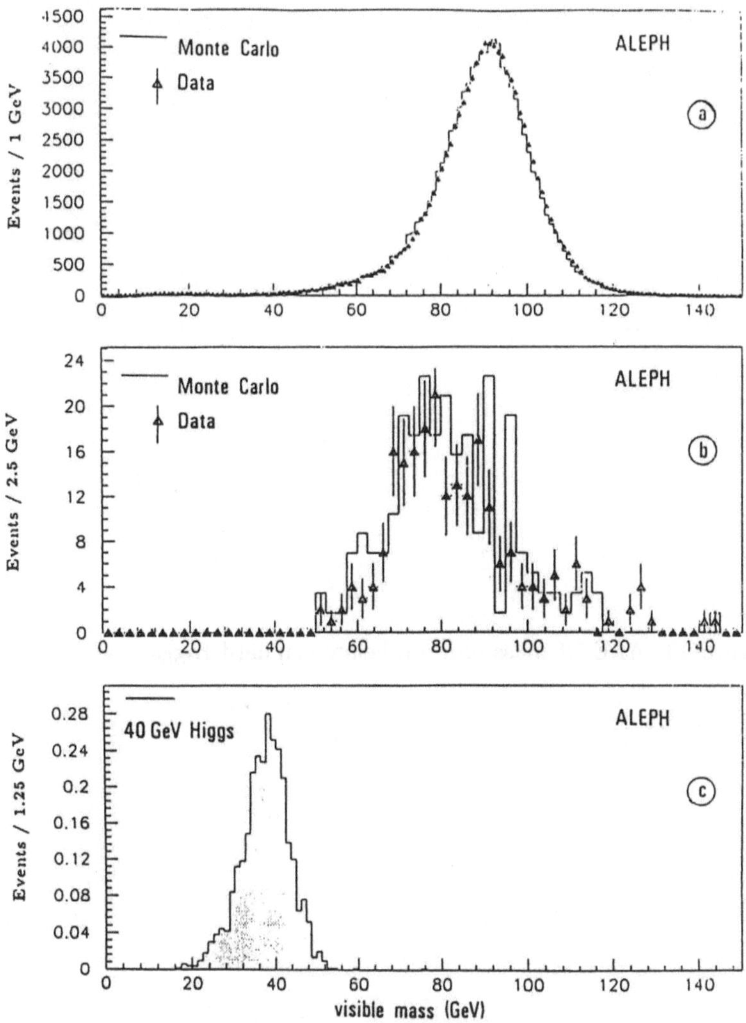

Figure–12: ALEPH calorimetry properties.

So, provided the detector is hermetic, one should be able with enough luminosity to push the Higgs search on the Z to higher masses. A factor ~ 10 in integrated luminosity is needed to go ~ 10 GeV higher in mass.

Above ~ 55 to 60 GeV, given the fact that $Z \rightarrow H\gamma$ is inaccessible, it will be more reasonable to move to LEP 200, provided the luminosity at high $\sqrt{s}$ is sufficient. As discussed in Appendix 3 one should then with enough available energy and integrated luminosity reach ~ 90 GeV.

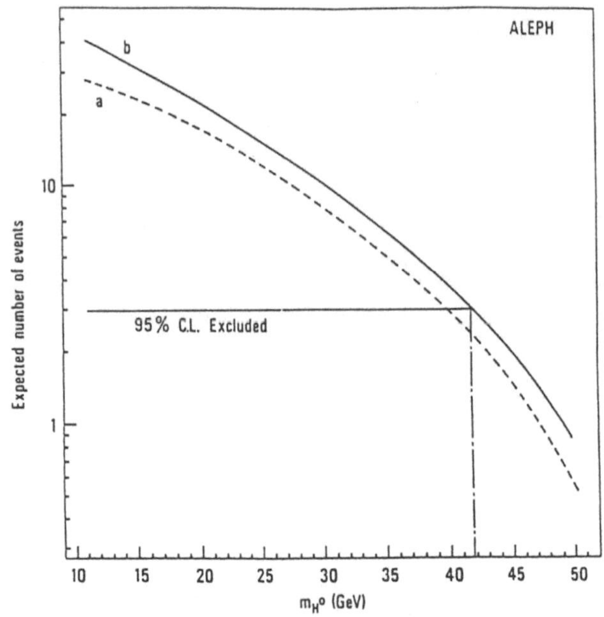

Figure–13: ALEPH mass limit on heavy standard Higgs.

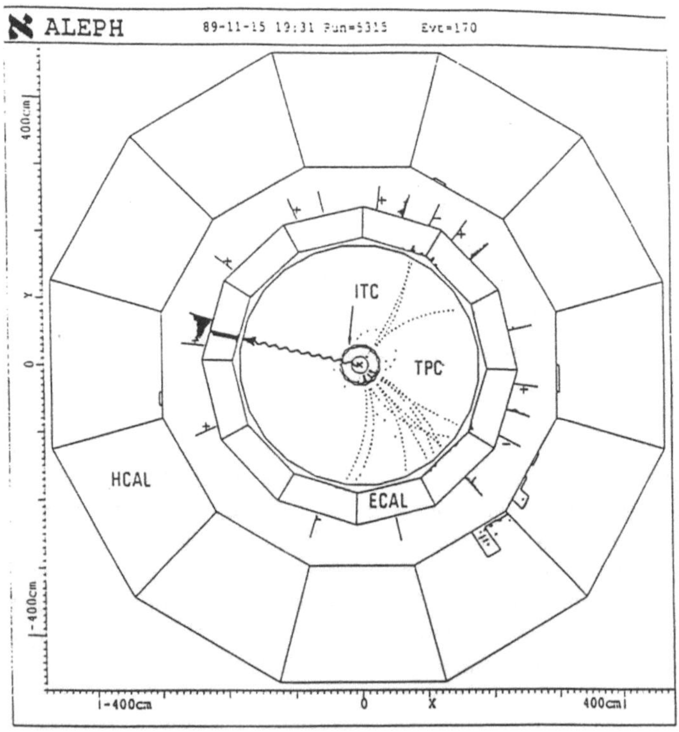

Figure–14: A $q\overline{q}$ radiative event.

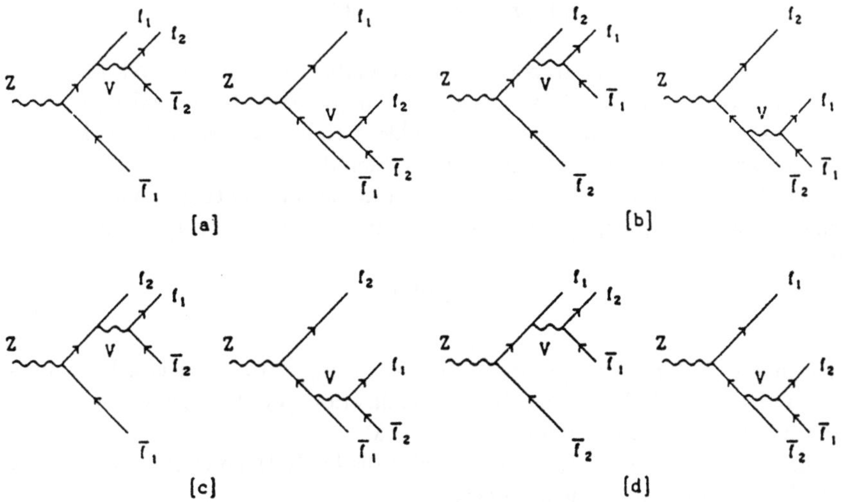

Figure–15: The four-fermion production mechanism.

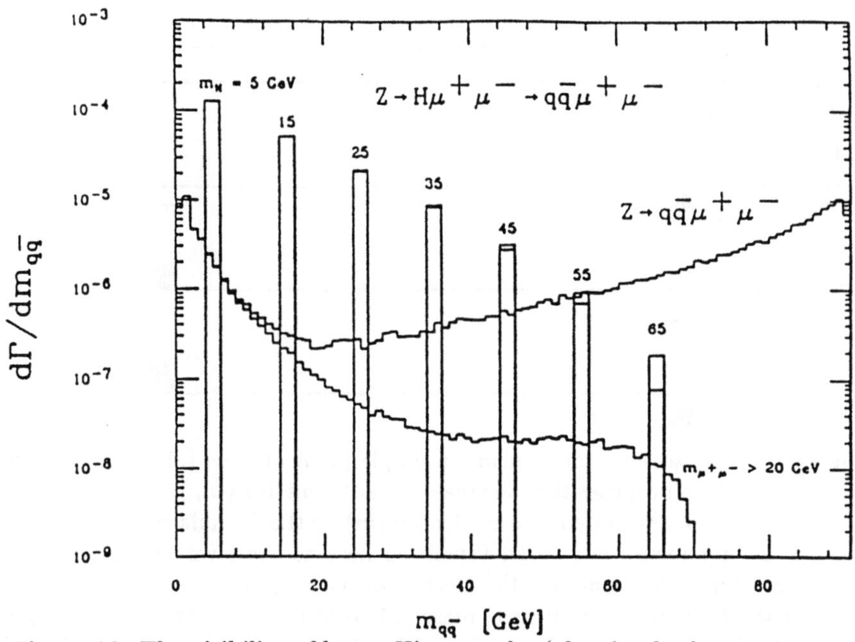

Figure–16: The visibility of heavy Higgs on the 4 fermion background.

6 Heavy neutral leptons (HNL)

Many reasons can push us to consider the possible existence of HNL [10].

In spite of the limits already set, one cannot exclude the existence of a further generation: one has then to provide a right handed field so that the neutrino can acquire a mass to respect the bound.

Right handed HNL would provide a natural mechanism to give a small mass to the normal neutrinos, by a see-saw mechanism (Fig. 17) [11]. From a mass matrix

$$\begin{pmatrix} 0 & m \\ m & M \end{pmatrix},$$

where m stands for a charged lepton mass ($m_e, m_\mu, ...$) and M is a high mass associated to the HNL one can by diagonalization get the eigenvalues m^2/M (identified to the small neutrino mass) and M.

The idea of Grand Unification by itself calls for HNL since it provides large multiplets with room for such objects.

The phenomenology of HNL depends of their nature. Sequential heavy leptons, members of an isospin doublet, are expected to be pair produced with normal strength. FCNC decays are absent. The decay is $L^o \to \ell W^*$ with an

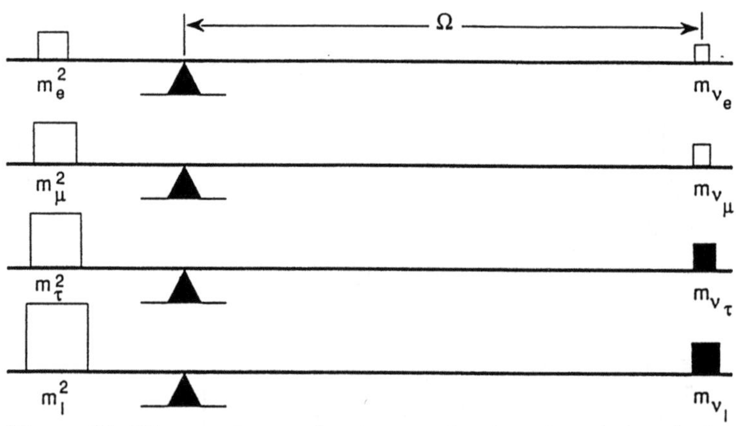

Figure–17: The see-saw mechanism: a simple point of view [11]. The see-saw mechanism proposes that the mass of each neutrino (m_ν) is related to the mass of the associated charged lepton (m) by the formula $m_\nu = m^2/\Omega$; Ω is an unknown mass scale, visualized here as the lever arm of the see-saw. Since, for example, the electron-neutrino mass is known to be less than 16 eV and the electron mass is known to be 0.5 MeV, the see-saw equation requires Ω to be at least 16 GeV. The tau mass is 1.8 GeV and cosmological limits on the tau-neutrino make it less than 65 eV. Running the see-saw equation with these values gives the stricter lower limit on Ω of 5×10^7 GeV. If Ω is related to a large fourth-lepton mass, the see-saw mechanism shows how this large mass could generate the very small neutrino masses.

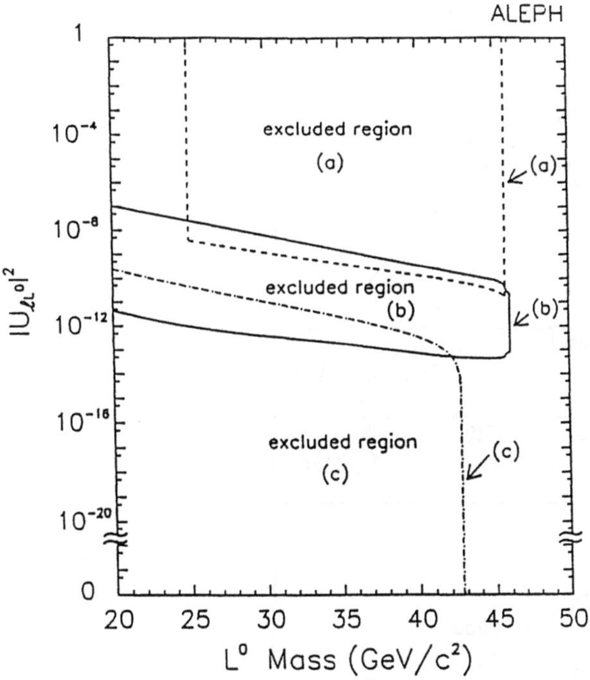

Figure–18: ALEPH neutral lepton exclusion.

amplitude $|U_{\ell L}|$ depending on the mixing matrix element with lower generations.

Extremely small $|U_{\ell L}|$ lead to stable invisible L^0 (that one may call ν_H proper, heavy stable neutrino); the only way of identifying their existence is through their influence on Γ_Z or Γ invisible. For some intermediate domain of $|U_{\ell L}|$ the lifetime is such that the L^0 can be found by its secondary vertex. For larger $|U_{\ell L}|$ normal $L^0 \to \ell W^\star$ decays occur leading to topologies A or B described previously.

Figure 18 from ALEPH [42] illustrates the result of the three methods, excluding most of the possibilities for sequential HNL up to ~ 43 GeV. Similar results come from other experiments. For instance L3 [13] gives (Fig. 19) limits on heavy neutrinos from Γ_Z and $\Gamma_{\rm inv}$.

Right handed HNL are produced in association with a neutrino

$$ee \to \nu N \ .$$

FCNC decay $N \to \nu Z^\star$ is competing with the charged current decay $N \to \ell W^\star$ (ratio NC/CC ~ 1/2 and matrix element $|U_{\ell N}| \sim m/M$). OPAL results [14] are shown in (Fig. 20) and we shall come back briefly to this topic in Appendix 3 since rare decay modes of the Z can be related to such HNL.

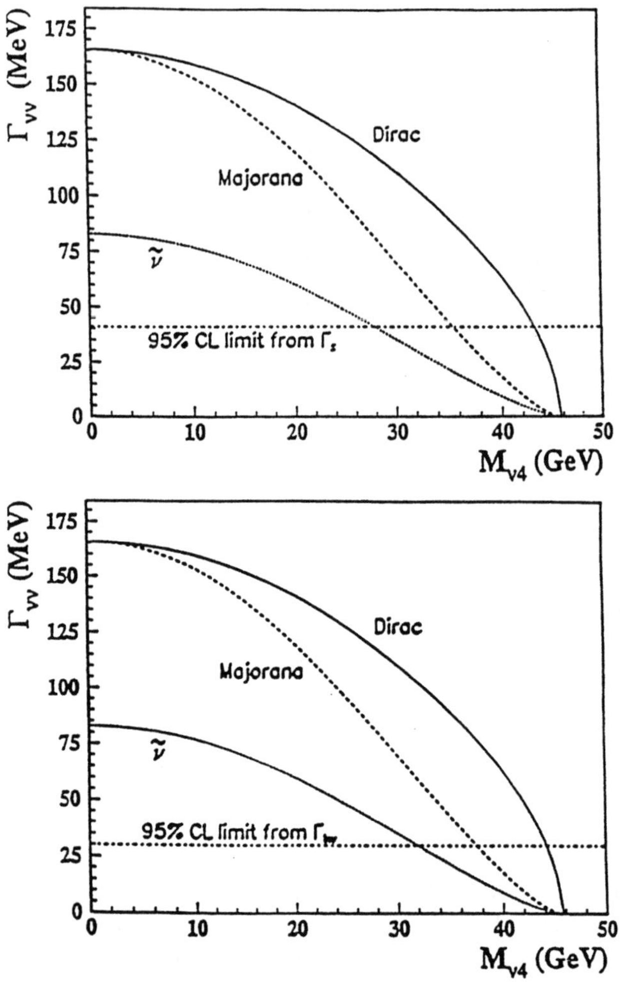

Figure–19: L3 exclusion of Dirac and Majorana L^0 from Γ_Z and $\Gamma_{\rm invisible}$.

Mirror leptons (i.e., RH doublets and LH singlets) can be considered as well. Their decay modes either to the normal standard world or to "sterile" mirror world channels depend on the details of the theory.

Charged leptons, stable or decaying, have been excluded similarly [15] (Fig. 21).

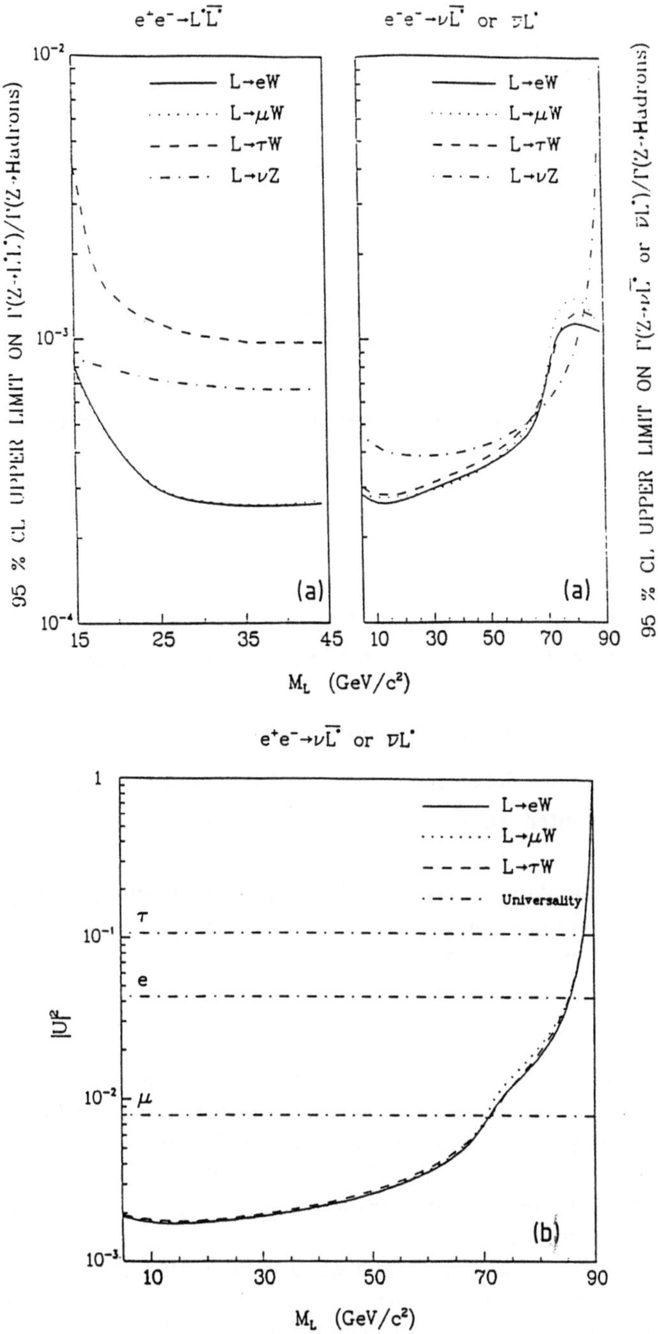

Figure–20: Exclusion curves by OPAL of heavy neutral leptons.

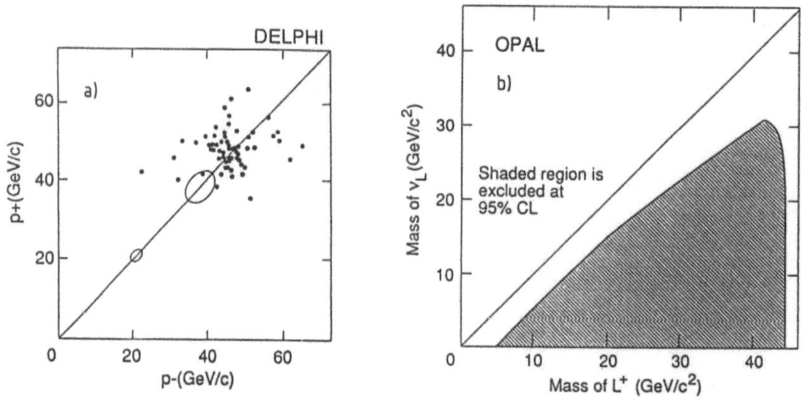

Figure–21: Exclusion of heavy charged leptons: a) stable, from DELPHI, b) from OPAL.

7 Supersymmetry

I think we need here some clarification about the meaning of the theory which is tested for. We will concentrate on a version of SUSY which stems from the following assumptions:

a) supergravity [16] provides the needed clue for supersymmetry breaking.

b) three simple assumptions define the scenery of SUSY at low energies:

– Minimal particle content supermultiplets of quarks, leptons and pair of Higgs doublets.

– Universality i.e., breaking of flavour and/or CP only through properly s-symmetrized Yukawa couplings at some grand mass M_x, universal mass term for all scalars, and universal gaugino mass term.

– R-parity unbroken. The number of Superpartners is conserved modulo 2 and the LSP is stable.

The set of these assumptions describe what is called the Minimal Supersymmetry Standard Model [17]. The important fact is that MSSM is a falsifiable theory: the constraints imposed on its components are such that experimentation can turn it wright or wrong, as expected from a physical theory. In other words, it cannot be saved for ever and be kept as an argument for an escalade in the available $\sqrt{s}$ of future machines.

Appendix 2 helps realizing the constraints which exist within the MSSM. The Supersymmetric Lagrangian valid at some high scale M_x (M_{Pl}?) has universal terms for scalars and gauginos, with mass terms m and M respectively. When energy is downscaled to present ones, the Lagrangian is renormalized and parameters like the top mass enter the game. In the gaugino sector the three terms correspond to gluinos, neutralinos and charginos. For neutralinos the mass matrix is 3×3, with parameters M_1, M_2, β, μ_R. For charginos, it is a 2×2 matrix, with parameters M_2, β, μ_R. The relationships between renormalized mass terms are

$$M_i = M\frac{\alpha_i(M_W)}{\alpha_i(M_x)}$$

with

$$M_{\tilde{\gamma}} = M_1\cos^2\theta_W + M_2\sin^2\theta_W \simeq 0.6M_2$$

being the mass of the LSP (for small values of M_2).

For the Higgs sector the Lagrangian is shown in Appendix 2. When the symmetry breaking mechanism has acted, one gets the masses of the surviving Higgs particles as functions of g, g', m, μ, M. Because of the presence of g, g' in the Lagrangian, Higgs masses and IVB masses are connected. In particular one arrives to the important inequality for the lightest Higgs

$$m_{h^0} \leq |\cos 2\beta|\, m_Z \ ,$$

valid at tree level.

Another concept, naturalness [17], can be described as follows: standard quantities like the Z mass depend on the parameters of SUSY. It is well known that, for SUSY to fulfil its role, the masses of superpartners cannot be too heavy. To keep at their observed values quantities like m_Z –or m_H– there must be fine tuning among the SUSY parameters. A natural theory is one which does not require too much fine tuning. A way to express it is to impose that the relative variation of an observable like M_Z does not depend too much on the relative variation of SUSY parameters a_i. Let us fix for instance:

$$\frac{dM_Z}{M_Z} / \frac{da_i}{a_i} < 10 \ .$$

Such a condition has the merit to set upper bounds to the masses of SUSY objects: because of the renormalization of couplings when one goes down from high scales to our world (Appendix 2) such bounds depend for instance on the top mass. Nevertheless SUSY partners as charginos and neutralinos, within the present limits on top, are bound to be fairly light objects ($\lesssim$ 200 GeV) if the above naturalness condition is required. This is at least an encouragement to look for such objects within the accessible energy domain at LEP.

MSUSY Higgs

As we saw much more restrictive are the bounds in the MSUSY Higgs sector. The phenomenology of MSUSY Higgses on the Z is described in Fig. 22 [18]. We call: h^0 the lighter scalar, H^0 the heavier scalar, A^0 the pseudoscalar. For low values of $\tan\beta \equiv v_2/v_1$, the ratio of expectation values of the two doublets, the mechanism ee $\rightarrow$ Zh is operative with appreciable strength (Fig. 22b) and searches for h^0 follow the pattern of standard Higgs ones. When $\tan\beta$ increases

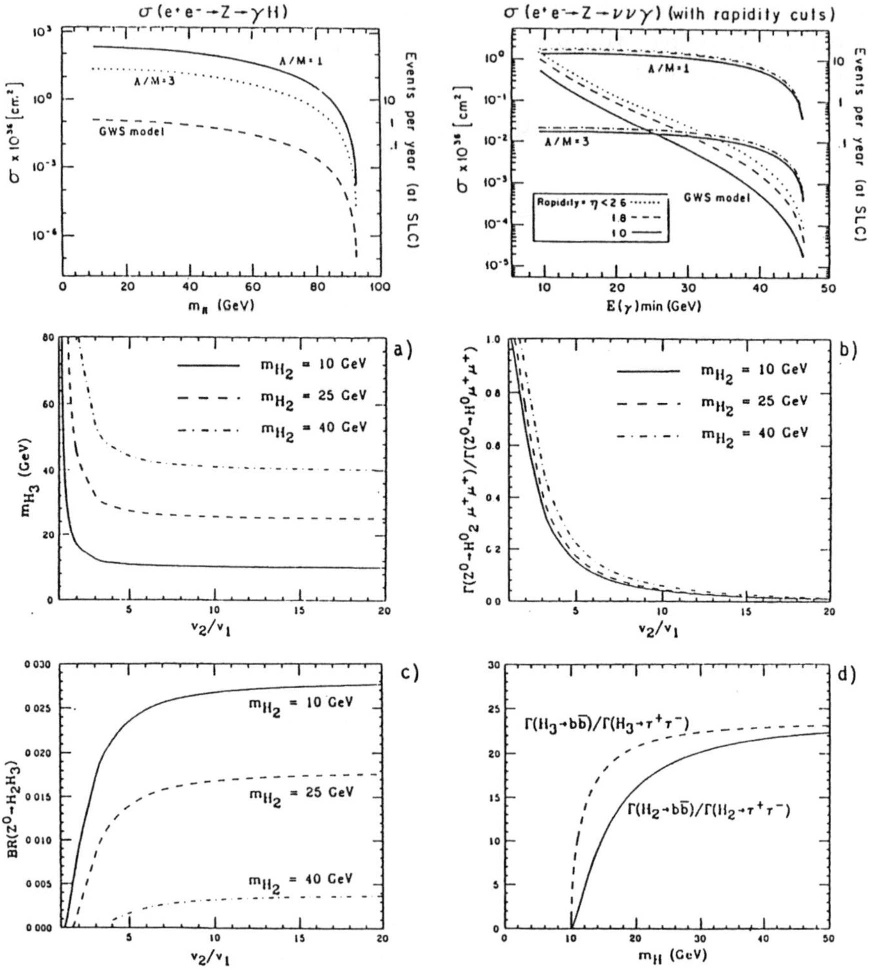

Figure–22: The phenomenology of SUSY Higgses.

(and $\tan\beta \gg 1$ is theoretically favoured because of the high top mass), this mechanism gradually disappears, but the associated production

$$ee \longrightarrow hA$$

of the lightest higgs with its then nearly mass degenerate pseudoscalar partner takes over (Fig. 22c), providing an extremely useful method of search through topologies A, B and C.

Decay patterns of h and A are governed as well by $\tan\beta$. For $\tan\beta > 1$ the branching ratio into $\tau\nu$, leading to A and B, is appreciable ($\geq 5\%$). For $\tan\beta < 1$ however one has to look to $h \rightarrow c\bar{c}$ modes. Either 4 jets (topology C) or special c tagging through $D^{\star}$ can then be used.

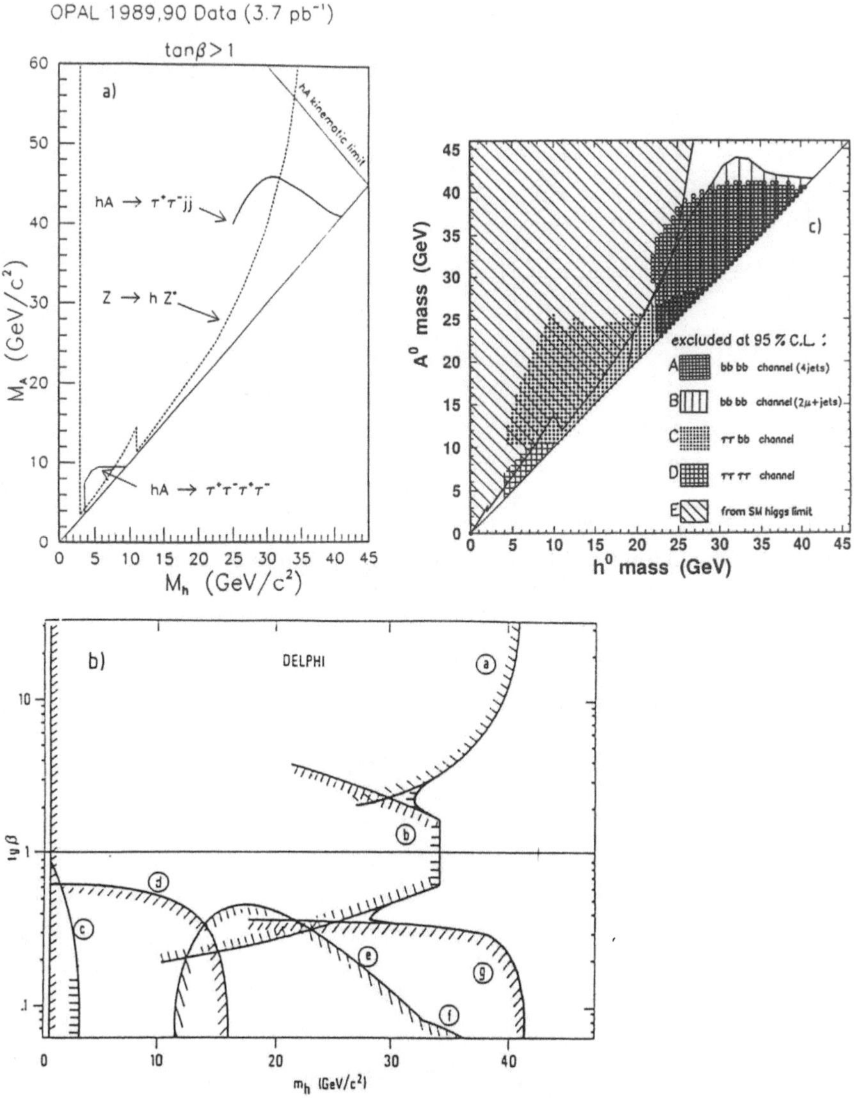

Figure–23: Limits on SUSY Higgses; a) OPAL, b) DELPHI, c) L3.

The LEP results [19] are shown in Figs. 23 and 24. The first ones give details, in the $\tan\beta$ versus m_h plane, on the strategies used, every experiment bringing one or more specific methods. The second one summarizes the global result up to now. The achievement is spectacular and mostly due to the mild ($\beta^3 \equiv (P/E)^3$) dependence on the associated cross section with m_h. However, the way to go to explore fully the domain allowed by MSSM ($m_h \leq m_z$) is still long as shown in Fig. 25. LEP I with increased L will close the gap up to

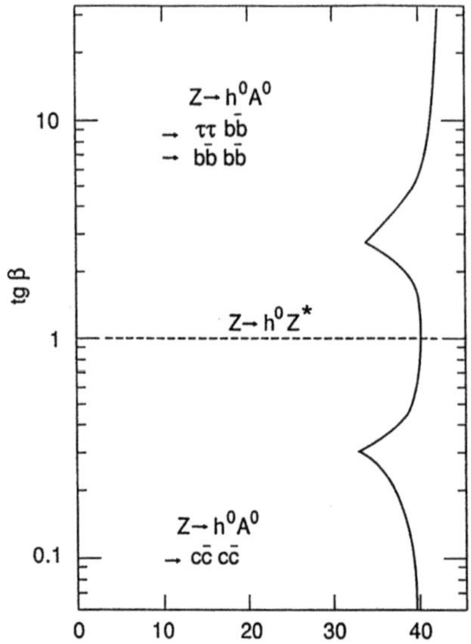

Figure–24: Global limit of LEP on SUSY Higgs (from M.J. Oreglia, talk at the CERN 90 Neutrino Conference, June 90).

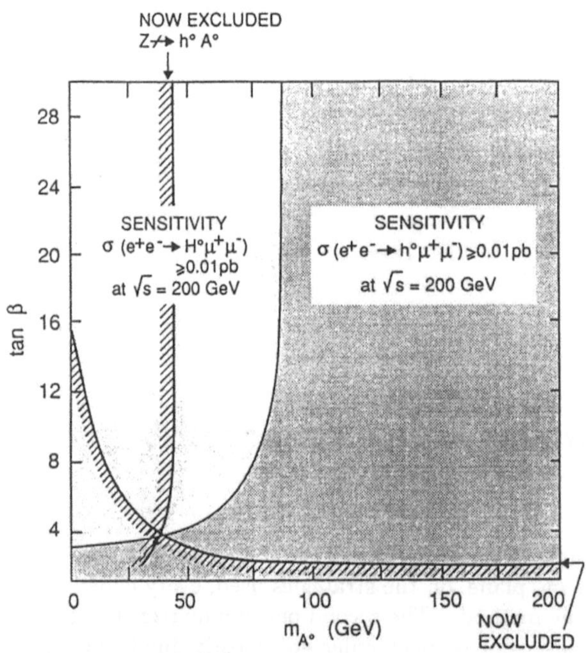

Figure–25: The MSUSY plane, seen in tg β versus M_A (courtesy of F. Zwirner and F. Pauss); LEP 200 with $\sqrt{s} \geq 190$ GeV, $L = 500$ pb^{-1} can exclude the whole plane.

$M_Z/2$, but LEP200 with much luminosity will be needed to complete the job (Fig. 25). We come back to this crucial issue in Appendix 3.

8 Neutralinos and charginos

These are the generic names of the particles resulting from mixing among (photino + zino + higgsinos) and (winos + higgsinos), respectively. The phenomenology is much dependent on the MSUSY parameters, but does not vary too rapidly with $\tan\beta$.

Exclusion of such objects can, as usual, rely on global methods like the inspection of Γ_Z or on specific channels. For these the only assumption is that the lightest neutral object, $\tilde{\chi}$, is stable and escapes detection. Signatures are therefore

$$ee \rightarrow \tilde{\chi}'\tilde{\chi} \; ,$$

where the heavier $\tilde{\chi}'$ cascades into the lightest one

$$\begin{aligned} \tilde{\chi}' \rightarrow \quad & \tilde{\chi}\, f\bar{f} \quad \text{(Zen one sided events)} \\ & \tilde{\chi}\,\gamma \quad \text{(single } \gamma \text{ on the } Z^0 \text{ peak)} \end{aligned}$$

with probabilities which again depend on the parameters.

Similarly ee $\rightarrow \tilde{\chi}^+\tilde{\chi}^-$ with $\tilde{\chi}^\pm \rightarrow \tilde{\chi}W^{*\pm}$ and obvious signatures.

The excluded regions are shown in the M, μ_R parameter plane (Fig. 26) [20]. Here M represents for small values the mass of the lightest $\tilde{\chi}$ (essentially a photino when μ is large and negative) while $|\mu|$ for small values again represents the Higgsino mass.

The most recent ALEPH exclusion region [21] for $\tan\beta = 2$ is shown in Fig. 27 where various exclusion methods are no more distinguished (to make the figure more reader friendly). One sees that the horizontal limit on M for negative μ is equivalent to a limit on $m_{\tilde{\chi}}$ (at $\sim$ 20 GeV) or a limit on $\tilde{g}$ at $m_{\tilde{g}} \sim$ 100 GeV (Fig. 28). This last limits, valid for $\tan\beta \geq 2$, are quite important since $\tilde{\chi}$ could be suspected to be a good candidate for non baryonic dark matter [22]. However, LEP data do not presently allow to set a bound on $m_{\tilde{\chi}}$ when $\tan\beta < 2$: the tree level limit $\tan\beta \geq 1.6$ quoted in the past from the non observation of h^0 (see above) is unfortunately not valid when the effect of loops are included (see Appendix 3). Calculations of this effect are under way.

Chargino exclusion [23] is illustrated by Fig. 29.

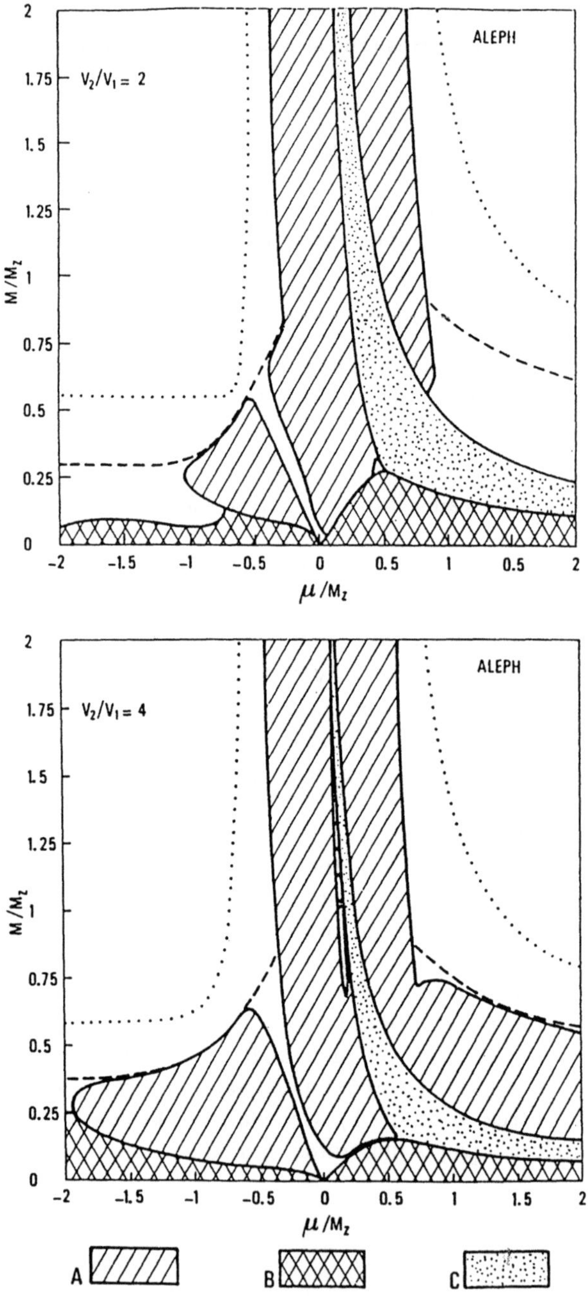

Figure–26: Neutralino limits from ALEPH.

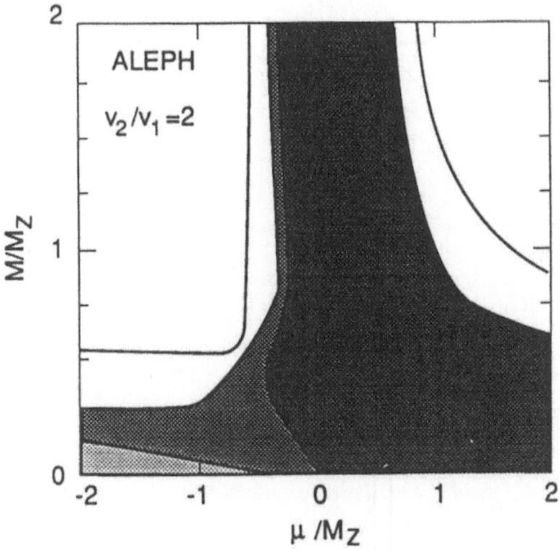

Figure–27: Last limit from ALEPH on neutralinos (Singapore).

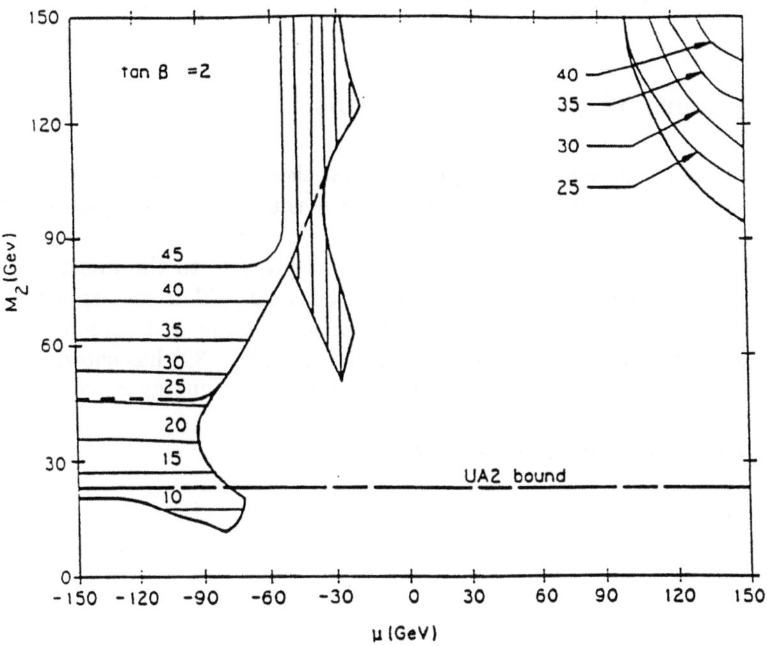

Figure–28: Iso-LSP mass curves in the M_2-μ plane.

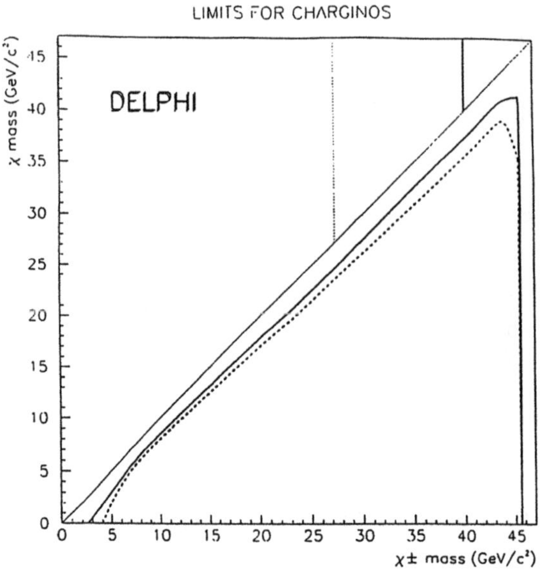

Figure–29: Limits on charginos (DELPHI).

Superpartners of quarks and leptons.

Even naturalness arguments do not promise that these objects should be light.

The exclusion of sleptons through topology A is a relatively simple matter [24] (Fig. 30).

For squarks one problem is to go as close as possible to the diagonal of Fig. 31, in case the LSP is nearly as heavy as the squark. This corresponds to events with low visible energy. $\gamma\gamma$ and beam gas interactions are then notorious backgrounds but non resonating ones. DELPHI [25] (Fig. 32) has shown that no resonant component is present among the low visible energy events, thus excluding the region quoted in Fig. 31.

Sneutrinos have been excluded as well.

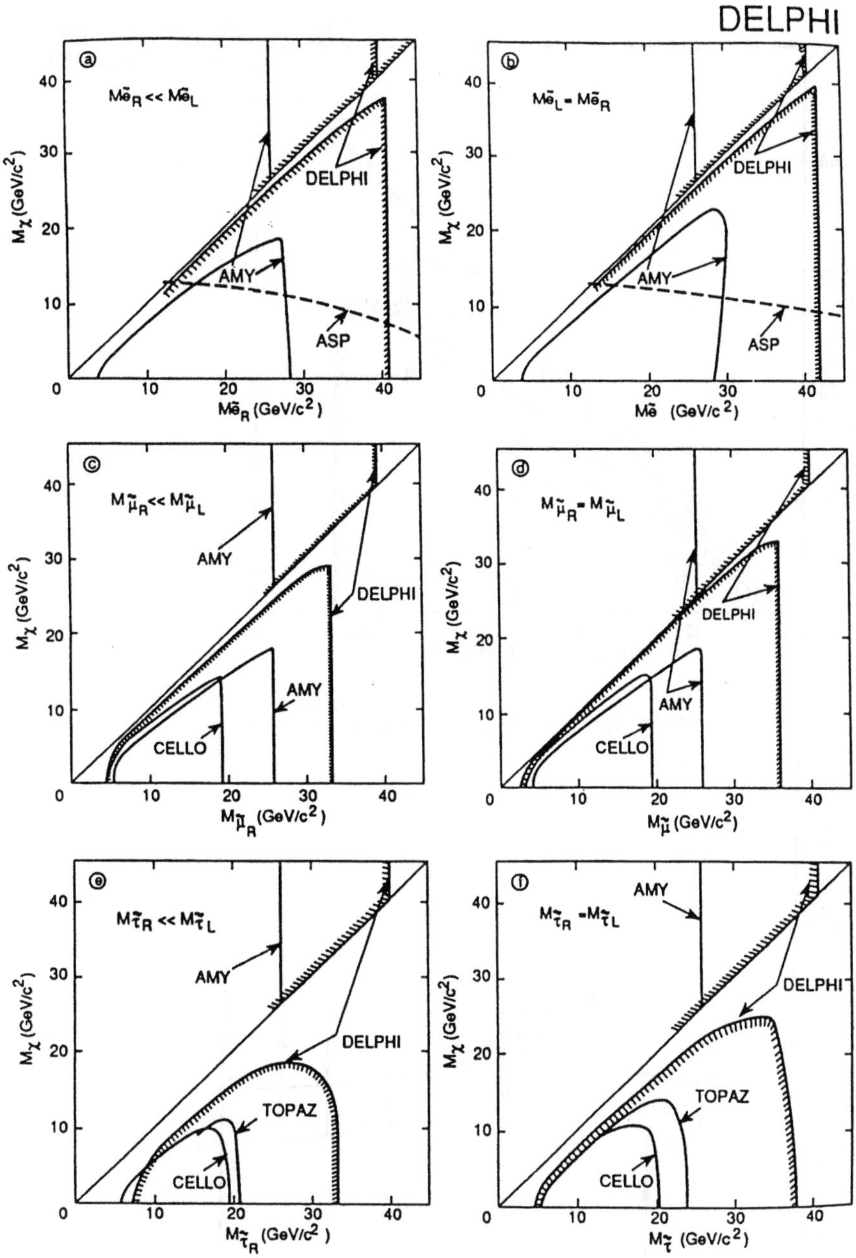

Figure–30: Limits on sleptons (DELPHI).

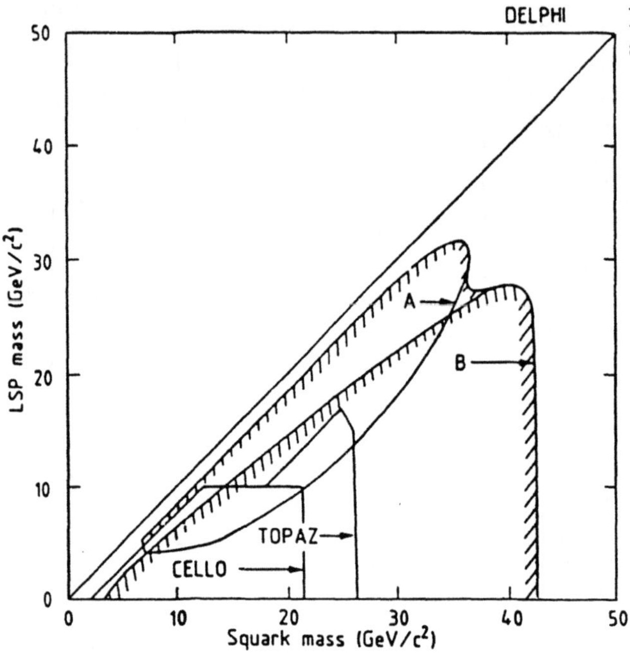

Figure–31: Limit on squarks (DELPHI).

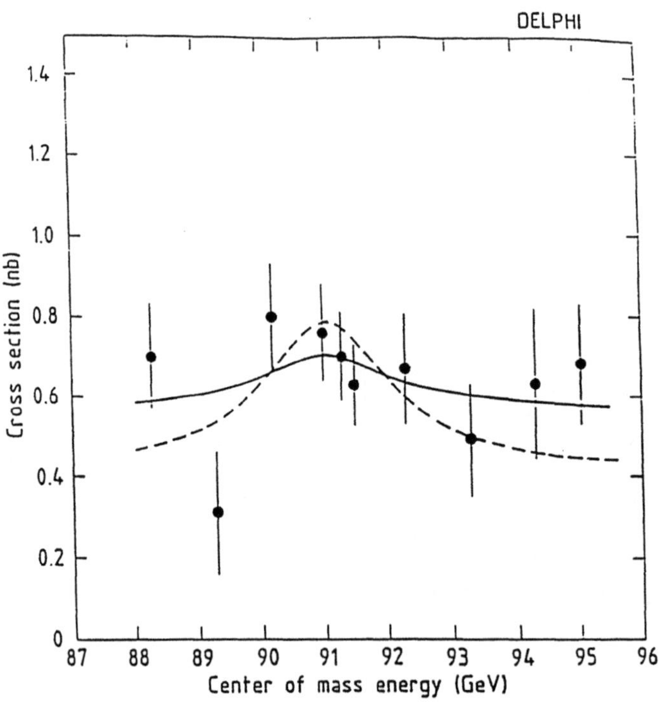

Figure–32: Absence of a resonant signal for low visible-energy events (DELPHI).

9 Compositeness

We come here to more speculative–and less theoretically grounded–issues to cope with the deficiencies of the SM. The fact that no theory ever managed to build nearly zero mass objects (light quarks and leptons) from heavy constituents (preons, rishons, ...) is not a reason to drop the idea of a further shell of constituents, explaining the strange features of the SM.

At a very high scale Λ we suppose that basic constituents manifest themselves: they can be created or exchanged, as partons are at present energies. For E $\ll \Lambda$ the only manifestation left is some kind of effective Lagrangian, giving rise to contact terms or excitation patterns. The ambiguity of the definition of Λ which requires that the strength of the interaction is defined as well has been emphasized in the past [26].

On the Z at LEP the manifestation of contact terms is minimal since their real contribution cannot interfere with a purely imaginary amplitude.

One should rather look for excited objects or specific decay modes of the Z manifesting its composite nature.

Excited leptons, foreseen by compositeness, can be pair produced with no ambiguity in the coupling. LEP easily sets a limit of $M_Z/2$ on their mass. They can also be singly produced

$$\mathrm{ee} \rightarrow \ell\ell^{\star} .$$

However the coupling is then an unknown magnetic one which has to be parametrized as follows

$$L_{eff} = \sum_{V=j,Z} e\frac{\lambda_V}{m_{e^\star}}\overline{\psi}_{e^\star}\sigma^{\mu\nu}\left(c_v - d_v\gamma_5\right)\psi_e\partial_\mu V_\nu \ .$$

c_V and d_V are determined by constraints from g-2 and electric dipole of the leptons. The unknown quantity λ can be related to the compositeness scale Λ:

$$\frac{1}{\Lambda_V^2} \simeq \alpha\left(\frac{\lambda_V}{m_{e^\star}}\right)^2$$

and stays as an open parameter. Therefore LEP results [27] can only be expressed in the plane: $\lambda - m_\ell^\star$ as shown in Fig. 33. L3 has up to now published the stronger limits.

Another manifestation of an excited e, e*, would be its t-channel exchange, in the process

$$\mathrm{ee} \rightarrow \gamma\gamma \ .$$

This ee $\rightarrow \gamma\gamma$ channel has been explored by all four experiments. Without any interference with an exotic Z component, its cross section can be written as

$$\frac{d\sigma}{d\Omega} = \frac{\alpha^2}{s}\frac{1+cos^2\theta}{\varepsilon - cos^2\theta} \text{ with } \varepsilon \simeq 1 + \frac{2m_e^2}{s}$$

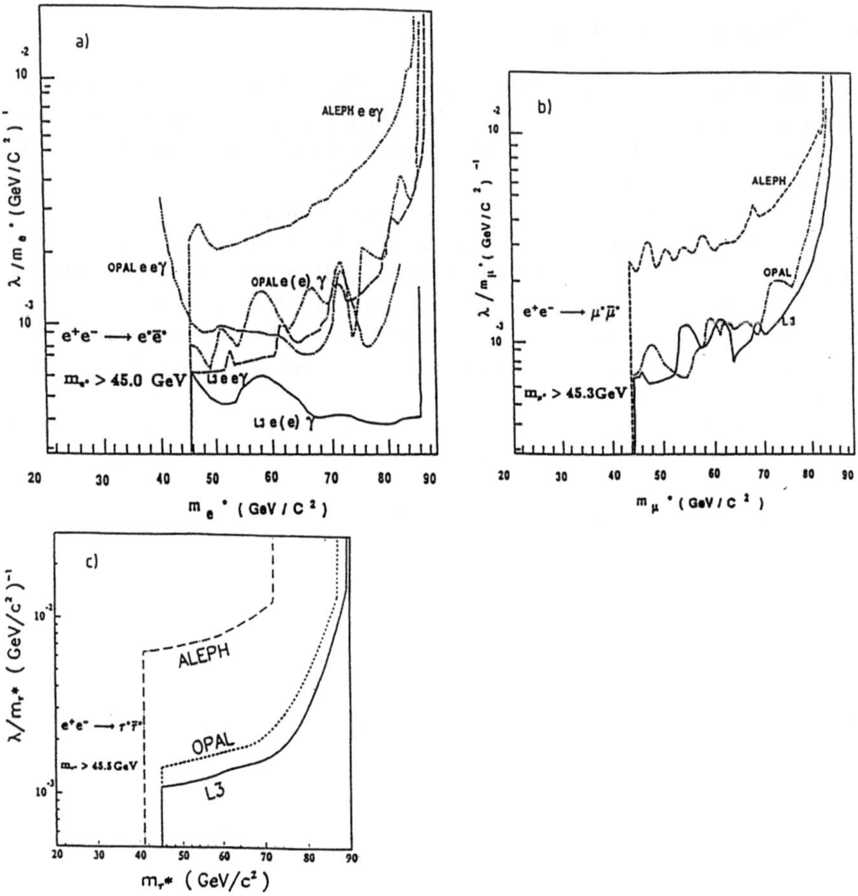

Figure–33: Limits on excited leptons, a) e*, b) μ*, c) τ*.

or

$$\frac{d\sigma}{d\Omega} / \frac{d\sigma}{d\Omega_{QED}} = 1 \pm \frac{s^2}{2\Lambda_\pm^4} \sin^2\theta H\left(\cos^2\theta\right)$$

when the explicit contribution of an e* is shown.

$$H\left(\cos^2\theta\right) \rightarrow 1 \text{ as } m_{e*} \gg \sqrt{s}\,.$$

$\Lambda^\pm$, the QED cut off parameter, is related to the m_{e*} and the compositeness scale Λ by

$$\frac{1}{\Lambda_\pm^4} \approx \frac{1}{\Lambda^2} \frac{1}{\alpha m_{e*}^2}\,.$$

The present e* mass limit, coming from the cross sections of Fig. 34 [28], is still rather poor (less than $\sqrt{s}/2$). A scaling law deduced from the above

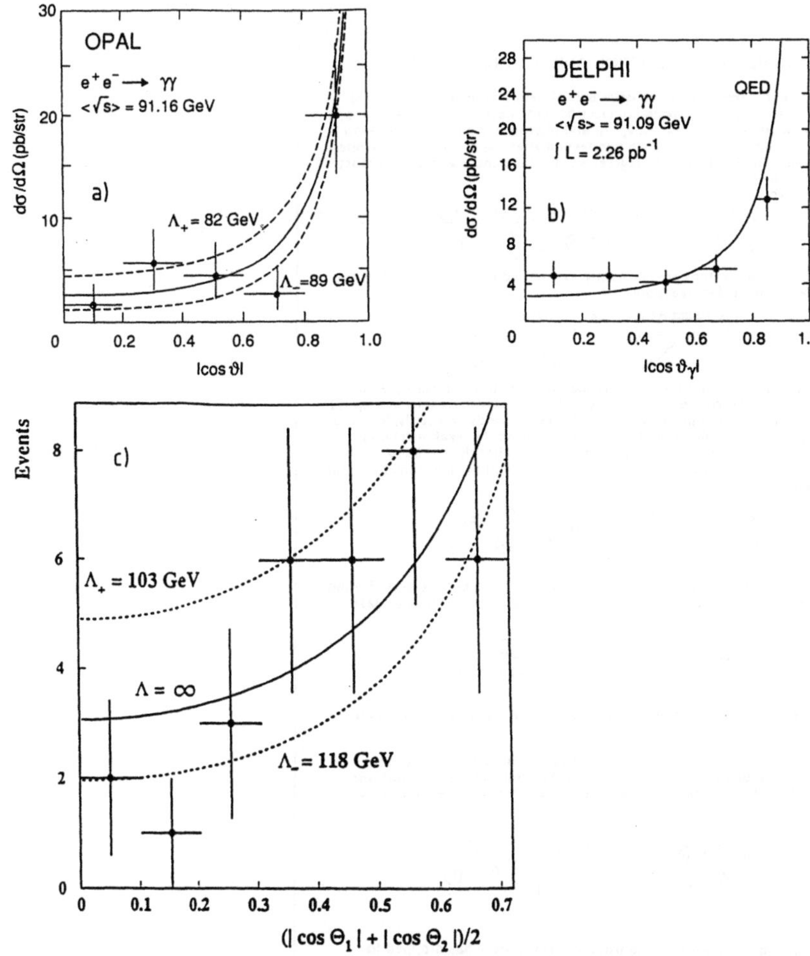

Figure–34: ee→ $\gamma\gamma$ cross-sections from; a) OPAL, b) DELPHI, c) L3.

formula reads as follows:

$$\frac{\alpha}{m_{e*}^4} s^{3/2} \sqrt{\int \mathrm{Ldt}} = \text{constant} \;,$$

where m_{e*} is the accessible mass for $\lambda = 1$. This shows that "patience does not pay" and that it is preferable to go to higher energies than insist on the Z.

A striking manifestation of the compositeness of the Z could be the observation of the decay $Z \to 3\gamma$. The expected branching ratio is $\sim 2\ 10^{-4} Q^6$, where Q is the mean electric charge of the preons. The present limit of OPAL [29] is $\sim 5.2\ 10^{-5}$. With more statistics one can improve it easily.

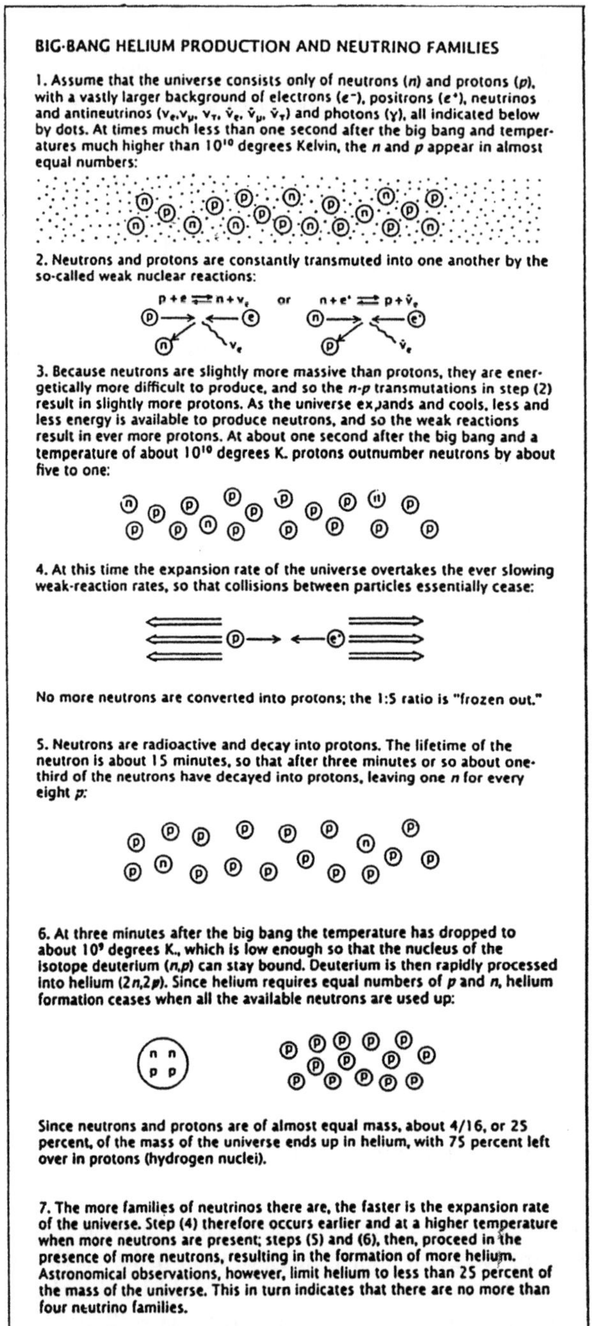

BIG-BANG HELIUM PRODUCTION AND NEUTRINO FAMILIES

1. Assume that the universe consists only of neutrons (n) and protons (p), with a vastly larger background of electrons (e^-), positrons (e^+), neutrinos and antineutrinos ($\nu_e, \nu_\mu, \nu_\tau, \bar{\nu}_e, \bar{\nu}_\mu, \bar{\nu}_\tau$) and photons ($\gamma$), all indicated below by dots. At times much less than one second after the big bang and temperatures much higher than 10^{10} degrees Kelvin, the n and p appear in almost equal numbers:

2. Neutrons and protons are constantly transmuted into one another by the so-called weak nuclear reactions:

$p + e \rightleftharpoons n + \nu_e$ or $n + e^+ \rightleftharpoons p + \bar{\nu}_e$

3. Because neutrons are slightly more massive than protons, they are energetically more difficult to produce, and so the n-p transmutations in step (2) result in slightly more protons. As the universe expands and cools, less and less energy is available to produce neutrons, and so the weak reactions result in ever more protons. At about one second after the big bang and a temperature of about 10^{10} degrees K. protons outnumber neutrons by about five to one:

4. At this time the expansion rate of the universe overtakes the ever slowing weak-reaction rates, so that collisions between particles essentially cease:

No more neutrons are converted into protons; the 1:5 ratio is "frozen out."

5. Neutrons are radioactive and decay into protons. The lifetime of the neutron is about 15 minutes, so that after three minutes or so about one-third of the neutrons have decayed into protons, leaving one n for every eight p:

6. At three minutes after the big bang the temperature has dropped to about 10^9 degrees K., which is low enough so that the nucleus of the isotope deuterium (n,p) can stay bound. Deuterium is then rapidly processed into helium ($2n,2p$). Since helium requires equal numbers of p and n, helium formation ceases when all the available neutrons are used up:

Since neutrons and protons are of almost equal mass, about 4/16, or 25 percent, of the mass of the universe ends up in helium, with 75 percent left over in protons (hydrogen nuclei).

7. The more families of neutrinos there are, the faster is the expansion rate of the universe. Step (4) therefore occurs earlier and at a higher temperature when more neutrons are present; steps (5) and (6), then, proceed in the presence of more neutrons, resulting in the formation of more helium. Astronomical observations, however, limit helium to less than 25 percent of the mass of the universe. This in turn indicates that there are no more than four neutrino families.

Figure–35: A simple way to understand the effect of N_ν on primordial element abundances (Ref.[11]).

10 Conclusion

Many other searches have been performed, generally by all four experiments. For the sake of completeness one can consult the talk of F. Dydak at Singapore which gives all results available.

In fact, the first result to come out from LEP, besides M_Z, was the number of light neutrinos. The present value is $N_\nu = 2.89 \pm 0.10$.

This accurate measurement is a good example of an indirect information on our universe. Further light neutrinos are excluded (and heavy ones for masses below $\sim$ 40 GeV as well through the width measurements); sneutrinos are excluded. This information has strong consequences on our view of the post big bang evolution: Figure 35 [30] recalls how it tightly constrains quantities like the He abundance in the Universe. The need for non baryonic dark matter is reinforced. But at the same time, as we saw, searches at LEP could in the future exclude up to $\sim$ 20 GeV one of the best candidates for such a dark matter, the LSP, making the situation still more exciting.

My own conviction is that searches with the guidance of the SM, SUSY, astrophysical ideas and in parallel with accurate measurements, should be a major concern at LEP until the end of its hopefully thorough exploitation.

Acknowledgements

I am grateful to many colleagues for fruitful discussions, in particular F. Richard and R. Barbieri. Previous reviews (G. Wormser, F. Richard, ...) were quite useful as well. I warmly thank the service of M. Jouhet for efficient typing and drawing.

After completion of this review the first numerical results on the effect of loops on SUSY Higgs masses came out [43]. This effect has indeed to be taken into account: it depends dramatically on the top mass.

References

[1] A. Djouadi et al., Preliminary report to the LEPC from the Working Group on High Luminosities at LEP (CERN, Geneva, May 1990), p.121.

[2] $H^\pm$ searches at LEP:
ALEPH, CERN-EP/90-34, Phys. Lett. **B241** (1990) 623,
DELPHI, CERN-EP/90-33 and Phys. Lett. **B241** (1990) 449,
L3, preprint # 18 (Sept. 90) to be published in Phys. Lett. **B**.
OPAL, CERN-EP/90-38 (March 90), and Phys. Lett. **B242** (1990) 299.

[3] M. Felcini, talk given at the DESY Theory Workshop, Hamburg (Oct. 90).

[4] OPAL, CERN-EP/89-154 (Nov. 89) and update for Singapore.

[5] Aachen Workshop on LEP 200, vol.2, p. 251.

[6] Light Higgs at LEP:
ALEPH CERN-EP/90-70 (May 90) Phys. Lett. **B245** (1990) 289
DELPHI CERN-EP/90-44
L3 Preprint # 19 (Sept. 90) to be published in Phys. Lett. **B**,
OPAL CERN-PPE/90-116 (Aug. 90) and Phys. Lett. **B251** (1990) 211.

[7] Higgs at LEP:
DELPHI CERN-EP/90-60 (May 90) updated for Singapore
L3 Preprint # 10 (June 90), Phys. Lett. **B248** (1990) 203,
OPAL EP/90-100 (July 90), PPE/90-150 (Oct. 90).

[8] ALEPH Higgs search: CERN-PPE/90-101 (July 90) Phys. Lett. **B246** (1990) 306.

[9] E.Glover et al., CERN TH 5584/89.

[10] F. Gilman, Comments Nucl. Part. Phys. 16 (1986) 231.

[11] D. Cline, Scient. American (Aug. 85).

[12] ALEPH, Phys. Lett. **B236** (March 90) 4.

[13] S. Ting, presented at Singapore conference, and L3 Preprint # 16 (Aug. 90), Phys. Lett. **B251** (1990) 321.

[14] OPAL results on Heavy Neutral Leptons, CERN-EP/90-72 (May 90) and Phys. Lett. **B247** (1990) 448.

[15] Charged leptons at LEP:
ALEPH Phys. Lett. **B236** (1990) 511.
DELPHI CERN-EP/90-80, 1990.
L3, Preprint # 16 (Aug. 90), Phys. Lett. **B251** (1990) 321.
OPAL preprint CERN-EP/90-09, PPE/90-132 (Sept. 90) and Phys. Lett. **B240** (1990) 250.

[16] H. Nilles, Phys. Rep. **C10** (1984) 1.
R. Barbieri, Riv. Nuovo Cimento **11** (1988).

[17] R. Barbieri, in Physics at LEP I, CERN 89-08, editors G. Altarelli, R. Kleiss, C. Verzegnassi vol. 2.

[18] G.F. Giudice, Phys. Lett. **B208** (1988) 315.

[19] LEP results on SUSY Higgs:
ALEPH CERN-EP/90-70 (May 90) Phys. Lett. **B237** (1990) 291,
L3 preprint # 15 (Aug. 90), Phys. Lett. **B251** (1990) 311,
OPAL CERN-EP/90-100 (July 90),
DELPHI CERN-EP/90-60 (May 90) updated.

[20] Neutralinos at LEP:
ALEPH CERN-EP/90-63 (May 90) Phys. Lett. **B244** (1990) 541,
DELPHI CERN-EP/90-80, 1990.
L3 Preprint # 15 (Aug. 90), Phys. Lett. **B251** (1990) 311,
OPAL PPE 90-95 (July 90) and Phys. Lett. **B248** (1990) 211.

[21] A. Roussarie, talk at Singapore for the ALEPH collaboration.

[22] LEP and the Universe: CERN TH 5709/90 (April 90).

[23] Charginos at LEP:
ALEPH CERN-EP/89-158 (Dec. 89) Phys. Lett. **B236** (1990) 86,
L3 preprint # 002, Phys. Lett. **B233** (1989) 530,
DELPHI CERN-EP/90-80 (June 90).
OPAL CERN-EP/89-176 (Dec. 89), Phys. Lett. **B240** (1990) 261.

[24] Sleptons at LEP see ref. [23].

[25] DELPHI, CERN EP/90-79 (June 90).

[26] Compositeness at LEP 2: preprint CERN-EP/87-50 (March 87).

[27] Excited leptons at LEP:
ALEPH Phys. Lett. **B236** (1990) 501, and PPE 90-107 (Aug. 90) Phys. Lett. **B250** (1990) 172,
L3 Preprint, # 007 (June 90) Phys. Lett. **B247** (1990) 177, L3 # 0014 (Aug. 90) Phys. Lett. **B250** (1990) 205, L3 # 0021 (Oct. 90) to be published in Phys. Lett. **B**,
OPAL CERN EP/90-49 (April 90) and Phys. Lett. **B244** (1990) 135.

[28] ALEPH Phys. Lett. **B241** (May 90) 635,
L3 preprint # 013 (Aug. 90) Phys. Lett. **B250** (1990) 199,
OPAL CERN EP/90-29 (Feb. 90) and Phys. Lett. **B241** (1990) 133.

[29] OPAL CERN-EP/90-29, updated.

[30] See Ref. [11].

[31] Physics at LEP 200, Aachen, CERN 87–08.

[32] Polarization at LEP, CERN 88–06.

[33] High luminosity at LEP, Preliminary report to the LEPC.

[34] D. Treille, preprint CERN–EP/90–30, March 1990.

[35] P. Mättig, preprint CERN–EP/90–71.

[36] M. Placidi, Int. Symposium on High Energy Spin Physics, Bonn, September 1990.

[37] R. Barbieri and F. Zwirner, private communication.

[38] C. Arnaud et al., preprint CERN/80–AF/90–06, presented at the EPAC Conference, Nice (1990).

[39] J. Jowett, More bunches in LEP, CERN LEP TH/89–17.

[40] J. Drees et al., CERN 88–06, Vol. 1, p. 317.

[41] E. Lieb et al., DELPHI 90–44, Phys. 71 (1990).
See also Ref. [5].

[42] P. Roudeau, LAL 89–21 (1989).
C. Defoix, DELPHI 90–40, Phys. 67 (1990).
H.G. Moser, MPI preprint Exp. 209.

[43] Y. Okada et al., TU-360, Oct. 1990 (+Erratum).
J. Ellis et al., CERN TH 5946, Nov. 1990,
R. Barbieri et al., IFUP-TH 46/90, Dec. 1990.

A Appendix 1

At LEP 1 Workshop (CERN 86-02, p. 297) a very general formula was given, which applies to the production of any spin 1/2 fermion–antifermion pair, or of a pair of charge-conjugate spin 0 bosons produced via γ or Z^0 exchange in the s-channel.

The cross-section reads as

$$\sigma_0^f(Q, T_{3L}T_{3R}, \beta, s) =$$

$$\frac{4\pi\alpha^2}{3s}\beta \times \left[Q^2\left\{\frac{3-\beta^2}{2}\right\} - 2QC_VC'_V\frac{s(s-m_Z^2)}{(s-m_Z^2)^2+m_Z^2\Gamma_Z^2}\left\{\frac{3-\beta^2}{2}\right\}\right.$$
$$\left. + \left(C_V^2+C_A^2\right)\frac{s^2}{(s-m_Z^2)^2+m_Z^2\Gamma_Z^2}\left\{C_V'^2\left\{\frac{3-\beta^2}{2}\right\} + C_A'^2\left\{\beta^2\right\}\right\}\right] \quad (1)$$

where

$$C_V = \frac{1-4\sin^2\theta_W}{4\sin\theta_W\cos\theta_W}, \quad C_A = \frac{1}{4\sin\theta_W\cos\theta_W} \quad (2)$$

are the vector and axial-vector couplings of the electron, and

$$C'_V = \frac{-2(T_{3L}+T_{3R})+4Q\sin^2\theta_W}{4\sin\theta_W\cos\theta_W}, \quad C'_A = \frac{2(T_{3L}+T_{3R})}{4\sin\theta_W\cos\theta_W} \quad (3)$$

are those of the produced fermion f, whose left- and right-handed components have third components of weak isospin T_{3L} and T_{3R}, respectively. The centre-of-mass energy $E_{cm} = \sqrt{s}$ and $\beta = \left(1-4m_f^2/s\right)^{1/2}$ is the centre-of-mass velocity of the produced fermion. The term $\propto C_A'^2$ has a different form from the others because the axial current can only generate fermions in a P-wave, not in an S-wave.

For s-channel pair production of spin-zero particles such as sleptons, squarks, Higgs bosons, we get the special case

$$\sigma_0^s(Q,T_3,\beta,s) = \frac{\beta^3}{4}\sigma_0^f(Q,T_3,T_3,1,s)\,. \tag{4}$$

The main corrections are radiative corrections. The initial-state electromagnetic one yields

$$\frac{\partial\sigma}{\partial x_\gamma} = \frac{\alpha}{\pi}\left(\ell n\frac{s}{m_e^2}-1\right)\frac{1+(1-x_\gamma)^2}{x_\gamma}\sigma_0\left[s\left(1-x_\gamma\right)\right], \tag{5}$$

where $x_\gamma \equiv 2E_\gamma/E_{cm} = E_\gamma/E_{beam}$ and σ_0 is the total leading-order cross-section (A1) or (A4).

The final-state QCD correction leads to

$$\sigma_{QCD} = N\sigma_0\left(Q,T_{3L},T_{3R},\beta,s\right)F_{QCD}, \tag{6}$$

N = 1 for lepton, 3 for colour triplets etc... $F_{QCD} = 1 + C\alpha_s(\beta)$ at leading order. Details can be found in the Yellow Book.

A Appendix 2

At very high scale (Planck scale?) M_x one assumes a Lagrangian

$$\begin{aligned} L &= L_{SUPERSYM} + L_{BREAKING}, \\ L_{SUPERSYM} &= L_{SUPER}(SU(3)\times SU(2)\times U(1); f), \end{aligned} \tag{7}$$

where

$$f = f_Y \;+ \mu\, H_1 H_2. \tag{8}$$

f_Y is the Yukawa part; μ governs the mass coupling between Higgses.

The symmetry breaking Lagrangian can be written as:

$$m^2\sum_i|\phi_i|^2 + M\sum_\alpha\overline{\lambda}_\alpha\lambda_\alpha + \left(Amf_Y + Bm\mu H_1H_2 + \text{h.c.}\right), \tag{9}$$

where m is a universal mass term for scalars and M a universal mass term for gauginos.

When one goes down to our energies, there are effects of renormalization (such $\mu \rightarrow \mu_R$) and the gaugino–higgsino sector becomes;

$$L = 1/2M_3\overline{\lambda}_a\lambda_a \quad + \quad 1/2\overline{\chi}M^{(0)}\chi \quad + \quad \left(\psi M^{(c)}\overline{\psi} + \text{h.c.}\right). \tag{10}$$

The first term describes 8 gluinos. The second term describes the neutralino sector: $M^{(o)}$ is a 4×4 matrix whose parameters are M_1, M_2, β, μ_R. The third term describes the chargino sector: $M^{(c)}$ is a 2×2 matrix whose parameters are M_2, β, μ_R.

We have defined

$$M_i = M\alpha_i(M_W)/\alpha_i(M_X), \tag{11}$$

$$\text{tg}\,\beta = v_2/v_1, \text{ and } \alpha_i = g_i^2/4\pi. \tag{12}$$

Indices 1, 2, 3 are for $U(1)_Y$, $SU(2)_L$, $SU(3)_C$, respectively. M_2 is the $SU(2)_L$ gaugino mass, M_1 is the $U(1)_Y$ gaugino mass, $m_{\tilde{g}}$ is the $SU(3)_C$ gaugino mass.

All three masses are related, from Grand Unification, by

$$M_1 = \frac{5}{3}\frac{\alpha_1}{\alpha_2}M_2 = \frac{5}{3}\text{tg}^2\theta_W M_2,$$

$$M_2 = \frac{\alpha_2}{\alpha_3}m_{\tilde{g}} = 0.3m_{\tilde{g}}\ .$$

Charginos are a mixing of Winos ($\tilde{W}^{\pm}$) and Higgsinos ($\tilde{H}^{\pm}$). The matrix $M^{(c)}$ reads as:

$$\begin{bmatrix} M_2 & \sqrt{2}M_w\sin\beta \\ \sqrt{2}M_w\cos\beta & \mu \end{bmatrix} \begin{matrix} \tilde{W} \\ \tilde{H}\ . \end{matrix}$$

Neutralinos are a mixing of photino $\tilde{\gamma}$, zino $\tilde{Z}^0$, and Higgsinos $\tilde{H}^0$, $\tilde{h}^0$. The matrix $M^{(o)}$ reads as

$$\left|\begin{matrix} M_1 & 0 & -M_Z Sc & -M_Z Ss \\ . & M_1 & M_Z Cc & M_Z Cs \\ . & . & 0 & -\mu \\ . & . & . & 0 \end{matrix}\right| \begin{matrix} \tilde{B} \\ \tilde{W}_3 \\ \tilde{H}_0 \\ \tilde{H}^0 \end{matrix}\ ,$$

where $S = \sin\theta_W$, $C = \cos\theta_W$, $s = \sin\beta$, $c = \cos\beta$.

A Appendix 3: Possibilities for the Future: a Summary

I would like to summarize here the main possibilities for the future of LEP. Details can be found in the Proceedings of various workshops [31]–[33] and other publications [34, 35].

One may go along three directions:

i) increasing the energy, an already approved option, whose objectives (max $\sqrt{s}$, $\int L\,dt$) need to be more precisely defined;

ii) increasing the luminosity, an option technically related to the previous one;

iii) increasing the accuracy of the SM tests, by getting longitudinal polarization in LEP [32]. This promising option, received a strong encouragement from the recent observation [36] of transverse polarization.

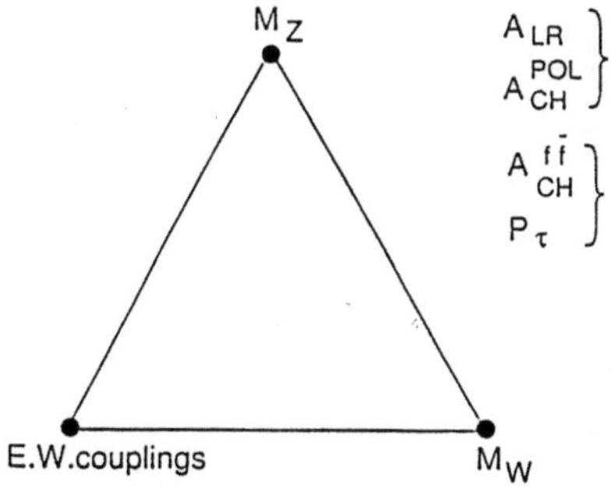

Figure–36: Symbolic representation of the impact of accurate measurements at LEP 200

Accurate measurements are the 'guaranteed' part of the LEP programme. Figure 36 summarizes the prospects: option (i) will give access to m_W with an accuracy at least equal to and probably better than what the hadron colliders will have achieved.

Options (ii) and (iii) will have major impacts on the third corner of the triangle (Fig. 36): the measurement of electroweak couplings.

The global potential physics output of the three options is shown in Table 1.

Table 1

		STAT.	SYST.
Polarization	$-A_{LR}$ $-A_{CH}^{POL}$	~ 1 MZ	OK from experimental point of view, problems are on the machine side
High L	–Accurate measur. $A_{CH}^{f\bar{f}}$, $\Gamma^{f\bar{f}}$,... –rare modes of Z, searches –B physics	several 10^7Z	heavy work ahead
LEP 200	–WW physics(M_w,...) –accurate measur. –searches	meager, luminosity as well as $\sqrt{s}$ are vital	

A.1 The Energy Increase

The most important fact to be kept in mind is the smallness of the cross-sections beyond the Z peak. At $\sqrt{s} = 190$ GeV the point-like cross-section is ~ 2 pb and the WW pair cross-section is ~ 17 pb. The main background is $ee \to Z\gamma$ ($\sim$ 70 pb); one should not forget either the quasi-symmetric radiative

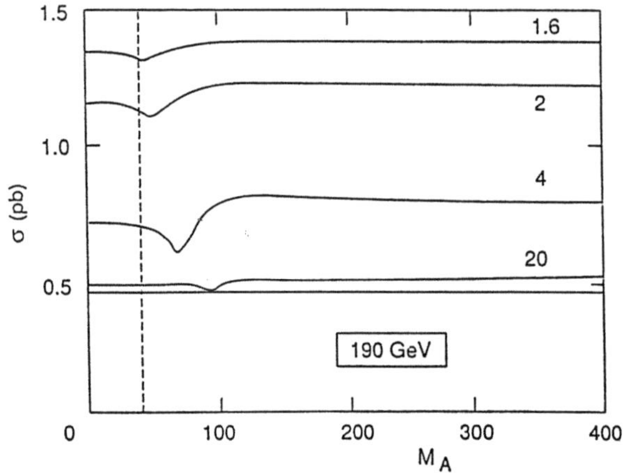

Figure–37: The possibility of excluding all MSUSY parameters plane at LEP 200 ($\sqrt{s} \geq 190$ GeV, $L \geq 500\ \mathrm{pb}^{-1}$). The curves show for various tg β the sum of the cross-sections for h^0 and H^0 at $\sqrt{s} = 190$ GeV. It is always higher than the cross-section of the standard Higgs ($m_H = m_Z$) at that energy. A 5σ signal is obtainable with $500\ \mathrm{pb}^{-1}$. (Courtesy of R. Barbieri.)

return to the Z, ee $\rightarrow$ Z($\gamma\gamma$) ($\sim$ 10 pb), which can be dangerous for some searches.

A standard Higgs of 90 GeV is produced with a cross-section of $\sim$ 0.45 pb at $\sqrt{s} = 190$ GeV and the threshold is at $\sqrt{s} \sim 180$ GeV. One can remember that to get the full production rate for a Higgs of m_H one needs

$$\sqrt{s} \cong m_H + m_Z + (10\,\mathrm{to}\,15\,\mathrm{GeV})\,.$$

Whilst a mass $m_H \sim m_Z$ has nothing magic for the Standard Higgs, for the neutral Higgses of Minimal Supersymmetry, the vicinity of m_Z is an 'accumulation point', where at least one of them *has to be*. Radiative corrections could in fact push the Higgs masses somewhat higher; this is under study. One would then need a corresponding increase in $\sqrt{s}$. This shift depends critically on the top mass. The anticipated total production rate of these scalars always exceeds the production rate of the would-be Standard Higgs of $m_H = m_Z$, provided that there is no kinematical suppression [37] (i.e. for $\sqrt{s} > 190$ GeV, see Fig. 37).

An energy of $\sqrt{s} \gtrsim m_h + 100$ GeV and an integrated luminosity $\int L\,dt \geq 500\ \mathrm{pb}^{-1}$ would allow us to achieve the following:

i) To bring important information on the WW pair physics: m_W, W properties and tests of the three-boson couplings. This has been studied in detail in Ref.[31] and the improvement of the test with increasing $\sqrt{s}$ is explicited by Fig. 38.

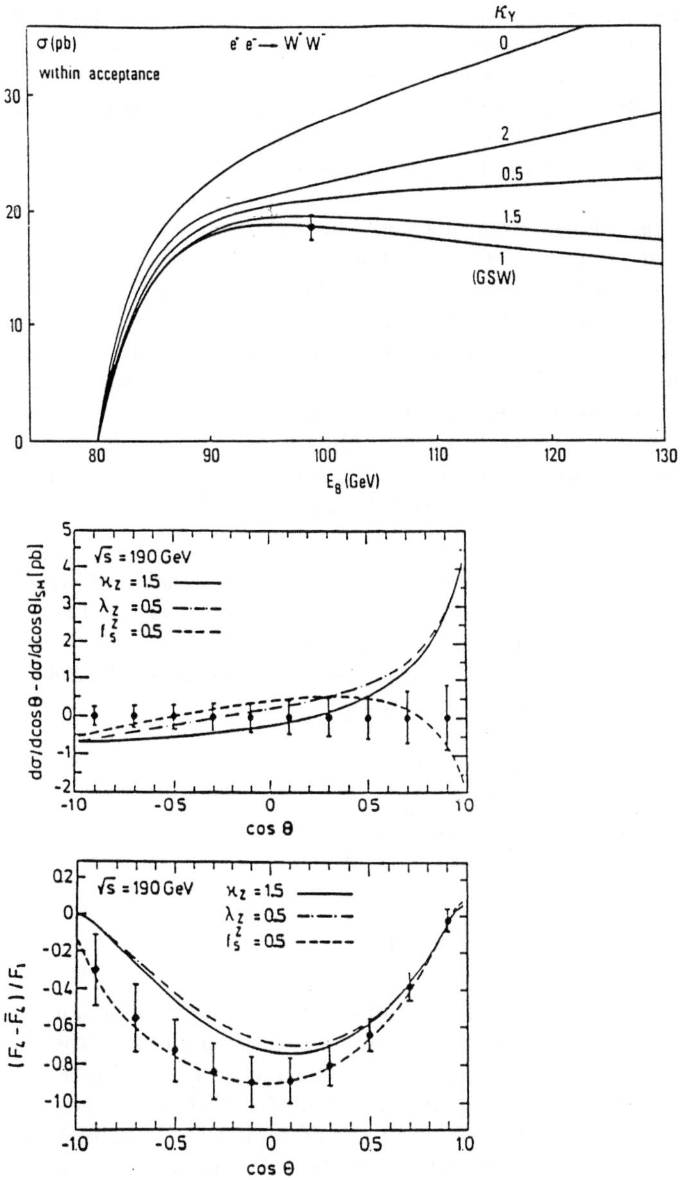

Figure–38: Tests of anomalous W couplings at LEP 200

ii) To explore the MSUSY neutral Higgs sector. MSUSY (the next theory after the SM in order of growing complexity) and most extensions from it can be tested there in a severe way, since, as we said, a light scalar should be present not far away from M_Z. A good b tagging is a key factor for such a programme. Similar tests at future hadron colliders, the detection of [H($\sim$ 90 GeV) $\rightarrow \gamma\gamma$] are, to say the least, not easy ones.

Other very interesting measurements would be performed, e.g., the fermion asymmetries $A^{\mathrm{ff}}_{\mathrm{CH}}$ at full energy which are major ingredients of the strategies of indirect searches. Various direct searches would push to $\sim \sqrt{s}/2$ the mass limit of pair-produced objects (charginos, sleptons).

Technically, the key point is the performance of a large set of superconducting cavities. Preliminary results [38] on a bunch of them are encouraging. Table 2 gives the energy achievable as a function of the number of cavities for various assumptions on their accelerating fields. Only under the most favourable conditions (7 MV/m, all cavities working altogether) can one approach (but not reach) the desirable $\sqrt{s}$ with the presently approved 192 cavities.

Table 2: Energy that can be achieved as a function of the number of SC cavities (warm cavities being removed); current that can be accelerated as a function of available power.
(* i.e. 8 bunches of 1.5 mA or more bunches; ** i.e. 8 bunches of 0.75 mA)

No. of	MV/m		
cavities	5	6	7
192	~ 84.1		~ 91.4
256	~ 89.6	~ 93.8	~ 98
Beam	Beam energy		
power	84	90	95
16 MW	~ 6 mA	~ 4 mA	~ 3 mA
32 MW	~ 12 mA*	~ 8 mA	~ 6 mA**

The luminosity is obviously of outmost importance. Figure 39 shows, under two assumptions on β^* (i.e., achievable beam focusing), the luminosity expected as a function of the available beam power. A safe way to gain a factor 2 in luminosity is to get 8 bunches in LEP instead of 4. With 8 bunches and the nominal current per bunch (0.75 mA) at $\sqrt{s}$ = 190 GeV, 32 MW are needed. The luminosity would be (5–10) $\times$ 10^{31} $\mathrm{cm}^{-2}\,\mathrm{s}^{-1}$ and the time needed to accumulate 500 pb^{-1} would then be $\sim$ 2 to 3 years.

A.2 The Luminosity Increase

Figure 39 shows that if a beam power P is available the luminosity achievable at a given beam energy E varies like

$$L \sim \frac{P}{E^3}\,.$$

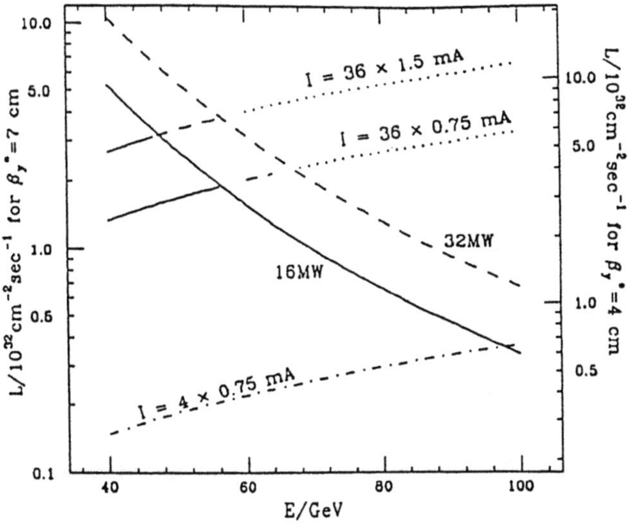

Figure–39: Luminosity versus $\sqrt{s}$ for given beam power values

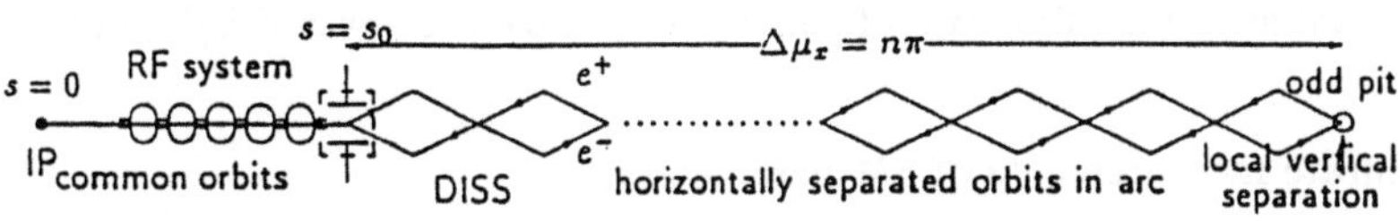

Figure–40: The pretzel scheme

This increase of L can be achieved by a multiplication of the number of bunches n_b in each beam. The pretzel scheme [39], with an induced separation of the beams (Fig. 40), offers such a possibility. Various conditions to be fulfilled lead to

$$n_b = 2, 4, 8, \ldots, 18, \ldots, 36, 40$$

as viable options. Taking $n_b = 8$ is an easy choice, which could be readily accepted by the experiments and requires the introduction in the machine of eight new separators. Since $n_b = 36$ is quite demanding from the experiments (although not impossible), $n_b = 18$ appears as the safe upper limit, likely to provide an increase of a factor of 4 in luminosity.

From the machine point of view, however, more studies and machine developments are needed before one can arrive to a real assessment and an optimized solution for a pretzel scheme.

Let us note that from the physics point of view the introduction of horizontal separators, because it leaves the door open to more bunches and because it is *a priori* harmless for polarization, is preferable to the mere mid-arc vertical separation, suggested for the eight-bunch scheme.

Such an increase in luminosity would make it possible to obtain an exposure of $N \simeq 25 \times 10^6$ Z in 2–3 years. Three main topics would greatly benefit from

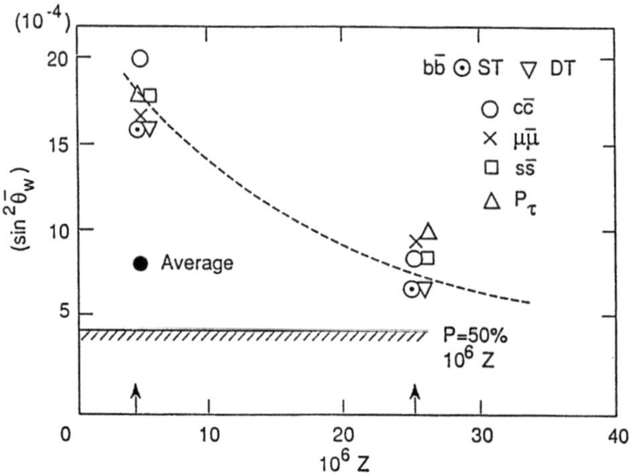

Figure–41: Accuracies on $\sin^2\theta_W$ from polarized and unpolarized measurements

this high statistics (the first two of them being missed for ever if LEP misses them):

a) accurate measurements of SM parameters;

b) rare decays of the Z;

c) fermion–antifermion physics, especially $b\bar{b}$.

a) SM parameters

In this domain we already said that a measurement of A_{LR} should give the most accurate determination of $\sin^2\theta_W$(Fig.40). However, the availability of longitudinal polarization at a sufficient level ($P \sim$ 40–50%) is not yet guaranteed and one may have to look for an alternative way by improving the measurements of $A_{CH}^{\ell^+\ell^-}$, P_τ, $A_{CH}^{q\bar{q}}$ and combining them. Furthermore, high statistics gives access to better determinations of other quantities such as $\Gamma^{f\bar{f}}$, i.e., information not contained in a polarization programme of 10^6 Z. If one day high luminosity *and* polarization are available (a polarized pretzel) so much the better!

In particular, high luminosity allows us to use, for quark tagging, besides the single-arm method [40] double-tag procedures [41], which are more demanding in statistics but less sensitive to systematic errors (although they still require a good measurement of the contamination). A strategy of flavour tagging with all steps based only on measurements (and not on Monte Carlos) can be devised. It is admitted that the combination of various measurements of unpolarized quantities, can bring with 25×10^6 Z an accuracy on $\sin^2\theta_W$ comparable to the one from the standard polarization programme. An accuracy of $\sim 2\%$ can also be reached on $\Gamma_{b\bar{b}}$.

b) Rare decays of the Z

The most outstanding classical decay involves the Higgs boson. While $Z \to H\gamma$ is inaccessible at its SM value because of $ee \to q\bar{q}\gamma$ background, one can possibly push the $Z \to Z^* H$ up to ~ 60 GeV: however, at some stage it is more effective to move to LEP 200 and search for the Higgs scalar in:

$$ee \to Z^* \to ZH\,.$$

Other possible rare decays, if they are observed at rates above the SM expectations, would reveal new physics. Their observation would be of the utmost importance, but we have no guarantee at all that they will show up. Among them, several have been demonstrated to be experimentally accessible down to very low branching ratios such as $Z \to 3\gamma$ (nearby compositeness), $Z \to e\tau, \mu\tau$ (FCNC), etc. [33].

c) Fermion–antifermion physics

Tau physics certainly deserves a special study: besides P_τ and $A_{\mathrm{CH}}^{\tau\tau}$ basic measurements, a general study of τ Lorentz parameters (Michel parameter, chirality parameters, etc.) can be envisaged through the complete measurement of the simple final state

$$ee \to \underset{\hookrightarrow A+\ldots}{\tau} \quad \underset{\hookrightarrow B+\ldots}{\tau}$$

$A, B = \pi, \ell, \rho$, etc.

The competition of eventual τ (and b) factories is obvious and for some aspects (τ properties, m_{ν_τ}, ...) completely overwhelming.

Beauty physics is by far the most promising sector. Table 3 summarizes the outstanding features of the Z as a beauty factory.

'One-arm' measurements (lifetimes, rare modes of the B, spectroscopy, ...) are quite accessible, but prone to competition (B factories, etc.). In particular,

Table 3: Properties of B production at LEP

Cross-section: $\sigma_{b\bar{b}} = 6.5$ nb.
Percentage, relative to the hadronic Z modes:
$\sigma_{b\bar{b}}/\sigma_{\mathrm{had}} = 0.22$.
Percentage, relative to the visible Z modes:
$\sigma_{b\bar{b}}/\sigma_{\mathrm{vis}} = 0.19$.
Population of various species, from 100 million Z:
15.5×10^6 B^0 $15.5 \times 10^6 B^+$
4.5×10^6 B_s^0 1.7×10^6 Λ_b
0.35×10^6 Ξ_b
Mean number of charged particles per B: ~ 5.
Mean number of charged particles at the primary vertex: ~ 10.
Mean flight path of B: ~ 2.2 mm.

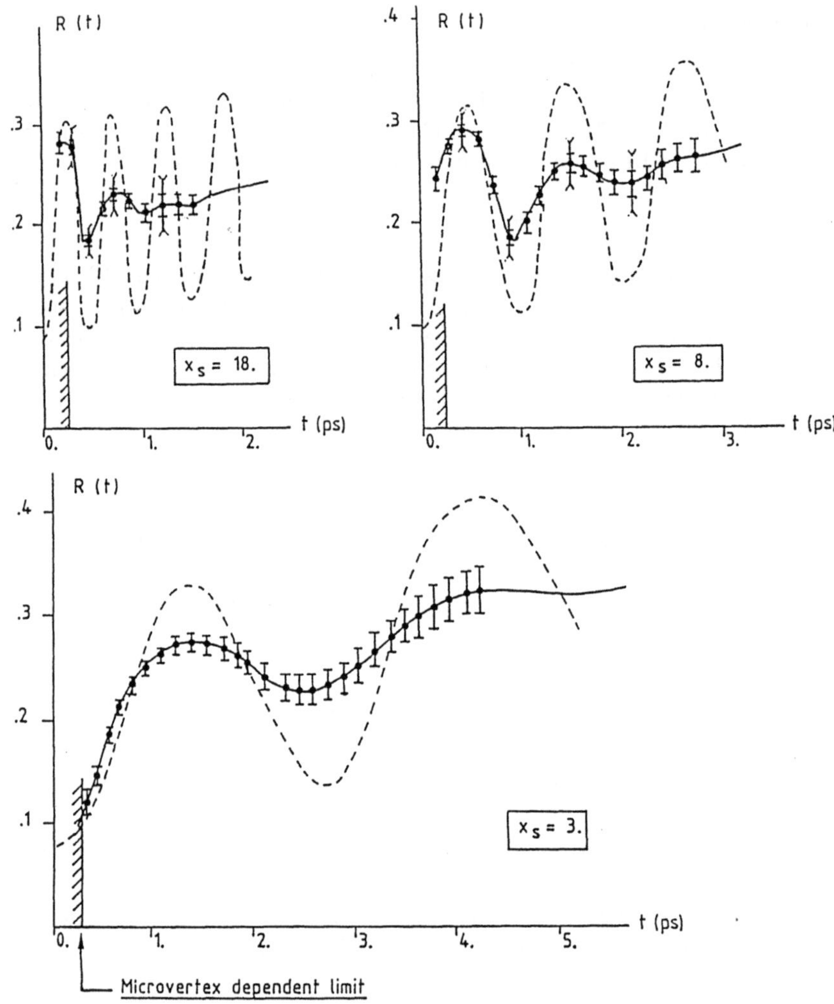

Figure–42: Oscillation pattern for B_{0s} mixing. The error bars are for 100 $\times 10^6$ Z; a few larger ones are indicated for 25 $\times$ 10^6 Z.

it is becoming clear that the Tevatron collider program under way can probably do a lot in this respect, especially through the B $\rightarrow \psi$ ... modes.

However, for 'double-arm' measurements where one B has to be reconstructed and the other one tagged (B oscillations, etc.) I think LEP has great advantages. At the collider the probability to have access to the other b is tiny; at LEP one knows exactly where to look and what to look for in a very clean environment.

One of the main concerns of LEP experiments should be (and indeed is) b tagging: high efficiency and purity should be very rewarding, both for searches ($H \underset{100\%}{\longrightarrow} b\bar{b}$) and for B physics itself. Good microvertices and a clever exploitation

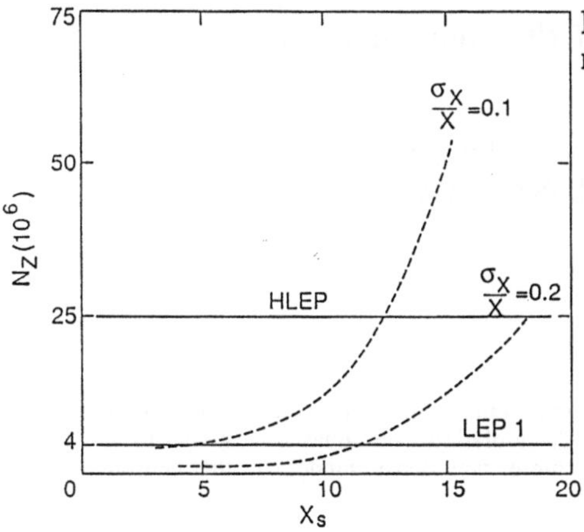

Figure–43: Number of Z needed to measure B_{0s} mixing

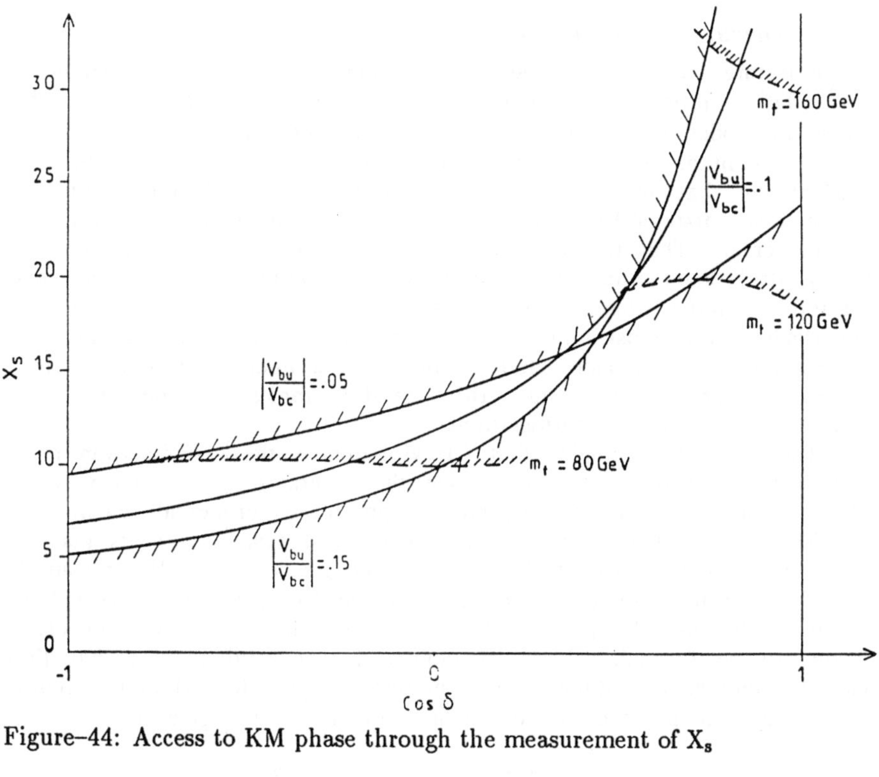

Figure–44: Access to KM phase through the measurement of X_s

of the special features of b fragmentation should make this programme possible with ~ 50% efficiency for $b\bar{b}$ tagging.

Figures 42 and 43 show what one could achieve for instance in the field of B_s mixing [42], an especially rewarding one since it gives access to the phase of the KM matrix, supposed to be the key to the CP problem (Fig. 44). In such a domain LEP with high luminosity should be better than any other competitor.

CP Violation in the Decays of the Neutral Kaon

B. Winstein

Enrico Fermi Institute, The University of Chicago,
5640 South Ellis Avenue, Chicago, IL 60637, USA

We treat current and future experiments that are likely to provide new information on the nature of CP violation. Emphasis is given to those involving the 2π as well as more rare decay modes of the neutral kaon.

1. General Introduction

These lectures will cover some topics of current interest in the study of CP violation, primarily in Kaon decays. We will be concerned with the 2π decay modes as well as with some much rarer ones.

The point of view adopted for these lectures, as well as in a broader context, is that we have just one confirmed manifestation of the phenomena, disregarding the apparent baryon number asymmetry in the universe. That manifestation is in an asymmetric mixing between the neutral kaon and its antiparticle. Although we have a very attractive model that relates this phenomenon to quark mixing, nevertheless it is most important to find other manifestations to give us more proof of the model. Indeed, there are a number of more or less definitive predictions that this model makes concerning the observability of new manifestations.

The possibility remains that CP violation is the manifestation of a completely new force in nature, one that operates at an energy scale many orders of magnitude beyond our present experimental reach.

The kaon system is a very well understood system with its masses, lifetimes, decay rates, and charge structure quite well measured. Probably the most important property of the K_L meson is its very long lifetime, allowing the production of relatively intense, pure, and clean beams to be formed. It is of course a wonderful feature that this particle contains an (almost) equal mixture of particle and anti-particle!

We recall that the rate for, in this case, $K^+ \to \mu\nu$ is given by

$$\Gamma(K\to\mu\nu) = \frac{G_F^2}{8\pi} f_K^2 |V_{us}|^2 m_\mu^2 \left(1-\frac{m_\mu^2}{m_K^2}\right)^2, \qquad (1)$$

where f_K is the kaon decay constant and V_{us} is the appropriate CKM matrix element. When we note that $G_F \propto M_W^{-2}$, we can see that a process with a branching ratio of 10^{-11}, the order of magnitude of the sensitivity of the best experiments at present, can probe a mass scale on the order of about 50 TeV.

Springer Proceedings in Physics, Vol. 65 **Present and Future of High-Energy Physics**
Editors: K.-I. Aoki and M. Kobayashi

2. The 2π decays of the neutral kaon.

We consider the experiments exploring the 2π decays of the neutral kaon, both those presently and future initiatives. First we will briefly review the relevant formalism.

2.1 ε'/ε

If the $\Delta I=1/2$ rule were exact in the 2π decays of the neutral kaon, then all manifestations of CP violation would be from the asymmetry in particle-antiparticle mixing and there would be no way to untangle the presence of a direct effect, namely a $\Delta S=1$ CP violating decay. However, since this rule is known to be violated at the 5% level (in amplitude), there is the possibility of observing a difference in the effect to $\pi^+\pi^-$ vs. $\pi^0\pi^0$ although it is suppressed.

The indirect, or mixing, effect is parameterized by ε which is well measured to be about 0.0023. The direct effect is parameterized by ε' and the size of this parameter has proven difficult to estimate. After we treat the current experiments on the subject, we will discuss the theoretical situation. Generally, it is thought that the value of ε'/ε is of the order of 10^{-3}, in the CKM model. There are, however, considerable uncertainties and in addition, the result will depend upon the mass of the top quark. It is possible that the value is accidentally even smaller as there are cancelling contributions.

Experimentally, one wants to measure the ratio

$$\left|\frac{\eta_{+-}}{\eta_{00}}\right|^2 = \left|\frac{\varepsilon+\varepsilon'}{\varepsilon-2\varepsilon'}\right|^2 \approx \left|1+3\frac{\varepsilon'}{\varepsilon}\right|^2 \approx 1+6\,\frac{\varepsilon'}{\varepsilon} \tag{2}$$

as systematically free as possible. Since

$$\eta_{+-} = \frac{\mathrm{amp}(K_L\to\pi^+\pi^-)}{\mathrm{amp}(K_S\to\pi^+\pi^-)}\,, \tag{3}$$

and

$$\eta_{00} = \frac{\mathrm{amp}(K_L\to\pi^0\pi^0)}{\mathrm{amp}(K_S\to\pi^0\pi^0)}\,, \tag{4}$$

the ratio in (2) is effectively a "double ratio". Experiments need to measure the four decay rates as systematically free as possible.

We will now discuss the experimental situation.

2.2 *The NA31 experiment at CERN*

This group reported a result [1] based upon their running in 1986 of

$$\left|\frac{\eta_{+-}}{\eta_{00}}\right|^2 = 0.980 \pm 0.004\ (\mathrm{stat.}) \pm 0.005\ (\mathrm{syst.}) \tag{5}$$

which, using (2), yields

$$\mathrm{Re}(\varepsilon'/\varepsilon) = (3.3 \pm 1.1) \times 10^{-3} \cdot \qquad (6)$$

The technique of the NA31 group is well known: both K_L decays are measured simultaneously and this running alternates with that where both K_S decays are then measured simultaneously. The K_S are produced directly from the target (*i.e.* no regenerator) and this K_S target moves throughout the decay volume in an attempt to mimic the K_L decay distribution. Even so, the data are binned in 10 regions of energy and 32 regions of vertex for the determination of the double ratio. The acceptance is very large and no magnet is used, the charged momenta being reconstructed with a hadron calorimeter. The neutral decays are reconstructed with a lead-liquid argon detector.

The group collected about 100K $K_L \to 2\pi^0$ decays and many more of the other three modes, accounting for the small statistical error. Systematic uncertainty arises from four different sources: acceptance corrections, background subtractions, energy scale errors, and accidental effects.

NA31 acceptance corrections. By the use of different targets, the two beams will have significantly different divergences. Also, when the K_S running is done, it is necessary to begin the decay region with a 7 mm lead sheet followed by a thin scintillator to cleanly eliminate both neutral and charged decays upstream of this point. Finally, the momentum spectra of the K_S and the K_L differ significantly. In combination, these effects amount to a correction of about 0.3%.

NA31 background subtractions. In the neutral mode, the major background arises from the $K_L \to 3\pi^0$ decays. These events masquerade as $2\pi^0$s when photons are missed and/or photons are fused in the electromagnetic detector. The liquid argon detector has excellent resolving power (position resolution) so that, for NA31, the residual background of 4.0% is almost exclusively a result of photons being missed.

For the charged mode, the hadron calorimeter has poor resolution resulting in a problem in the rejection of semileptonic backgrounds. This background is reduced to the 0.65% level only by cutting very hard on the energy distributions of the candidate pions in the calorimeters, so hard in fact that about 50% of the otherwise good events do not satisfy the cuts.

NA31 energy scales. It is important to determine the relative energy scale between charged and neutral decays: one is measuring the decay rates in the laboratory and what matters is the ratio of decay rates in the center of mass. Since the events are not monochromatic, it is the particle energies that tell us how to make this transformation.

For the charged modes, even though the energy resolution is poor, the kaon energy can be determined with reasonable precision from the ratio of the π^+ and π^- energies:

$$E^2 = \frac{(M_K{}^2 - M_\pi{}^2 R)R}{\Theta^2} , \qquad (7)$$

where Θ is the opening angle of the particles and $R = 2 + E_1/E_2 + E_2/E_1$. By restricting the energies so that R lies between 0.4 and 2.5, the energy resolution becomes about 1% and the calorimeter constants are adjusted to give the known K^0 mass. One of course needs to worry about non-Gaussian tails, non linearities and spatial variations in the calorimeter response.

For the neutral modes, the 4-gamma invariant mass is given by

$$M_{4\gamma} = \frac{\sum_{i>j}^{4} E_i \sum_{j=1}^{4} E_j r_{ij}{}^2}{z^2} , \qquad (8)$$

where E_i are the photon energies, r_{ij} are the distances between photons, and z is the decay vertex, measured from the calorimeter. It is seen that the overall energy scale is closely tied with the vertex scale. It is set by making use of those neutral decays that occur close to the lead/anti-counter where a sharp edge in production is smeared by experimental resolution. Here one knows the mass and the z of the decay and since the distances between showers is well measured, the overall energy scale can be determined. Good understanding of the smearing (i.e. the detector response) is needed to extract the exact position of the anti-counter. Also, the energy scale (or overall calibration constant) drifted by 10 to 15% over the run. The final uncertainty is only 0.1% but this comprises a large contribution to the systematic error.

NA31 accidental effects. The issue of "accidentals" is probably the most serious for the NA31 technique. One collects charged and neutral decays at the same time but these are reconstructed using essentially different detectors: the chambers and hadron calorimeter for the charged modes and the electromagnetic calorimeter for the neutral modes. Thus one has to be sure that changes in gas gain, wire efficiencies, PM gains, etc., which occur in every experiment are understood well enough that a systematic uncertainty doesn't creep in. This is exacerbated by the fact that the detector sees different conditions when running with the close-by target for K_S and that these conditions are changing as the target is further moved. Moreover, as the cuts are relatively tight, one has to be sure of understanding relatively subtle pile-up effects as they effect both the trigger at several stages and the resultant event reconstruction. The losses are at the 3% level and are similar for all modes but this problem is still getting lots of attention.

NA31 future. The NA31 group also ran in 1988 and 1989 and collected a total of about three times as much data. Improvements were made such as the addition of a transition radiation detector so that they could relax the selection criteria for $\pi\pi$ events without any increase in background. Thus a result with reduced error can be expected in the near future.

2.3 *The E731 Experiment*

The E731 group has reported a result [2] based upon 20% of its data. That result is:

$$\mathrm{Re}(\varepsilon'/\varepsilon) = (-4 \pm 14(\mathrm{stat.}) \pm 6(\mathrm{syst.})) \times 10^{-4}. \qquad (9)$$

The technique of E731 is also well known. Two nearly parallel K_L beams enter the detector region, one passing through a thick B_4C regenerator to provide K_S decays. In this manner, K_S and K_L decays are viewed simultaneously by the same detector. The regenerator alternates between the beams on a pulse-by-pulse basis (roughly every minute) so that any small difference in beam intensity or detector acceptance is essentially eliminated. There are different triggers for charged and neutral decays; however, the trigger in no way distinguishes between K_S and K_L. Thus the effect of the inevitable changes in chamber gas gain or drifts in calorimeter response over the course of a run will be reduced to high order. The same is the case for "noisy electronics" and similar effects. Moreover, changes in accelerator performance which affect the instantaneous rates in detector elements have little consequence with this technique.

With these advantages comes one principal disadvantage: the acceptance of the detector must be precisely known as a function of decay vertex because the lifetimes of the K_S and K_L are quite different.

The rates in the vacuum (R_V) and regenerated (R_R) beams are given (for example for charged decays) by:

$$R_V \propto |\eta_{+-}|^2 \qquad (10)$$

and

$$R_R \propto |\rho e^{-t/2\tau_S + i\Delta m t} + \eta_{+-}|^2, \qquad (11)$$

where ρ is the regeneration amplitude for the material. Since $|\rho| \approx 10|\eta|$, the ratio of the integrated numbers of regenerated to vacuum decays (defined as R_{+-}) is given approximately by:

$$R_{+-} \propto |\rho/\eta_{+-}|^2. \qquad (12)$$

Similarly,

$$R_{00} \propto |\rho/\eta_{00}|^2 \qquad (13)$$

and thus for the double ratio, we have

$$R \approx R_{00}/R_{+-}. \qquad (14)$$

The time dependence of the decay rate downstream of the regenerator, (11) above, can also be used to fit for τ_s, the K_S lifetime, Δm, the K_L-K_S mass difference, and Φ_η , the phase of the CP violating decay amplitudes (relative to the phase of ρ).

The charged decays are reconstructed by means of magnetic analysis using four precision drift chambers. A large lead-glass array serves to detect the neutral decays and permit off-line rejection of the $\pi e \nu$ decay mode; $\pi\mu\nu$ decays are eliminated with a muon filter. A background in the neutral mode arises from $K_L \rightarrow 3\pi^0$ decays where photons miss the lead-glass; these are largely eliminated by means of many planes of "photon vetoes" situated outside the solid angle of the lead-glass array.

Data collection and triggers. The experiment took data from August of 1987 to February of 1988. About 75% of the data was taken with either the charged or neutral triggers in operation. For neutral mode running, a thin lead sheet was placed at the "trigger plane" in order to, about 25% of the time, convert (usually) one photon. For the other 25% of the data, all four modes were collected simultaneously (with no lead sheet) and it is from the bulk of this latter data set that a result has been presented.

The charged trigger required a two-track topology in the scintillator hodoscopes with no muon signal. In addition to the two pion decay, $\pi\varepsilon\nu$ and $\pi^+\pi^-\pi^0$ events useful for calibration, alignment, and detector studies were collected.

The neutral trigger required either four or six clusters and greater than about 30 GeV energy deposit in the lead glass; $3\pi^0$ decays were thus also collected.

The analyzed data set consisted of approximately 20,000 spills (over an 18 day period) at the Fermilab Tevatron; each spill lasted for about 22 sec and the repetition rate was about one per minute. During each spill, a light flasher which illuminated each phototube of the lead-glass array was pulsed every second, determining the tube gains to about 1% per flash. Pedestal events were taken as frequently and an accidental trigger was taken about 70 times per spill to provide an unbiased sample of the underlying activity in the detector. Care was taken so that both triggers were "loose" and unbiased with respect to K_L vs K_S decays.

An event was retained only if its reconstructed z position lay between 120m and 137m from the production target and its reconstructed momentum lay between 40 GeV and 150 GeV; this selection was made to reduce the systematic error arising from acceptance and lead-glass energy scale uncertainties.

The event totals for this sample together with the magnitude of the corrections (to be discussed) is shown in the following Table.

Table 1. Event Totals and Corrections

	Neutral	Charged	R [from exp'n (14)]
Uncorrected Event Totals			
Vacuum	52,226	43,357	
Regen.	201,332	178,803	1.0698(78)
Background Fractions			
Vacuum	0.0515	0.0042	
Regen.	0.0257	0.0012	1.0445
Acceptance			
Vacuum	0.1885	0.5041	
Regen.	0.1813	0.5064	1.0001

We will now discuss briefly the various sources of systematic uncertainty.

E731 acceptance corrections. As mentioned above, it is important for this technique that the acceptance be properly understood. From Table 1, it is seen, for the chosen fiducial region, that difference in the sizes of the acceptance correction to be made is on the order of a few percent.

Possible uncertainties in the acceptance corrections are the most important source of systematic error for the measurement as E731 has chosen to perform it. The issue is of course the acceptance vs z-vertex; the agreement between data and monte-carlo for the 2 pion modes is quite good and that, for there to be an effect at the percent level, the disagreements would be rather obvious . In order to estimate the possible systematic error, the high statistics modes have been used. Although their decay kinematics are not identical to the 2π decays, monte-carlo studies show that if there is a problem in the acceptance for a 2π mode (such as an aperture that is in the wrong place or a partially inefficient chamber wire, etc), it is readily apparent in the $\pi e \nu$ or $3\pi^0$ modes.

The z-vertex distribution for $K_L \rightarrow \pi e \nu$ decays collected simultaneously is useful for limiting the uncertainty in the acceptance; the agreement over the fiducial decay volume for a sample of 16M reconstructed events is excellent and with only a small disagreement in the upstream region. The similar distribution for $3\pi^0$ decays also shows excellent agreement. The acceptance is in fact easier to determine for the neutral mode: photons travel in straight lines and there is essentially only one aperture to model, the lead-glass array.

For the charged mode where the z resolution is about 15cm, the result is consistent when fitting in 2m z-vertex bins, thereby nearly eliminating the dependence upon acceptance. (This is less reliable in the neutral mode where the z resolution is about 1m.) The effect that the acceptance disagreement observed in the $\pi \varepsilon \nu$ decays would have on the result has been systematically determined and the expected shift in R_{+-} is -0.05%. From this information and the stability of R_{+-} and R_{00} when apertures, beam shapes, cuts, and detector efficiencies are varied, a systematic uncertainty of $\leq$ 0.18% for each mode is assigned.

Fits for other physical parameters. The extraction of Re(ε'/ε) from the data is an intricate procedure and, as the acceptance error is the most important for E731, it is desirable to have independent checks for its determination. To give confidence in the results, the values of other parameters of kaon decay from the data using the same procedure were extracted. The $K_L - K_S$ mass difference Δm, the K_S lifetime τ_S, and the phase difference between η_{00} and η_{+-}, $\Delta\phi$ were all determined. The extraction of $\Delta\phi$, the phase difference between η_{+-} and η_{00}, has been published elsewhere [3]; its value was $-0.3° \pm 2.4°$ (stat.) $\pm$ 1.2° (syst.), consistent with expectations from *CPT* conservation.

Like $\Delta\phi$, the value of Δm depended on the shape of the decay distribution downstream of the regenerator. In the neutral and charged mode fits the values found were $(0.532 \pm 0.013) \times 10^{10}$ ħ sec^{-1} and $(0.535 \pm 0.013) \times 10^{10}$ ħ sec^{-1} respectively, both consistent with the PDG value [4] of $(0.5349 \pm 0.0022) \times 10^{10}$ ħ sec^{-1}.

The final fit was for the K_S lifetime. Of all the fits, that for τ_S was most sensitive to acceptance corrections. The fits gave $(0.8913 \pm 0.0027) \times 10^{-10}$ sec from the neutral and $(0.8891 \pm 0.0029) \times 10^{-10}$ sec from the charged fits, again consistent with one another and with the PDG value of $(0.8922 \pm 0.0020) \times 10^{-10}$ sec. The statistical uncertainties of the fits were equivalent to a precision of 0.05%/m on the kaon loss due to decay. Within this error the result was consistent with the PDG value, indicating that the error on the acceptance correction was of the order of 0.05%/m in each mode. Since the mean difference in the accepted vertex positions for K_L vs K_S decays is less than 2m, the possible systematic uncertainty in the acceptance corrections as estimated in this fashion are even smaller than the assigned uncertainties of 0.18%.

E731 background subtractions. The background summary and the estimated uncertainties are given in Table 2 below.

The largest background, and the one with the most uncertainty, is due to inelastic regeneration where a kaon from the Ks beam is reconstructed in the K_L beam. This background can be exactly simulated (the relevant scattering amplitudes are well enough known or are separately measured) and the simulation can be checked by using the measured P_T distribution of charged mode events where the cross-over probability will be identical. The total error in the double ratio due to uncertainties in the backgrounds is 0.18%.

Table 2. Summary of E731 Backgrounds

Mode	Process	Correction [%]	Error [%]
$K_L \to \pi^+\pi^-$	$\pi e \nu$	0.31	0.06
$K_S \to \pi^+\pi^-$	incoherent regeneration	0.13	0.01
$K_L \to 2\pi^0$	"cross-over"	4.66	0.14
$K_L \to 2\pi^0$	$3\pi^0$	0.37	0.07
$K_S \to 2\pi^0$	incoherent regeneration	2.58	0.07

E731 energy scale. The "energy scale" in the charged mode is easily determined: one simply adjusts the overall magnetic field so that the K^0 and Λ masses agree with the accepted values. It is important that the non-uniformities in the field be well understood. For the neutral scale, the lead-glass is calibrated using momentum-analyzed electrons (using of course the same magnetic field map as for the charged mode analysis) so that in principle the energy scales will coincide. In practice, this is only valid to within a few tenths of one percent as a result of uncertainties in the difference in response of the lead-glass for photons and electrons. The overall neutral energy scale can be adjusted by aligning the observed distributions, but there remains a residual uncertainty on the order of 0.1%. The effects of this uncertainty of the ratio R_{00} are largely eliminated by the choice of the fiducial region (in z-vertex): the region, which includes essentially all of the K_S events, is chosen so that roughly the same number of K_L events leave (enter) at the downstream boundary as enter (leave) at the upstream boundary when the gamma energies are decreased (increased). Empirically R_{00} changes very little with lead-glass energy scale. However the energy resolution itself is not perfectly understood and as a result an 0.2% uncertainty in R_{00} is assigned.

E731 accidental effects. In principle, the presence of accidental activity in the detector at the time of the event could result in a bias between K_S and K_L decays: the accidental activity tends to cluster near the vacuum beam and this could cause a systematic lowering of the efficiency for decays from that beam relative to decays from the other beam. Such a bias would obviously not cancel upon alternation of the regenerator position. The accidental activity, however, is at a low level: events have an extra cluster in the lead glass only 2.7% of the time (and only those which land on top of other clusters can cause a bias) and they have only about 8.5 extra chamber hits throughout the spectrometer (again only those which land on top of the true hits can cause a bias). Such biases have been searched for, primarily using the high statistics modes; indeed, distributions in track quality have a very slight broadening in the vacuum beam, but the cuts are sufficiently loose that this is inconsequential. The technique of overlaying accidental events upon monte carlo events to investigate this possible bias has been used. This procedure correctly reproduces a 3% loss in charged reconstruction efficiency (common to both modes) over the intensity range of the data (a factor of 4). There is, however, no asymmetry in K_L vs K_S for either mode within 0.07% so that the combined possible bias from accidental activity is taken to be 0.10%.

E731 future. The E731 group has about five times more data which has been fully processed. The delicate processes of alignment and calibration are taking place and, as with NA31, a result with better precision is expected soon. In particular, though the bulk of this remaining data was taken with charged and neutral modes collected separately, the systematic uncertainty does not appear to be any greater.

2.4 *Comparisons with theory based on the CKM Model*

As is well known, in the standard model, CP violation arises from a phase in the charged current coupling matrix between the up and down quarks. With the inclusion of a class of diagrams known as electroweak penguins, there is a contribution which grows as the top mass increases and is in the direction to reduce the expected size of ε'/ε. We show in Figure 1 the results of a calculation [5] of ε'/ε, where the several

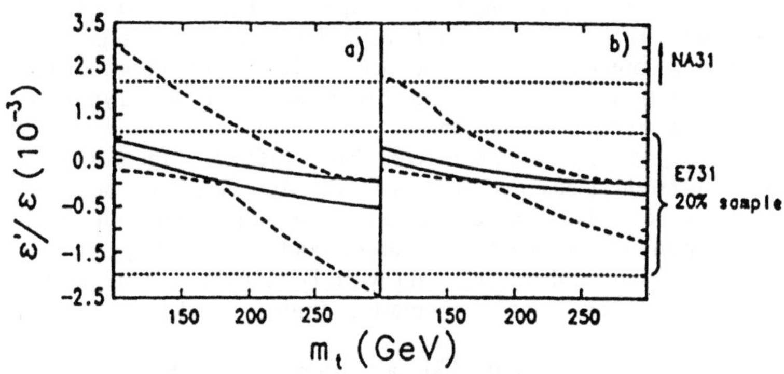

Figure 1 ε'/ε vs. M_t from Buchalla, Buras, and Harlander. The solid curves represent the "standard" range of ε'/ε values while the dashed curves represent the allowed range when all parameters (R, s_{23}, Λ_{QCD}, B_K, and M_s) take their extreme values. It should be noted that the allowed range could be event greater than that shown because of other assumptions inherent in the methods of Ref. 11. Also shown are the measurements of E731, and NA31. a) (b)) is for δ, the phase of V_{ub}, in the first (second) quadrant.

unknowns are allowed to vary within their allowed ranges, and the latest experimental results.

We note that there is at present considerable uncertainty as to the size of ε'/ε but that this will be reduced as we learn more, both experimentally and theoretically, about the weak couplings. The NA31 and E731 values are not in good agreement with the former favoring lower top-quark masses. When both groups report on their full data samples, we will learn more.

2.5 *New initiatives on* ε'/ε

It is clear that one has to make ever more precise measurements on ε'/ε. We will see later that there is a possibility of observing the effect in the system of B mesons but that this will take a new facility whereas one can make substantial improvements in the kaon system with existing facilities. We will describe new initiatives by NA31, by E731, and then the idea of a Φ factory.

The new NA31 experiment. This group has submitted a proposal [6] for a new experiment aimed at a determination with a precision of about 2×10^{-4}. A new beam line and detector are required for this effort which does not use the "NA31 technique". The group will no longer attempt to normalize charged and neutral decays but will now have two beams, one K_L and one K_S, running simultaneously. They will no longer have a "train" to move the K_S through the decay region but will have a fixed source of K_S and use a weighting scheme in the ratio calculations. They will no longer rely upon hadron calorimetry for hadron energy determination but will have a large aperture magnet and will construct new precision drift chambers. They will improve their electromagnetic calorimeter from liquid Argon to Xenon or Krypton for significantly better energy resolution.

The K_S and K_L beams will be derived from different targets and will therefore have different sizes and divergences. But they will be nearly overlapping in space; this is accomplished by having a reduced proton flux incident upon the downstream target and surrounding it with tagging counters. Thus a detected event will be assigned as a short or long candidate depending upon the relative timing of the nearest tag with the event time.

This idea is interesting, but, like all good ideas, there are some problems with it which the collaboration is addressing. First, the tagging counters are running at very high rates ($\approx$ 10MHz) and as a result there is a significant problem with accidental tags. Secondly, the event timing needs to be accurately and identically determined for charged and neutral decays and the group hopes to do this with a lead converter placed in front of the electromagnetic calorimeter which surely compromises its performance. Nevertheless, if the energy response of the calorimeter can be understood sufficiently well, the group should be able to make a determination with significantly reduced systematic uncertainty than with their present method.

The new experiment by this group which will likely be approved would run in 1994, 1995, and 1996.

The new E731 experiment. The E731 group has proposed [7] a new measurement at the Tevatron, using the same technique. The goal is to achieve a measurement in the range of 1.0×10^{-4} statistical error and smaller systematic uncertainty.

In order to execute this experiment, many improvements are required These will benefit the search for rare decays [8] that the same group is performing, E799.

Presently, muons dominate the rate responsible for chamber current and a factor of 10 improvement is desired. The beam halo and profile should be improved. Up to 5×10^{12} protons per pulse are needed, a factor of about 2.5 increase over what we have run at before.

The new experiment (P832) will run with charged and neutral modes taken simultaneously and the rate estimates are reliably based upon the 20% data of E731. The 20% statistical levels together with the assumed relative yields for P832 are shown below.

TABLE 3 Statistics for new ε'/ε experiment.

Mode	20% Statistics	P832 Stat's
$K_L \to 2\pi^0$	55K	N
$K_S \to 2\pi^0$	215K	2N
$K_L \to \pi^+\pi^-$	45K	5N
$K_S \to \pi^+\pi^-$	170K	10N

The basis of the assumptions is the following. The ratio of 2/1 for regenerated to vacuum will result from an increase in the decay volume (for K_L decays). The ratio of 5/1 for charged to neutral events arises from removing prescaling and the assumption, roughly true in E731, that the ratio of acceptance for charged vs. neutral will be about 2.5. The error, then, on the double ratio will be $1.34/\sqrt{N}$. About 0.00055 precision on the double ratio is needed to obtain 10^{-4} precision on ε'/ε where some "dilution" from interference is included. The statistical goal becomes 6×10^6 $K_L \to 2\pi^0$ events.

The attainment of this goal has large implications on the rest of the experiment and these will be mentioned here. First we will treat questions related to what is needed to be able to collect this amount of data and secondly we will treat the systematic issues.

Improvements to the apparatus. The major improvements to the apparatus that are foreseen are:

a) A new calorimeter, probably made from about 3000 CsI crystals.

b) A crude hadron calorimeter

c) A better photon veto system to handle an enlarged decay region and provide more hermetic coverage.

d) New higher gain front-end amplifiers for the drift chambers.

e) TDCs and crude flash ADCs on each calorimeter channel.

f) A new data acquisition system with parallel processing (Level 3).

P832 will also make good use of improvements already underway: a track processor, a system of TRD's, and a totally active regenerator.

The new calorimeter is required to have: (1) energy resolution of order $1\%/\sqrt{E(\mathrm{GeV})}$ or better, (2) no significant radiation damage at 1000 Rad, (3) narrow output pulse and good timing resolution to reject accidental hits. Both BaF_2 and CsI (pure) are under active study.

The literature on radiation hardness of BaF_2 and CsI are somewhat confusing due primarily to different grades of crystals used for tests. BaF_2, however, seems to be hard enough for the next round of experiments. More study is needed on the radiation hardness of CsI. One significant drawback of BaF_2 is that it is about three times as expensive as CsI.

Improvements to the trigger. As in E731, the majority of events written to tape would be of the high statistics modes: a great deal of these are needed for systematic checks.

About 2×10^4 events per spill will be written to tape where the 2π candidate triggers contribute about equally to the high statistics ones. A factor of 10 reduction in the four cluster trigger rate, presumably from an invariant mass calculation, is required. A factor of 50 reduction in the charged trigger rate, using the track processor and Level 3, is also needed.

Systematics. We now turn to the achievement of a systematic error below 5×10^{-4} in the double ratio. We first list in Table 4 the contributions to the systematic uncertainty for the 20% data set analysis together with the goals for P832. In order to achieve the required level of acceptance understanding for P832, another factor of about 2 or 3 improvement is needed. It is likely that this can be achieved even with E731 data. The remaining challenges, which all are tractable, are the following:

-- Control of accidental effects to the 10^{-4} level.

-- Reduction in the uncertainty in the neutral energy calibration by a factor of 10 (this is the major reason for a high precision calorimeter).

-- Reduction in the size of the incoherent regeneration effects by about a factor of 10 (this will be attainable with a totally active regenerator).

This new experiment, P832, will likely run in the late 1993 fixed-target run, along with the experiment concentrating upon rare decays (E799).

Table 4: Systematic Errors for ε'/ε

Effect	E731 Correction [%]	E731 Syst. Error (published) [%]	P832 Correction [%] Goal	P832 Syst. Error [%] Goal
πev	0.31	0.06	0.03	0.005
incoherent regeneration (charged)	0.13	0.01	0.025	0.005
3π0	0.31	0.06	0.10	0.01
"cross-over"	4.66	0.14	0.45	0.01
incoherent regeneration (neutral)	2.58	0.07	1.0	0.01
accidental effects (charged)	-	0.07	-	0.01
accidental effects (neutral)	-	0.07	-	0.01
energy scale resolution non-linearity	-	0.20	-	0.02
acceptance charged	0.45	0.18	≈2.0	0.02
acceptance neutral	3.97	0.18	≈8.0	0.02
TOTAL		0.38		0.04

Φ Factories. Another possibility for the study of ε'/ε is to use the Φ meson decay to K_L K_S as a source of Kaons. The largest advantage of this technique is that one has K_L and K_S decays simultaneously and in the same event.

We briefly discuss the prospects for determining ε'/ε at a Φ factory. The Φ's are produced by e^+e^- collisions and the final state of $K_LK_S \to \pi^+\pi^-$ 4γ is studied. One needs very high statistics and a detector which is large enough to collect as many K_L decays as is practical, given that the K_L lifetime is about 340cm. The overall acceptance (acc) must be high and its variation over the fiducial radius (R) of the detector must be very well understood. Very good electromagnetic calorimetry is required. From the known value of the Φ production cross-section, a straightforward calculation shows that the (statistical) error on ε'/ε can be expressed as:

$$\sigma_{\varepsilon'/\varepsilon} = 6 \times 10^{-4} \times \sqrt{10^{40}/\int Ldt} \times \sqrt{25\%/acc} \times \sqrt{20cm/R}, \qquad (15)$$

where $\int Ldt$ is the integrated luminosity in cm^{-2}. The acceptance could perhaps be made 50% and one proponent believes that a fiducial radius of 1m is possible. Since 6×10^{-4} is already achieved by two experiments, it is clear that integrated luminosities in excess of 10^{40} must be assured before one would embark upon such a project. We note that a luminosity of 10^{32} $cm^{-2}sec^{-1}$ has been proposed for the first phase of the approved project at Frascati for the mid 1990's, with a planned upgrade to follow.

The following Table, adopted from on [9] prepared by H. Nelson, shows the characteristics of the many proposed rings. While it will be a struggle to attain a sensitivity of 10^{-4}, such a measurement by a completely independent technique would nicely complement the determinations made at accelerators.

Table 5: Φ Factory Projects

Accel. Center	Principle	Beam	πD [m]	β^*_x β^*_z [cm]	ε_x ε_z	$\mathcal{L}$ [$cm^{-2}s^{-1}$]	Status
NIKEF	Double Ring, Wigglers, Multibunch	Flat	71.2	1.0 20.0	0.04 0.04	10^{32}	Proposed
Frascati	Double Ring, Wigglers, Multibunch	Flat	100.7	4.5 450	0.04 0.04	10^{32}	Going ahead 1995
UCLA	Double Ring and Asymmetric collider (0.3 x 0.7)	Flat				10^{32} 10^{33}	Proposed
KEK	Double Ring, Wigglers, Multibunch	Flat	≥96	0.3 30	0.03 0.03	$3x10^{33}$	Design Studies
PSI/Mainz/ Switzerland / Germany	Double Ring, Multibunch	Round or Flat	28.8	1.2 1.2	0.1 0.1	10^{33}	Design Studies
Novosibirsk	Single ring, 1 or 3 bunches	round	30.9	1.0 1.0	0.1 0.1	10^{33} $3x10^{33}$	1994

3. CP Violation in rare kaon decays

Partly because of the somewhat muddled theoretical situation in ε'/ε, there has been a lot of attention of late to the subject of the detection of a direct CP violating effect in very rare kaon decays, ones where the $\Delta I=1/2$ rule is inoperative. In this section, we will review this field, concentrating upon the decays of $K_L \rightarrow \pi^0 \ell\bar{\ell}$.

3.1 *Constraints on the Unitarity Triangle*

In the Standard Model, CP non-conservation arises from the complex nature of the CKM matrix. In this section, we will see how to estimate the size of the expected effect in the rare K decays, as well as in the B meson system. We adopt the Wolfenstein convention, where

$$V_{ub} = \lambda^3(\rho - i\eta) \text{ and,} \qquad (16)$$
$$V_{td} = \lambda^3(1-\rho-i\eta). \qquad (17)$$

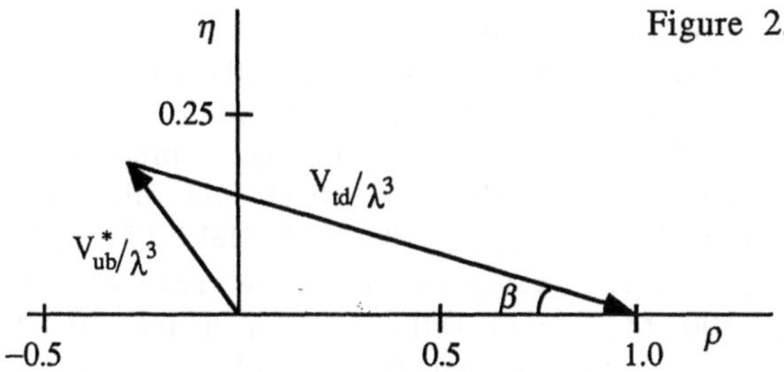

Figure 2. The Unitarity Triangle, scaled by λ^3.

All CP violating amplitudes are proportional to the parameter η. We have the following expressions [10] for the mixing in the neutral B system and for the CP violation parameter ε:

$$\Delta m/\Gamma \propto |V_{td}|^2 M_t F(M_t^2/M_W^2) f_B^2 B_B, \qquad (18)$$

and

$$|\varepsilon| \propto \mathrm{Im}(V_{td}^*)^2 M_t F(M_t^2/M_W^2) B_K. \qquad (19)$$

These expressions assume that the top-quark is heavy enough that it dominates both processes. In practice, for the conclusions that we will draw, it is sufficient that $M_t \geq 100$ GeV/c^2.

From the ratio of (3) and (4), we can find the angle β in the Unitarity triangle shown in Figure 2. It is easy to show that

$$\mathrm{Im}(V_{td}^*)^2 / |V_{td}|^2 = \sin(2\beta) \qquad (20)$$

where β is the angle that the "V_{td} leg" of the triangle makes with the real axis. This is useful in that the asymmetry in the ΨK_S class [11] of modes in B^0 decays is in fact sin(2β) and β is essentially independent of the top mass. Current estimates [12] on the parameter η, with some reasonable assumptions (but not employing measurements of ε'/ε) place it roughly in the range of $0.05 \leq \eta \leq 0.70$. The range of variation in β is largely due to the uncertainty in f_B.

Using expressions (3) and (4), we can estimate the size of β, and hence of the predicted level of asymmetries in the Standard Model, from the following expression where |ε| and Δm/Γ are quite well determined:

$$\sin(2\beta) = \text{const.} \cdot f_B^2 \cdot B_B/B_K \cdot |\varepsilon|/\Delta m/\Gamma. \qquad (21)$$

In fitting for the likely range for β, we use the experimental values for the mixing in the B system and CP impurity in the kaon system together with the latest values [13] of the bag factors from lattice calculations. The latter are $B_B = 0.85 \pm 0.10$ and $B_K = 0.80 \pm 0.20$. The values chosen for f_B deserve a separate discussion.

There are solid reasons to expect that, for a bound meson of one heavy and one light quark, the meson decay constant will scale inversely with the square root of the mass of the heavy quark. Thus one can use determinations in the charm system to make reliable estimates in the B system. Experimentally (from Mark III) one knows that $f_D < 290$ MeV (90% confidence); this implies $f_B < 165$ MeV. This we take for the likely upper value. One can also make theoretical estimates; from a recent review of Martinelli using QCD sum rules, the central values of three determinations of f_D are 180 ± 30 MeV, 174 ± 52 MeV, and 215 ± 60 Mev; these give a value for f_B in the range of 110 MeV. A value of about 100 MeV also follows from quark model arguments (Isgur and Godfrey) which say that f_D should approximate f_K. Thus we take 100 MeV as the lower value, recognizing that there may very well be unforeseen uncertainties. This then leads to the likely [14] range $3^0 < \beta < 14^0$. We are now in a position to estimate the size of the CP violating effects in the rare kaon decays.

3.2 $K_L \to \pi^0 e^+ e^-$

Closely coupled with the issue of a non-zero ε'/ε is the branching ratio for the $K_L \to \pi^0 e^+ e^-$ mode which is expected to be of the order of 10^{-11}. The decay is complicated in that, in addition to the direct CP violating term, there is likely a contribution from the K_1 (the indirect piece), from the K_2 through a 2γ transition (the CP conserving piece), and because there is a significant background.

The direct piece. A substantial fraction of this decay should be direct CP violation, arising from contributions with virtual top quarks. The direct branching ratio has been calculated [15] to be

$$BR(K_2 \to \pi^0 e^+ e^-) = 1.0 \times 10^{-5} (s_2 s_3 s_\delta)^2 G(M_t)\,, \tag{22}$$

where G is a function of the top quark mass of order unity and the original CKM matrix element notation is used. It is easy to show that one can use the constraint on the CKM mixing angles above to express this branching ratio in terms of β, B_B, the bag factor for the B_u meson system, and f_B, the B meson decay constant as well as another function of M_t of order unity:

$$BR(K_2 \to \pi^0 e^+ e^-) = \frac{5 \times 10^{-13} \sin^2\beta}{(B_B f_B{}^2 F(M_t))}\,. \tag{23}$$

Here we have a manifestly CP violating transition in that the branching ratio is directly proportional to $\sin\beta$. Given what we know about the unknowns in the above expression, the value for the direct branching ratio could range from about 10^{-12} to 10^{-11} with a central value of about 3×10^{-12}. We note again the significant dependence upon f_B.

Even if one could accumulate enough statistics at this level, there are, in addition to the direct CP violating term, three other contributions which need to be untangled. These are an indirect term, coming from the $K_1 \to \pi^0 e^+ e^-$ transition, a CP conserving term, coming from the $K_2 \to$

$\pi^0\gamma\gamma$ intermediate state, and a background coming from the $K_L \rightarrow e^+e^-\gamma\gamma$ radiative decay. These have been discussed extensively in the literature.

The indirect piece. There is a prediction [16] for the size of the indirect term. Based upon the use of chiral perturbation theory with some additional assumptions, the prediction for the branching ratio has a two-fold ambiguity: the value should be either 1.5×10^{-12} or 2.4×10^{-11}. This should be directly determined but for the time being, the ambiguity can be broken by a study of the similar $K^+ \rightarrow \pi^+ ee$ rate. Experiment E777 at Brookhaven has about 700 of these events with relatively high ee invariant mass and the spectrum favors [17] a rather stiff distribution for the e^+e^-; this suggests the lower value for the corresponding K_1 transition which is good for the mode in question.

The CP conserving piece. For the CP conserving transition, there are competing theories [18] which give values between 10^{-14} and 10^{-11} for the two photon (CP conserving) K_2 transition to $\pi^0 ee$. There are now two observations [19] of the $K_L \rightarrow \pi^0\gamma\gamma$ branching ratio and high values for the $\gamma\gamma$ invariant mass are strongly favored in both. This again favors the chiral perturbation theory prediction for the lower branching ratio [20]. Again, this mode needs to be better measured.

The background. To reach the level of direct CP violation in this mode, it will be necessary to run the experiment in a much higher rate environment than has been attempted before. Many backgrounds [21] are understood for this mode, including a whole variety of accidental effects.

The most severe background [22] appears to arise from the $K_L \rightarrow e^+e^-\gamma\gamma$ decay; its branching ratio has recently been determined to be at the level of 5×10^{-7} (depending upon cutoff). These decays tend to have one very low energy gamma and a very low mass ee pair. However, after reasonable cuts on these quantities, still a sizable background remains and one has only the π^0 mass as a final constraint. With the high precision calorimeter, this background can be reduced to the level of about 10^{-11} and one will probably have to live with it at this level.

Experimental status and prospects. Clearly to be able to explore this mode, a very high flux facility is needed. The $K_L \rightarrow \pi^0 e^+e^-$ sensitivity is now in the 10^{-9} range as a result of a combination of BNL845 [23] and FNAL731 [24] and it will be pushed to nearly the 10^{-11} level in E799 at the Tevatron. There is also a dedicated experiment [25] at KEK pursuing the $K_L \rightarrow \pi^0 e^+e^-$ mode. A very attractive facility [26] for future studies would be an extracted beam from the 120 GeV Main Injector likely to be constructed at Fermilab. Beams of 3×10^{13} protons per pulse, with a repetition rate of 2.9 sec, provide year round a very intense neutral kaon beam with desirable features. There the flux necessary to permit sensitivities to this and other modes in the range of 10^{-10} per hour of running are obtainable. Studies show that the detector acceptance for the $\pi^0e^+e^-$ mode is about 15% with the requirement that both photons exceed 1 GeV. The decay rate for kaons greater than 10 GeV is about 33×10^6 per spill in a 20m decay region. After subtracting the estimated background, a three sigma sensitivity for a residual (presumably direct CP violating) branching ratio of about 5×10^{-13} results for a one year run. For this and the other high-rate running conditions, the singles rates are about 100MHz in the largest drift chamber and the maximum rate on a single wire (3mm pitch) is about 600kHz. The time frame for this facility would be in the latter half of the 1990's. Similar studies could be made at KAON in TRIUMF.

3.3 $K_L \to \pi^0 \nu\bar{\nu}$

An important related decay [27] is $K_L \to \pi^0 \nu\bar{\nu}$. This decay, in the standard model is essentially pure direct CP violating: in principle, the clean observation of just a single unambiguous event would establish the long-sought for effect! Also, the expected branching ratio [28] is about six times greater than for the $\pi^0e^+e^-$ case: roughly, a factor of 2 comes because one has both vector and axial vector couplings and a factor of 3 is for three types of neutrinos. Thus the central value is expected to be about 2×10^{-11}. A very good vacuum, on the order of 10^{-6} Torr, is needed so that the interactions in the residual gas do not fake the topology. The most significant background [29] will be the $K_L \to 2\pi^0$ decay where two photons are missed so that a very elaborate γ veto system is required. From the work of the BNL E787 collaboration together with further design studies to optimize the veto efficiency for the required energy range, it looks as though an experiment can be mounted with residual background below 10^{-11}; however, tests must clearly be performed. While the background and instrumental problems are challenging, the flux to do the measurement is available at the Main Injector and the relatively higher photon energies are easier to detect and to veto.

4. Summary

Presently, the field of precise and sensitive kaon decay studies is flourishing. There are ongoing, dedicated efforts at BNL, FNAL, CERN and KEK with significant detector upgrades in the works. Future facilities include the Φ factory at Frascati, the kaon beam at the FNAL Main Injector, and possibly KAON in Canada.

The quantity ε'/ε can probably be measured in the next generation experiments at the level of 10^{-4} with extreme care in dealing with systematic issues. This effort is likely the best bet to establish a direct CP non-conserving effect in the next several years: if the standard model is correct, then an effect on the order of 10^{-3} is predicted and it would take a particularly unlucky cancellation to reduce the effect to 10^{-4} or below.

The $K_L \rightarrow \pi^0 e^+ e^-$ mode and the related mode with a muon pair can be sought to the 10^{-11} level with current facilities. The effective ratio of direct to indirect CP violation is of order unity so that one needs excellent sensitivity and background rejection rather than excellent control of systematics. This represents a 2-order-of-magnitude advance over the current level of sensitivity and a non-Standard Model effect could show up. However, to go much beyond this level, a new facility is required.

The $K_L \rightarrow \pi^0 \nu\bar{\nu}$ mode is the cleanest theoretically but perhaps the dirtiest experimentally. More design work and tests of prototype veto systems will hopefully establish that a search at the level of the Standard Model prediction is viable. A new facility is probably required.

Finally, we call for continued theoretical investigations into the ranges for the parameters f_B, B_B, and B_K which are so important is estimating (and interpreting the eventual determinations of), CP violating asymmetries in the Standard Model.

ACKNOWLEDGEMENT

The author would like to thank the organizers of the 5th Nishinomiya Yukawa-Memorial Symposium for their warm hospitality, particularly M. Kobayashi and Ken-Ichi Aoki.

References

1. H. Burkhart et al., Phys. Lett. **B206** (1988) 169.
2. J.R. Patterson et al., Phys. Rev. Lett. **64** (1990) 1491.
3. M. Karlsson et al., Phys. Rev. Lett. **64** (1990) 2976.
4. Particle Data Group, G.P. Yost et al., Phys. Lett. **B204** (1988) 1.
5. G. Buchalla, A.J. Buras, M. Harlander, MPI-PAE-Pth-30/90, July 1990.
6. G. Barr et al., Proposal to CERN for a new measurement of ε'/ε.
7. P832, K. Arisaka et al.
8. E799 proposal, A. Barker et al.
9. N. Nelson, International Conference on High Energy Physics, Singapore, 1990.
10. See, for example, C.S. Kim, J.L. Rosner, and C.P. Yuan, Phys. Rev. **D42**, (1990) 96. See also the contribution from M. Wise, these proceedings.
11. See, for example, SLAC and Cornell B-factory proposals.
12. C.S. Kim et al., op cit.
13. G.W. Kilcup et al., Phys. Rev Lett. **64** (1990) 25; Ref. 9 and references therein.

14. The situation is probably not as clean as this paragraph sounds. In these proceedings, G. Martinelli has said that f_B might be as large as 300 MeV; if true, then β will be significantly greater than 14^0.
15. C.O. Dib, I. Dunietz, F.J. Gilman, Phys. Rev. **D39** (1989) 2639.
16. G. Ecker, A. Pich and E. de Raphael, Nuc. Phys. **B291** (1987) 692.
17. M. Zeller and P. Cooper, private communications.
18. Chiral perturbation theory and Vector dominance; see Ref. 16 and references therein.
19. G. Barr et al., Phys. Lett. **242B** (1990) 523; A. Barker et al., EFI 91-01.
20. However, since the observed branching ratio is about a factor of three higher than the prediction, there could be higher order terms present which could complicate the interpretation here.
21. E799 proposal; proceedings of the Breckenridge conference, Physics at Fermilab in the 1990's, edited by D. Green and H. Lubatti.
22. H. Greenlee, Phys. Rev. **D42** (1990) 3724.
23. K.E. Ohl et al., Phys. Rev. Lett. **64** (1990) 2755.
24. A. Barker et al., Phys. Rev. **D41** (1990) 3546.
25. KEK E162.
26. KAMI Conceptual Design Report, G. Bock et al., Feb. 1991.
27. L. Littenberg, Phys. Rev. **D39** (1989) 3322.
28. C.O. Dib, I. Dunietz, F.J. Gilman, Phys. Lett. **218B** (1989) 487.
29. "Last year's sensation is this year's calibration....and tomorrow's background." V.L. Telegdi, private communication. Perhaps "year" should be modified to "decade."

CP Violation

M.B. Wise

California Institute of Technology, Pasadena, CA 91125, USA

Abstract. The possibility of determining the unitarity triangle from measurements of CP conserving quantities is discussed. The predictions of the minimal standard model for CP violation in the Kaon system and $\bar{B}$ decays are reviewed. Electric dipole moments of the neutron and the electron in models with an extended Higgs sector are discussed.

I. Introduction

At the present time CP violation has only been observed in kaon decays. The quantities

$$\eta_{+-} = \frac{<\pi^+\pi^-|H_{\rm eff}^{|\Delta s|=1}|K_L>}{<\pi^+\pi^-|H_{\rm eff}^{|\Delta s|=1}|K_S>}, \quad \eta_{00} = \frac{<\pi^0\pi^0|H_{\rm eff}^{|\Delta s|=1}|K_L>}{<\pi^0\pi^0|H_{\rm eff}^{|\Delta s|=1}|K_S>} \tag{1}$$

are CP violating. If CP was conserved $|K_L>$ would be a CP odd state and it could not decay to $|\pi^+\pi^->$ or $|\pi^0\pi^0>$ which (in an S-wave) are CP even states. Experimentally η_{+-} and η_{00} are small but non zero; $\eta_{+-} \simeq \eta_{00} \simeq 2.2 \times 10^{-3} e^{i\pi/4}$ [1]. In the standard model with minimal particle content CP violation can only enter in two ways. The QCD vacuum angle $\theta_{\rm QCD}$ occurs in a coupling [2]

$$\mathcal{L} = \frac{g^2}{16\pi^2}\theta_{\rm QCD} Tr G\tilde{G} \tag{2}$$

that violates both P and CP. In eq (2) g is the strong SU(3) gauge coupling, G is the gluon field strength and $\tilde{G}$ its dual . The stringent bound [1] on the electric dipole-moment of the neutron $d_n \leq 10^{-25} e - cm$ implies that [3] $\theta_{\rm QCD} \leq 10^{-8}$. Such a small value of $\theta_{\rm QCD}$ means that strong interaction CP violation is negligible for CP violating quantities that can be measured in the weak kaon and B-meson decays. In the minimal standard model the other source of CP violation is in the coupling of the W-bosons to the quarks. It has the form

$$\mathcal{L}_{int} = -\frac{g_2}{\sqrt{2}}\bar{u}_L^j \gamma_\mu V^{jk} d_L^k W_{(+)}^\mu + h.c. \ . \tag{3}$$

Here the repeated generation induces j and k are summed over 1,2,3 and g_2 is the weak SU(2) gauge coupling. V is a 3×3 unitary matrix that arises from diagonalization of the quark mass matrices. By redefining the phases of the quark fields it is possible to write V in terms of four angles $\theta_1, \theta_2, \theta_3$ and δ (For N_g generations there are $(N_g - 1)^2$ angles). The θ_j are analogous to the Euler angles and δ is a phase that gives rise to CP violation. Explicitly [4]

$$V = \begin{pmatrix} c_1 & -s_1c_3 & -s_1s_3 \\ s_1c_2 & c_1c_2c_3 - s_2s_3e^{i\delta} & c_1c_2s_3 + s_2c_3e^{i\delta} \\ s_1s_2 & c_1s_2c_3 + c_2s_3e^{i\delta} & c_1s_2s_3 - c_2c_3e^{i\delta} \end{pmatrix}, \tag{4}$$

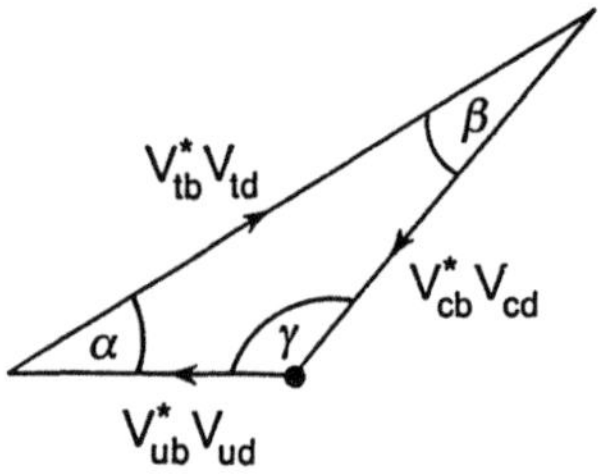

*Fig.*1 The Unitarity Triangle

where $c_i \equiv \cos\theta_i$ and $s_i \equiv \sin\theta_i$. It is possible to choose the θ_j to lie in the first quadrant. Then the quadrant of δ has physical significance and cannot be chosen by convention. A value of δ not equal to zero or π gives rise to CP violation.

Experimental information on nuclear β decay and the weak decays of kaons, hyperons and B mesons shows that all the Euler like angles θ_j are small (but different from zero). The angle θ_1 is essentially the Cabibbo angle. It is by far the best known of the angles [1]

$$\sin\theta_1 = 0.22 \tag{5}$$

(with an error at the percent level). The other angles and δ are not known nearly as well.

Unitarity of the Kobayashi–Maskawa Matrix V gives that

$$V_{ud}V^*_{ub} + V_{cd}V^*_{cb} + V_{td}V^*_{tb} = 0 \ . \tag{6}$$

We can think of each of the 3 complex numbers ($V_{td}V^*_{tb}$, etc.) on the left-hand side of eq. (6) as vectors in the complex plane. Then eq (6) gives that these vectors add to zero and so by translating them they form the sides of a triangle, usually called the unitarity triangle. With the parametrization of the Kobayashi–Maskawa matrix in eq (4) we have

$$V_{ud}V^*_{ub} \simeq -s_1 s_3 \qquad , \tag{7a}$$
$$V_{td}V^*_{tb} \simeq -s_1 s_2 e^{-i\delta} \qquad , \tag{7b}$$
$$V_{cd}V^*_{cb} \simeq s_1(s_3 + s_2 e^{-i\delta}) \ . \tag{7c}$$

Fig. (1) shows the unitarity triangle in a case when $s_\delta > 0, c_\delta < 0$, and $s_2 > s_3$.

The unitarity triangle specifies the angles θ_2, θ_3, and δ. From eq (7) it is clear that the length of two of the sides give θ_2 and θ_3 while the angle α is equal to $\pi - \delta$.

The orientation of the triangle in the complex plane depends on the phase convention for the Cabibbo–Kobayashi–Maskawa matrix. Fig. 1 corresponds to the choice in eq. (4) where $V^*_{ub}V_{ud}$ is real. The angles α, β and γ, however, are physical quantities independent of the phase convention. When there is no CP violation the unitarity triangle collapses to a line. There has become, in the literature, a commonly used orientation for the triangle; it has $V_{cd}V^*_{cb}$ lying on the real axis instead of $V_{ud}V^*_{ub}$. This orientation is shown in Fig. 2. It is conventional to rescale the side on the real axis to unit length and locate the corner associated with β at the origin of the complex plane. With this convention the unitarity triangle is sometimes specified by the coordinates (in the complex plane) of the corner associated with α.

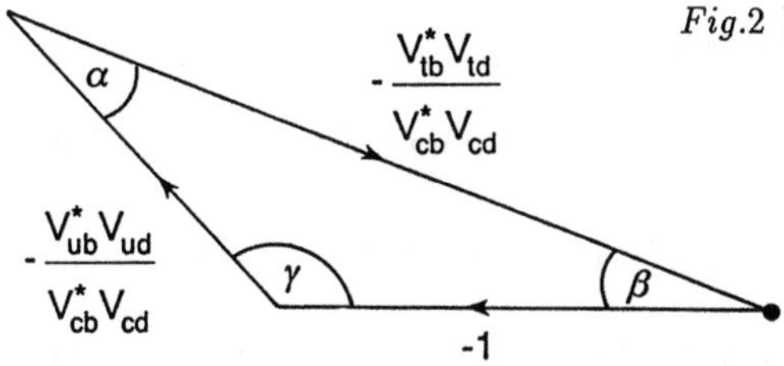

*Fig.*2 Rescaled Unitarity Triangle

II. Weak Mixing Angles

The unitarity triangle may eventually be determined by measurements of CP conserving quantities. Then CP violating quantities can be predicted. There are several approaches available to determine the magnitude of the Kobayashi–Maskawa matrix elements V_{ub} and V_{cb}. Since the b-quark is heavy compared with the QCD scale inclusive semileptonic $\bar{B}$ meson decay can be approximated by b-quark decay. Corrections to this picture are presumably suppressed by factors of Λ_{QCD}/m_b. The b-quark decay either proceeds through the $b \to c$ or $b \to u$ weak W-boson couplings so:

$$\Gamma(\bar{B} \to Xe\bar{\nu}_e) \simeq \Gamma(b \to ce\bar{\nu}_e) + \Gamma(b \to ue\bar{\nu}_e) \ , \tag{8}$$

where

$$\Gamma(b \to ce\bar{\nu}_e) \simeq \frac{|V_{cb}|^2 G_F^2 m_b^5 g(m_c/m_b)}{192\pi^3} \ , \tag{9a}$$

$$\Gamma(b \to ue\bar{\nu}_e) \simeq \frac{|V_{ub}|^2 G_F^2 m_b^5}{192\pi^3} \ . \tag{9b}$$

In eq. (9) g is a function that takes into account the effects of the charm quark mass (it is about equal to one-half) and G_F is the Fermi constant

$$\frac{G_F}{\sqrt{2}} = \frac{g_2^2}{8M_W^2} \ . \tag{10}$$

Experiments reveal that the $b \to c$ transition dominates the rate. From the measured $\bar{B}$-meson lifetime $\tau_B \simeq 10^{-12}$ and the semileptonic branching ratio $\sim 10\%$ eq. (9a) implies that

$$|V_{cb}|^2 \simeq 0.2 \times 10^{-2} \ . \tag{11}$$

Eq. (11) used $m_b = 5.2 GeV$. Because of the sensitive dependence of eq. (9) on the mass of the bottom quark there is a large uncertainty ($\sim 20\%$) associated with the value of $|V_{cb}|^2$ extracted this way. Fortunately there is another complimentary approach for determining $|V_{cb}|$. Experimentally it is observed that the rate for semileptonic $\bar{B}$ decay is dominated by two processes $\bar{B} \to De\bar{\nu}_e$ and $\bar{B} \to D^* e\bar{\nu}_e$. The rates for these processes are determined by the magnitude of V_{cb} and by the matrix elements

$$< D(v')|\bar{c}\gamma_\mu b|\bar{B}(v) >, \ < D^*(v',\epsilon)|\bar{c}\gamma_\mu\gamma_5 b|\bar{B}(v) >, \ < D^*(v',\epsilon)|\bar{c}\gamma_\mu b|\bar{B}(v) > \ .$$

Recently it has been realized that in the limit of QCD where heavy quark masses go to infinity [5-9] new spin and flavor symmetries appear and that these symmetries can be used to predict the needed matrix elements [6]. Explicitly

$$< D(v')|\bar{c}\gamma_\mu b|\bar{B}(v) >= \sqrt{m_B m_D} C_{cb}\xi(v\cdot v')(v+v')_\mu \ , \tag{12a}$$

$$< D^*(v',\epsilon)|\bar{c}\gamma_\mu\gamma_5 b|\bar{B}(v) >= \sqrt{m_B m_{D^*}} C_{cb}\xi(v\cdot v')[(1+v\cdot v')\epsilon^*_\mu - (\epsilon^*\cdot v)v'_\mu] \ , \tag{12b}$$

$$< D^*(v',\epsilon)|\bar{c}\gamma_\mu b|\bar{B}(v) >= i\sqrt{m_B m_{D^*}} C_{cb}\xi(v\cdot v')\epsilon_{\mu\nu_\alpha\beta}\epsilon^{*\nu}v'^\alpha v^\beta \ . \tag{12c}$$

In eq. (12) C_{cb} is a calculable factor that sums, in the leading logarithmic approximation, dependence on the heavy bottom and charm quark masses from perturbative (high momentum) QCD effects [7,8]. ξ is a function of $v\cdot v'$ and is independent of the heavy quark masses. It is not possible to calculate $\xi(v\cdot v')$ since nonperturbative strong interactions determine it. However, at zero recoil where $v = v'$ the heavy quark flavor symmetry implies that

$$\xi(1) = 1 \ . \tag{13}$$

These results allow one to determine $|V_{cb}|$ from data on exclusive semileptonic $\bar{B}$ decays. The weakest feature of this approach is treating the charm quark as heavy compared with the QCD scale. The results in eq. (12) receive corrections of order $\alpha_s(m_Q)/\pi$ and Λ_{QCD}/m_Q where $Q = b$ or c is a heavy quark. Most of the perturbative $\alpha,(m_Q)/\pi$ have been calculated [8,9]; these do not diminish the predictive power and are not large. The most troublesome source of corrections is the Λ_{QCD}/m_c effects. These involve new nonperturbative matrix elements that cannot be computed. However, it has recently been shown [10] that at $v = v'$ there are no Λ_{QCD}/m_c corrections to eq.(12). This result (which is analogous to the Ademollo–Gato theorem) means that exclusive $\bar{B}$ decays (particularly $\bar{B}\to D^* e\bar{\nu}_e$ since the matrix element for $\bar{B}\to De\bar{\nu}_e$ at zero recoil is proportional to m_e) may eventually allow a determination of $|V_{cb}|$ comparable in precision to that $\sin\theta_1$. An important aspect of this is that the heavy-quark symmetries give information on several form factors so that it will be possible to study experimentally the size of corrections to (12). Furthermore there are other very different exclusive decays where an independent determination of $|V_{cb}|$ is possible. For example, the rate for the semileptonic decay $\Lambda_b\to\Lambda_c e\bar{\nu}_e$ is determined by $|V_{cb}|$ and the matrix elements; $< \Lambda_c(v',s')|\bar{c}\gamma_\mu b|\,\Lambda_b\,(v,s) >$ and $< \Lambda_c(v',s')|\bar{c}\gamma_\mu\gamma_5 b|\,\Lambda_b\,(v,s) >$. Using the heavy quark methods one can show that [11]

$$< \Lambda_c(v',s')|\bar{c}\gamma_\mu b|\,\Lambda_b\,(v,s) >= C_{cb}\eta(v\cdot v')\bar{u}(v',s')\gamma_\mu u(v,s) \ , \tag{14a}$$

$$< \Lambda_c(v',s')|\bar{c}\gamma_\mu\gamma_5 b|\,\Lambda_b\,(v,s) >= C_{ab}\eta(v\cdot v')\bar{u}(v',s')\gamma_\mu\gamma_5 u(v,s) \ , \tag{14b}$$

where C_{ab} is the same factor as in $\bar{B}$ decays and η is a new function of $v\cdot v'$ that the heavy-quark symmetries normalize to unity at zero recoil, i.e.,

$$\eta(1) = 1 \ . \tag{15}$$

Again it has been shown that there are no Λ_{QCD}/m_c corrections of (14) at zero recoil. In fact in this case Λ_{QCD}/m_c corrections introduce no new unknown functions of $v\cdot v'$ but rather are expressed as a known function of $v\cdot v'$ times the quantity $\bar{\Lambda} = m_{\Lambda_b} - m_b = m_{\Lambda_c} - m_c$ [12].

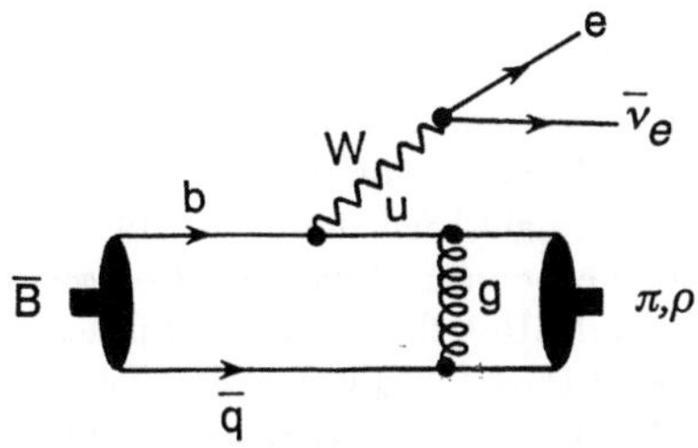

*Fig.*3 Quark line diagram for $\bar{B} \to (\pi \text{ or } \rho)e\bar{\nu}_e$

Nonleptonic $\bar{B}$ decays may also be useful in assessing the size of Λ_{QCD}/m_c corrections to eqs (12). Because $m_b - m_c$ is large compared with the QCD scale the amplitudes for $\bar{B} \to D\pi$ and $\bar{B} \to D^*\pi$ factorize and their rates are related to semileptonic form factors near maximum recoil [13].

Although the $b \to u$ coupling is responsible for only a small fraction of the semileptonic $\bar{B}$ decays, it is possible to isolate this contribution, without reconstructing hadronic final states, by examining the electron spectrum

$$\frac{d\Gamma}{dE_e}(\bar{B} \to X e\bar{\nu}_e) \ . \tag{16}$$

Kinematically the production of final hadronic states X with mass m_X is forbidden for electron energies greater than

$$(E_e)_{\max} = (m_B^2 - m_X^2)/2m_B \ . \tag{17}$$

The lowest mass particle containing a charm quark is a D meson with a mass of $1.9GeV$. So by examining the endpoint region $E_e > 2.3GeV$ of the electron spectrum one is sure to be focusing on the $b \to u$ contribution. What is needed theoretically is a method for normalizing the electron spectrum in this region. Unfortunately one cannot justify the use of the b-quark decay picture for the production of low mass hadronic final states [14] like the π and the ρ. Figure 3 shows a diagram contributing to this decay. Here for large recoils the gluon is transferring a large momentum to the light spectator quark so that the final hadronic state can have a low invariant mass.

This means, for low mass hadronic final states, large recoils are suppressed by hadronic form factors. This effect is not included in the b-quark decay picture. Comparison of experimental information on the endpoint region of the electron spectrum with various phenomenological models gives [15]

$$|V_{ub}/V_{cb}|^2 \sim (0.2)s_1^2 \ . \tag{18}$$

It is difficult to assess the uncertainties in (18), however, in my opinion, there is no reason to believe that they are less than about a factor of two.

The recently developed heavy quark methods open another avenue to determining $|V_{ub}|$. Ordinary isospin symmetry plus heavy quark flavor symmetry implies, for example, that [6, 16]

$$< \rho(k,\epsilon)|\bar{u}\gamma_\mu(1-\gamma_5)b|\bar{B}(v) >= \left(\frac{m_B}{m_D}\right)^{1/2} \left[\frac{\alpha_s(m_b)}{\alpha_s(m_c)}\right]^{-6/25}$$

$$\cdot < \rho(k,\epsilon)|\bar{d}\gamma_\mu(1-\gamma_5)c|D(v)> \ . \tag{19}$$

Eq. (19) is valid (in the rest frame of the $\bar{B}$ and D) for light four momenta k small compared with the heavy quark masses. Since in the Cabibbo suppressed semileptonic decay $D \to \rho\bar{e}\nu_e$ the weak mixing angles are known, the right-hand side of eq. (19) can be determined experimentally. With this information experimental data on $\bar{B} \to \rho e\bar{\nu}_e$ will allow a determination of $|V_{ub}|$. If one uses light quark SU(3) flavor symmetry instead of isospin then the Cabibbo allowed semileptonic decay $D \to K^*\bar{e}\nu_e$ can be used for the right-hand side of eq. (19). The form factors for this decay have already been studied experimentally [17]. The weakest feature of this program is the neglect of Λ_{QCD}/m_c corrections. In this case there is no theorem that forces these to vanish. However, since eq. (19) provides relations between several form factors and similar relations hold for other final states (e.g. the pion) there will be many experimental clues to the importance of Λ_{QCD}/m_c effects. It is not unreasonable to imagine that, with a little luck, this method will eventually allow a determination of $|V_{ub}|$ to an uncertainty of about 10%.

So far I have described the determination of the length of two sides of the unitarity triangle. We still need the determination of $|V_{td}|$. Experimental information on it comes from $B^0 - \bar{B}^0$ mixing. In the minimal standard model $B^0 - \bar{B}^0$ mixing arises from a box diagram with two W-bosons in the loop. Using the experimental value $(\Delta m/\Gamma)_B \simeq 0.7$ one finds from a calculation of the box diagram that

$$1 \simeq f(m_t^2/M_W^2)\left(\frac{m_t}{180GeV}\right)^2 \frac{|V_{td}|^2}{(1.4\ 10^{-4})}\left(\frac{f_B^2 B_B}{10^{-2}GeV^2}\right), \tag{20}$$

where [18]

$$f(x) = \frac{x^2 - 11x + 4}{4(x-1)^2} + \frac{3}{2}\frac{x^2 lnx}{(x-1)^3} \ . \tag{21}$$

The function $f(x)$ is typically close to unity. For example $f(0) = 1$ and $f(4) = 0.6$. In eq.

$$< \bar{B}^0|(\bar{b}_{L\alpha}\gamma_\mu d_{L\alpha})(\bar{b}_{L\beta}\gamma^\mu d_{L\beta})|B^0> = \frac{2}{3}B_B f_B^2 m_B^2 \tag{22}$$

with the four fermion operators evaluated at a subtraction point equal to m_b. In order for eq. (20) to provide a reliable determination of $|V_{td}|^2$ we need to know the top quark mass m_t and the nonperturbative strong interaction matrix element (22). The heavy quark flavor methods relate f_B to f_D [7]

$$f_B = \left(\frac{m_D}{m_B}\right)^{1/2}\left[\frac{\alpha_s(m_b)}{\alpha_s(m_c)}\right]^{-6/25} f_D \tag{23}$$

and it may be possible to measure f_D in the next generation of D decay experiments (currently there is a limit $f_D < 290MeV$) [19]. Some recent lattice Monte Carlo calculations suggest, however, that there are very large Λ_{QCD}/m_Q corrections to eq. (23). Their calculations give a very large value, of about 300 MeV, for f_B [20].

Another feature of the determination of $|V_{td}|$ from (20) is its reliance on the assumption of a minimal Higgs sector for the standard model. While the extraction of $|V_{ub}|$ and $|V_{cd}|$ does not strongly rely on this, $B^0 - \bar{B}^0$ mixing is particularly sensitive to the nature of the Higgs sector because it arises from loop effects with a top quark in the loop. In the minimal standard model which has a single Higgs doublet there is only one physical (neutral) Higgs scalar. Since the same transformation, which diagonalize the quark mass matrices diagonalize the couplings of the Higgs scalar there are no flavor changing Higgs couplings at tree level. There is no direct experimental information on the Higgs sector, and even the simplest extension of the Higgs sector can dramatically effect $B^0 - \bar{B}^0$ mixing. For example, in models with several Higgs doublets there are physical charged scalars. It is possible to build models where there are no flavor changing couplings of the neutral scalars at tree level, although this is less automatic then in the minimal standard model [21]. In these models the Higgs scalars couplings to quarks q are proportional to (m_q/M_W). Consequently in low energy physics it is processes where a virtual top quark plays a prominent role that are most easily affected by the additional Higgs scalars. In $B^0 - \bar{B}^0$ mixing charged Higgs can replace the W-bosons in the box diagram. In a model with several Higgs doublets the value of $|V_{td}|^2$ would be less than that implied by eq. (20) [22].

One can remove somewhat our dependence on nonperturbative physics and our sensitivity to extensions of the Higgs sector by measuring $B_s^0 - \bar{B}_s^0$ mixing. In the minimal standard model and in the class of multi-Higgs doublet models discussed above

$$\left(\frac{\Delta m}{\Gamma}\right)_{B_s} = \frac{|V_{cb}|^2}{|V_{td}|^2}\left(\frac{\Delta m}{\Gamma}\right)_B \hat{\eta} \ . \tag{24}$$

$\hat{\eta}$ is a factor that would be unity in the limit of SU(3) light quark flavor symmetry,

$$\hat{\eta} = \frac{< \bar{B}_s^0|(\bar{b}_{L\alpha}\gamma_\mu s_{L\alpha})(\bar{b}_{L\beta}\gamma^\mu s_{L\beta})|B_s^0 >}{< \bar{B}^0|(\bar{b}_{L\alpha}\gamma^\mu d_{L\alpha})(\bar{b}_{L\beta}\gamma_\mu d_{L\beta})|B^0 >} \ . \tag{25}$$

Since $\hat{\eta}$ is one in the SU(3) limit ones reliance on nonperturbative methods is limited to estimating deviations from light quark SU(3) symmetry, which is presumably easier than estimating the whole matrix element in eq. (22). Eventually it may be possible to measure both f_D and f_{D_s}. Since the heavy quark flavor symmetry implies that $(f_{D_s}/f_D) = (f_{B_s}/f_B)$ this would help in estimating $\hat{\eta}$. If $(\Delta m/\Gamma)_{B_s}$ is large then it is very difficult to measure it. Large values of f_B like those reported in some recent Lattice Monte Carlo Calculations decrease $|V_{td}|$ (for a given value of m_t and B_B) and hence lead to larger expected values of $B_s^0 - \bar{B}_s^0$ mixing. Also in the multi-Higgs doublet models described above the additional contributions to $B^0 - \bar{B}^0$ mixing lower the value of $|V_{td}|$ expected, for a given value of m_t and $B_B f_B^2$.

Kaon decays can also provide valuable information on the weak mixing angles [23]. The effective Hamiltonian for the decay $K^+ \to \pi^+\nu\bar{\nu}$ arises from short distance physics and its matrix element is free of uncertainties associated with our inability to handle nonperturbative strong interaction physics. Using the notation of Dib et.al., the branching ratio for this process (for three neutrino species) is [24]

$$Br(K^+ \to \pi^+\nu\bar{\nu}) = 1.5 \times 10^{-5}|V_{cb}|^2|\tilde{C}_{\nu,t}|^2\left| -\left(\frac{\tilde{C}_{\nu,c}}{|V_{cb}|^2\tilde{C}_{\nu,t}}\right) + \frac{V_{tb}^*V_{td}}{V_{cd}V_{cb}^*}\right|^2 . \tag{26}$$

In eq. (26) $\tilde{C}_{\nu,c}$ and $\tilde{C}_{\nu,t}$ are calculable functions of the charm and top quark masses. From eq. (26) it is evident that an accurate measurement of this branching ratio would restrict the "α-corner" of the unitarity triangle in figure two to lie on a circle centered about $\tilde{C}_{\nu,c}/|V_{cb}|^2\tilde{C}_{\nu,t}$. This process, like $B^0 - \bar{B}^0$ mixing, is sensitive to extensions of the

Higgs sector. A comparison of the unitarity triangle extracted from $B^0 - \bar{B}^0$ mixing with that extracted from the branching ratio for $K^+ \to \pi^+ \nu\bar{\nu}$ would provide a powerful probe of the Higgs sector.

III. CP Violation in Kaon Decays

The CP violating quantities η_{+-} and η_{00} get contributions from CP violation in second order weak $K^0 - \bar{K}^0$ mixing and CP violation in the first order weak Kaon decay amplitudes A_I defined by

$$< \pi\pi(I)|H_{\text{eff}}^{|\Delta s|=1}|K^0 >= iA_I e^{i\delta_I} \; . \tag{27}$$

In eq. (27) I is the isospin of the final two pion state and δ_I is the $\pi\pi(I)$ phase shift. One usually adopts a phase convention where A_0 is real. Then

$$\eta_{+-} \simeq \epsilon + \epsilon' \; , \tag{28a}$$

$$\eta_{+-} \simeq \epsilon - 2\epsilon' \tag{28b}$$

with ϵ characterizing the CP violation in $K^0 - \bar{K}^0$ mixing

$$\epsilon \simeq \frac{1}{\sqrt{2}} e^{i\pi/4} \frac{ImM_{12}}{M_L - M_S} \tag{29a}$$

and ϵ' characterizing the CP violation in $K \to \pi\pi$ decay amplitudes

$$\epsilon' = \frac{1}{\sqrt{2}} e^{i\pi/4} \frac{ImA_2}{A_0} \; . \tag{29b}$$

In eq. (29a) M_{12} is the $K^0 - \bar{K}^0$ mixing element. The phases $\pi/4$ are approximate following from various measured quantities (e.g. pion phase shifts in eq. (29b)). In the minimal standard model the effective Hamiltonian for $|\Delta s| = 2$ $K^0 - \bar{K}^0$ mixing is derived from the familiar box diagram with two W-bosons and charge 2/3 quarks in the loop. It gives [25, 26]

$$\epsilon \simeq \frac{s_1^2 B_K G_F^2 f_K^2 m_K m_c^2}{6\sqrt{2}\pi^2 (M_L - M_S)} s_2 s_3 s_\delta \left[-\eta_1 + \eta_3 \ell n(m_t^2/m_c^2) \right.$$
$$\left. + \eta_2 f(m_t^2/M_W^2)(m_t^2/m_c^2)(s_2^2 + s_2 s_3 c_\delta) \right] e^{i\pi/4} \; . \tag{30}$$

In Eq. (30), η_1, η_3 and η_2 are calculated QCD correction factors. The three terms in the square bracket of eq. (30) successively reflect the contribution of virtual momentum scales of order the charm quark mass, between the charm quark and top quark mass, and of order the top quark mass. The quantity B_K involves low momentum nonperturbative strong interaction physics. It is defined by

$$< \bar{K}^0|(\bar{s}_{L\alpha}\gamma^\mu d_{L\alpha})(\bar{s}_{L\beta}\gamma_\mu d_{L\beta})|K^0 >= \frac{2B_K f_K^2 m_K^2}{3} \; , \tag{31}$$

where the operator is evaluated at a subtraction point μ which is the same as in η_1, η_2 and η_3. There is a prediction for B_K based on chiral perturbation theory [27]. Under chiral SU(3)$_L\times$ SU(3)$_R$ the operator $(\bar{s}_{L\alpha}\gamma^\mu d_{L\alpha})(\bar{s}_{L\beta}\gamma_\mu d_{L\beta})$ transforms as $(27_L, 1_R)$. In chiral perturbation theory the pseudo Goldstone bosons are put in a 3×3 unitary matrix

$$\sum = \exp\frac{2iM}{f} \quad , \tag{32a}$$

where

$$M = \begin{bmatrix} \pi^0/\sqrt{2}+\eta/\sqrt{6} & \pi^+ & K^+ \\ \pi^- & -\pi^0/\sqrt{2}+\eta/\sqrt{6} & K^0 \\ K^- & \bar{K}^0 & -\sqrt{2/3}\eta \end{bmatrix} \tag{32b}$$

and f is the pseudoscalar decay constant ($\sim$ 135 MeV). Under chiral SU(3)$_L\times$ SU(3)$_R$ transformations

$$\sum \to L\sum R^\dagger \quad , \tag{33}$$

where $L\epsilon\ SU(3)_L$ and $R\epsilon\ SU(3)_R$. For matrix elements of $(\bar{s}_{L\alpha}\gamma^\mu d_{L\alpha})(\bar{s}_{L\beta}\gamma_\mu d_{L\beta})$ involving the pseudo-Goldstone bosons we construct an operator that transforms as this component of $(27_L, 1_R)$ involving $\sum$ with the least number of derivatives or insertions of the light quark mass matrix m (which transforms as $(\bar{3}_L, 3_R)$ or $(3_L, \bar{3}_R)$). This procedure gives

$$(\bar{s}_{L\alpha}\gamma^\mu d_{L\alpha})(\bar{s}_{L\beta}\gamma_\mu d_{L\beta}) = \alpha T^{ij}_{k\ell}\left(\sum\partial_\mu\sum^\dagger\right)^k_i\left(\sum\partial^\mu\sum^\dagger\right)^\ell_j \quad , \tag{34}$$

where $T^{ij}_{k\ell}$ is the traceless symmetric tensor with nonzero components

$$T^{33}_{22} = 1 \tag{35}$$

and α is an unknown constant. The decay $K^+\to\pi^0\pi^+$ proceeds through the $\Delta I = 3/2$ part of the effective Hamiltonian for $\Delta s = 1$ weak nonleptonic decays

$$H^{|\Delta s|=1,|\Delta I|=3/2}_{\text{eff}} = -\frac{4G_F}{\sqrt{2}}s_1\hat{C}(\mu)\hat{O}(\mu) \quad , \tag{36}$$

where

$$\hat{O} = (\bar{s}_{L\alpha}\gamma^\mu d_{L\alpha})(\bar{u}_{L\beta}\gamma_\mu u_{L\beta}) + (\bar{s}_{L\alpha}\gamma^\mu u_{L\alpha})(\bar{u}_{L\beta}\gamma_\mu d_{L\beta})$$
$$-(\bar{s}_{L\alpha}\gamma^\mu d_{L\alpha})(\bar{d}_{L\beta}\gamma_\mu d_{L\beta}) \tag{37}$$

and $\hat{C}$ is a calculable coefficient. In the leading logarithmic approximation, at a subtraction point where $\alpha_s(\mu) = 1$, it is about equal to 0.2. The operator $\hat{O}(\mu)$ is also $(27_L, 1_R)$. It is part of the same representation as $(\bar{s}_{L\alpha}\gamma^\mu d_{L\beta})(\bar{s}_{L\beta}\gamma_\mu d_{L\beta})$. So for its low momentum matrix elements we can use

$$\hat{O} = \alpha\hat{T}^{ij}_{k\ell}\left(\sum\partial_\mu\sum^\dagger\right)^k_i\left(\sum\partial^\mu\sum^\dagger\right)^\ell_j \tag{38}$$

when $\hat{T}^{ij}_{k\ell}$ is the symmetric traceless tensor with nonzero components

$$\hat{T}^{13}_{12} = \hat{T}^{31}_{21} = \hat{T}^{31}_{12} = \hat{T}^{13}_{21} = 1/2 \quad , \tag{39a}$$

$$\hat{T}^{23}_{22} = T^{32}_{22} = -1/2 \quad , \tag{39b}$$

and α is the same constant as in eq. (34). Expanding (34) and (38) gives

$$< \bar{K}^0|(\bar{s}_{L\alpha}\gamma^\mu d_{L\alpha})(\bar{s}_{L\beta}\gamma_\mu d_{L\beta})|K^0 >= -8\alpha m_K^2/f^2 \ , \tag{40a}$$

$$< \pi^+\pi^0|\hat{O}|K^+ >= -\frac{12i\alpha}{\sqrt{2}f^3}m_K^2 \quad . \tag{40b}$$

So

$$|B_K| = |A(K^+ \to \pi^+\pi^0)/2G_F\hat{C}(\mu)s_1 f m_K^2| \quad . \tag{41}$$

Using the measured $K^+ \to \pi^+\pi^0$ decay rate one finds, for a subtraction point where $\alpha_s(\mu) = 1$, that

$$|B_K| \simeq 0.4 \quad . \tag{42}$$

Unfortunately it is not possible to have complete confidence in this result because the strange quark mass is not small enough to ensure that the corrections to (42) are small. Formally the leading corrections to eqs. (40) are of order $m_K^4 \ell n m_K^2$. These "chiral logarithms" have been computed and are very large [28]. Of course it is still possible that (42) will survive since there are order m_K^4 corrections which can cancel the effects of the chiral logarithms.

There are other approaches to calculating B_K. In the large N_c limit matrix elements factorize and so $B_K = 3/4$ (similarly in the large N_c limit $B_B = 3/4$). Lattice Monte Carlo calculations have typically found B_K near unity [30]. However, some very recent results suggest that the lattice spacing is not small enough to be in the continuum limit and that B_K is actually considerably smaller than this [30].

The parameter ϵ measures CP violation in second order weak (i.e. order G_F^2) $K^0 - \bar{K}^0$ mixing while ϵ' measures CP violation in first order weak (i.e. order G_F) decay amplitudes. The quantity ϵ is much more sensitive than ϵ' to new physics beyond the standard model. For example suppose that δ was very small and the CP violation parameter ϵ is due to new physics associated with a very large mass scale $M_{\rm NEW}$. At low energies this new physics manifests itself as SU(3) × SU(2) × SU(1) invariant nonrenormalizable terms in the Lagrangian density like

$$\mathcal{L}_{\rm NEW} = \frac{1}{M_{\rm NEW}^2} g^{ijk\ell}(\bar{Q}_L^i d_R^j)(\bar{d}_R^k Q_L^\ell) \ , \tag{43}$$

where $Q_L = \binom{u_L}{d_L}$ is an SU(2) doublet. For complex g of order unity such terms can accommodate the measured value of ϵ when $M_{\rm NEW} \simeq 10^8 GeV$. With such a large mass scale associated with the new physics its contribution to ϵ' is negligible. This scenario for CP violation is called the superweak model [31].

In the minimal standard model where the angle δ is responsible for ϵ we expect a non-negligible value for ϵ'. The contribution to ϵ' comes, to leading order in electroweak interactions from strong interaction Penguin-type Feynman diagrams with a heavy charm and top quark in the loop [32,33]. The effects of these diagrams are reproduced by the matrix elements of four fermion operators in the effective Hamiltonian for $|\Delta s| = 1$ weak nonleptonic decays. The most important of these is thought to be

$$Q_6 = (\bar{s}_{L\alpha}\gamma_\mu d_{L\beta})[(\bar{u}_{R\beta}\gamma^\mu u_{R\alpha}) + (\bar{d}_{R\beta}\gamma^\mu d_{R\alpha}) + (\bar{s}_{R\beta}\gamma^\mu s_{R\alpha})] \quad . \tag{44}$$

This operator is $I = 1/2$ and transforms under chiral $SU(3)_L \times SU(3)_R$ as $(8_L, 1_R)$. Its $K \to \pi\pi$ matrix element are proportional to m_K^2 and hence are suppressed by a factor of the strange quark mass. Because it is $I = 1/2 Q_6$ contributes an imaginary part to A_0. Rotating the Kaon fields to preserve the phase convention that A_0 is real the strong interaction penguin type diagrams give a contribution to ImA_2. Neglecting isospin violation we write it as

$$\left(\frac{ImA_2}{A_0}\right) \simeq s_2 s_3 s_6 f_s \left(\frac{A_2}{A_0}\right) . \tag{45}$$

In eq. (45) f_s arises from the $K \to \pi\pi(I=0)$ matrix element of the term in the effective Hamiltonian involving Q_6. If Q_6 dominated this decay amplitude then f_s is of order unity. Isospin violation due to the difference between m_u and m_d allows the matrix elements of $I = 1/2$ operators like Q_6 to contribute directly to ImA_2. This gives a contribution [34]

$$\left(\frac{ImA_2}{A_0}\right) \simeq \left(\frac{m_d - m_u}{3\sqrt{2} m_s}\right) s_2 s_3 s_6 f_s . \tag{46}$$

Note that although (46) is suppressed by $(m_d - m_u)/m_s$ it is enhanced compared with (45) by a factor of $A_0/A_2 \sim 20$ since it contributes directly to the imaginary part of A_2. There are also important contributions to ϵ' coming from electroweak Penguin-type diagrams (and box diagrams). These Feynman diagrams give rise to local four fermion operators that are $I = 3/2$ and transform under chiral $SU(3)_L \times SU(3)_R$ as $(8_L, 8_R)$. Explicitly

$$Q_7 = (\bar{s}_{L\alpha}\gamma_\mu d_{L\alpha})[(\bar{u}_{R\beta}\gamma_\mu u_{R\beta}) - \frac{1}{2}(\bar{d}_{R\beta}\gamma_\mu d_{R\beta}) - \frac{1}{2}(\bar{s}_{R\beta}\gamma_\mu s_{R\beta})] , \tag{47}$$

$$Q_8 = (\bar{s}_{L\alpha}\gamma_\mu d_{L\beta})[(\bar{u}_{R\beta}\gamma_\mu u_{R\alpha}) - \frac{1}{2}(\bar{d}_{R\beta}\gamma_\mu d_{R\alpha}) - \frac{1}{2}(\bar{s}_{R\beta}\gamma_\mu s_{R\alpha})] . \tag{48}$$

Although the coefficients of these operators are suppressed by $(\alpha_{em}/\alpha_s(m_t))$, compared with Q_6, their $K \to \pi\pi$ matrix elements are enhanced by Λ_{QCD}/m_s because of the $(8_L, 8_R)$ chiral structure [35]. Furthermore in ϵ' their effects are enhanced over those of Q_6 by a factor of 20 because they contribute directly to A_2's imaginary part. We write the electroweak contribution as

$$ImA_2 \simeq s_2 s_3 s_6 f_w A_2 . \tag{49}$$

Putting these contributions together

$$\epsilon' \simeq \frac{e^{i\pi/4}}{20\sqrt{2}}[f_w + (1 - \Omega) f_s] s_2 s_3 s_6 , \tag{50}$$

where

$$\Omega = 20 \left(\frac{m_d - m_u}{3\sqrt{2} m_s}\right) . \tag{51}$$

Phenomenological models [36] for the matrix elements of the four quark operators Q_6, Q_7 and Q_8 suggest that for smaller values of the top quark mass both f_w and f_s are positive (with f_s somewhat larger) and that f_w partially cancels against $-\Omega f_s$ in the expression for ϵ'. Then we expect ϵ'/ϵ positive and of order 10^{-3}. On the other hand for large values of m_t (around 200 GeV) the situation is dramatically different. In this regime the Z-Penguin dominates f_w driving it negative. For large m_t f_w is comparable in magnitude with f_s and so ϵ'/ϵ can be negative [37].

The theoretical estimate for ϵ'/ϵ in the minimal standard model is very difficult. For large m_t there are contributing terms of order 10^{-3} which have different signs and partially cancel against each other. Despite these problems an uncontroversial measurement [38] of a nonzero value for ϵ'/ϵ is very important. It would rule out the superweak model showing that at least part of the CP violation observed in kaon decays is not due to very short distance physics.

There are other interesting possibilities for observing CP violation in kaon decay amplitudes. These include the processes $K_L \to \pi^0 e^+ e^-$, and $K_L \to \pi\nu\bar{\nu}$. For $K_L \to \pi^0 e^+ e^-$ there is a CP conserving "background" that arises from a two photon intermediate state. The cleanest theoretically is $K_L \to \pi^0 \nu\bar{\nu}$ which has no hadronic uncertainties. Of course it would be a very difficult measurement.

IV. CP Violation in $\bar{B}$ decay

As we have just observed, apart from the decay $K_L \to \pi^0 \nu\bar{\nu}$ the interpretation of CP violation in kaon decays is clouded by theoretical uncertainties. The situation for $\bar{B}$ meson decays, however, is much better. There are CP violating quantities that, in the minimal standard model, are free of dependence on nonperturbative strong interaction physics [39]. Consider, for example, the asymmetry

$$A_{\psi K_S}(t) = \frac{\Gamma(\bar{B}^0(t) \to \psi K_S) - \Gamma(B^0(t) \to \psi K_S)}{\Gamma(\bar{B}^0(t) \to \psi K_S) + \Gamma(B^0(t) \to \psi K_S)} \quad . \tag{52}$$

In eq. (51) $\bar{B}^0(t)$ and $B^0(t)$ are time evolving states that initially, i.e., at $t = 0$, are $\bar{B}^0$ and B^0 respectively. $A(t)$ is a CP violating quantity since the final ψK_S must have one unit of orbital angular momentum and is therefore has definite CP. (i.e., -1). In computing A we note that for the $B^0 - \bar{B}^0$ system $|\Gamma_{12}| << |M_{12}|$ and so we can neglect the width mixing. It is also convenient to redefine the phase of the b-quark field so M_{12} is real. Then a $\bar{B}^0$ or B^0 state produced at $t = 0$ has the following time evolution

$$|\bar{B}^0(t)> = \frac{e^{-im_Bt-\Gamma t/2}}{2}\{(1+e^{-i\Delta mt})|\bar{B}^0> - (1-e^{-i\Delta mt})|B^0>\} \ , \tag{53a}$$

$$|B^0(t)> = \frac{e^{-im_Bt-\Gamma t/2}}{2}\{-(1-e^{-i\Delta mt})|\bar{B}^0> + (1+e^{-i\Delta mt})|B^0>\} \ . \tag{53b}$$

Here the mass eigenstates are $1/\sqrt{2}\{|\bar{B}^0> + |B^0>\}$ and $1/\sqrt{2}\{|\bar{B}^0> - |B^0>\}$ with masses $m_B + \Delta m$ and m_B and $CP = -1$ and $+1$ respectively. Neglecting the small CP violation in $K^0 - \bar{K}^0$ mixing

$$<\psi K_s|H_{\text{eff}}^{|\Delta b|=1}|\bar{B}^0> = a(\epsilon_\psi^* \cdot p_B)e^{-i\beta} \quad , \tag{54a}$$

$$<\psi K_s|H_{\text{eff}}^{|\Delta b|=1}|B^0> = a(\epsilon_\psi^* \cdot p_B)e^{+i\beta} \quad , \tag{54b}$$

where β is the angle in the unitarity triangle of Fig, 2. It arises from a combination of the phase from the phase of $B^0 - \bar{B}^0$ mixing (i.e., arg $(V_{tb}^* V_{td})$) with the phase of the decay amplitude (i.e. arg $(V_{cb} V_{cs}^*)$.) It follows from eqs. (52), (53) and (54) that

$$|<\psi K_S|H_{\text{eff}}^{|\Delta b|=1}|\bar{B}^0(t)>|^2 = e^{-\Gamma t}|\epsilon_\psi \cdot p_B|^2|a|^2[1 + \sin 2\beta \sin \Delta mt] \tag{55a}$$

and

$$|<\psi K_S|H_{\text{eff}}^{|\Delta b|=1}|B^0(t)>|^2 = e^{-\Gamma t}|\epsilon_\psi \cdot p_B|^2|a|^2[1 - \sin 2\beta \sin \Delta mt] \ . \tag{55b}$$

This yields

$$A_{\psi K_S}(t) = \sin 2\beta \sin \Delta m t \tag{56}$$

free from nonperturbative hadronic uncertainties which are parametrized by a. Note that we have not assumed that final state interactions are small, they are included by having a be complex. Also we have not assumed that Penguin type diagrams are negligible. Penguin type diagrams with a top in the loop have (to leading order in small angles) the same combination of weak mixing angles as the "tree-level" $b \to c\bar{c}s$ processes.

$A_{\psi K_S}(t)$ arises from the interference of a phase from $B^0 - \bar{B}^0$ mixing with a phase in the decay amplitude. To isolate CP violation in first order weak decay amplitudes one must measure the analog of $A_{\psi K_S}(t)$ for another process. For example the analogous asymmetry $A_{\pi\pi}(t)$ is given by

$$A_{\pi\pi}(t) = \sin 2\alpha \sin \Delta m t \tag{57}$$

and the deviation of $A_{\pi\pi}(t)/A_{\psi K_S}(t)$ from unity is a measurement of CP violation in first order weak decay amplitudes. Even if $A_{\pi\pi}$ is not measured, if eventually the shape of the unitarity triangle is determined by measuring CP conserving quantities and $A_{\psi K_S}$ is found to agree with eq (56) then there would be very powerful evidence in support of the standard Kobayashi-Maskawa model where CP violation arises from δ.

The prediction for the asymmetry $A_{\pi\pi}(t)$ in eq (57) is not quite on the same footing as that for $A_{\psi K_S}(t)$ in eq (56). Penguin type diagrams do destroy eq (57), however, they are expected to be a small effect. Also the penguin type diagrams are $I = 1/2$ while the "tree level" W-exchange has both $I = 1/2$ and $I = 3/2$ pieces. For the isospin two $\pi\pi$ final state eq. (57) holds without corrections from Penguin type diagrams [40].

V. CP Violation from an Extended Higgs Sector

As was mentioned in the introduction the standard model with three generations of quarks and leptons and a single Higgs doublet has only two sources of CP violation; the vacuum angle for the strong interactions $\theta_{\rm QCD}$ and the phase δ in the Cabibbo–Kobayashi–Maskawa matrix. The stringent experimental limit on the electric dipole moment of neutron $d_n \leq 10^{-25} e - cm$ gives rise to the bound $\theta_{\rm QCD} \leq 10^{-8}$ on the vacuum angle. The phase δ, however, is not restricted by the present limit on the electric dipole moment of the neutron (it is expected to contribute at the $10^{-30} e - cm$ level) and in the minimal standard model it must be the source of the CP violation observed in Kaon decays.

Understanding the conservation of CP by the strong interactions is an important problem in particle physics. Speculative explanations for the smallness of $\theta_{\rm QCD}$ do exist. For example, there could be a U(1) Peccei–Quinn symmetry that is spontaneously broken at a large energy scale [41]. This converts $\theta_{\rm QCD}$ to a dynamical variable which is determined to be near zero by minimizing the vacuum energy. Alternatively if quantum fluctuations in the topology of spacetime occur, the wave function of the universe may be infinitely strongly peaked on the subspace of universes where $\theta_{\rm QCD}$ is near zero [42].

In the standard model only one Higgs doublet is required to spontaneously break SU(3) × SU(2) × U(1) to the low energy gauge group $SU(3)_c\times U(1)_{e.m.}$ and to give mass to the quarks and leptons. However, there is no compelling reason for such a minimal Higgs sector. Extensions of the standard model motivated by the hierarchy puzzle typically have a much more complicated Higgs sector. The simplest extension of the Higgs sector is the addition of more doublets. With one doublet the Higgs sector contributes a single real scalar to the spectrum of theory. It automatically has no flavor changing couplings to quarks; its coupling to quarks is proportional to their mass matrices so the same transformation which diagonalize the quark mass matrices diagonalizes its couplings. With n doublets the physical degrees of freedom arising from the Higgs sector are (2n-1) neutral scalars and (n-1) charged scalars. It is no longer automatic that the tree level couplings of the neutral scalars are flavor diagonal. However, flavor changing tree level couplings of the neutral scalars are absent if the up type quark Yukawa couplings involve only one of the doublets and the down type quark Yukawa couplings also only involve on of the doublets [21]. For definiteness consider a model of this type with n-doublets $H_j, j = 1, ..n$, where H_1 gives mass to the up type quarks, H_2 gives mass to the down type quarks and the remaining $n-2$ doublets do not Yukawa couple to the quarks. For three or more doublets there are new phases in the Higgs sector which contribute to CP violation [43]. In this model the coupling of the charged Higgs $H_j^{(\pm)}$ to the quarks is given by

$$\mathcal{L} = -\frac{1}{2v_1^*} H_1^{(+)} (\bar{u}, \bar{c}, \bar{t}) M_U V (1-\gamma_s) \begin{pmatrix} d \\ s \\ b \end{pmatrix} ,$$

$$+ \frac{1}{2v_2^*} H_2^{(+)} (\bar{u}, \bar{c}, \bar{t}) V M_D (1+\gamma_5) \begin{pmatrix} d \\ s \\ b \end{pmatrix} + h.c. \quad , \tag{58}$$

where

$$H_j = \begin{pmatrix} H_j^{(0)} \\ H_j^{(-)} \end{pmatrix} , H_j^{(-)\dagger} = H_j^{(+)} \tag{59}$$

and $< H_j^{(0)} >= v_j$. In eq. (1) V is the Cabibbo–Kobayashi–Maskawa matrix and

$$M_U = \begin{pmatrix} m_u & 0 & 0 \\ 0 & m_c & 0 \\ 0 & 0 & m_t \end{pmatrix} , M_D = \begin{pmatrix} m_d & 0 & 0 \\ 0 & m_s & 0 \\ 0 & 0 & m_b \end{pmatrix} . \tag{60}$$

The quark fields in eq (58) are mass eigenstate fields but that is not true of the Higgs fields. The mass eigenstate charged scalars $\phi_j^{(+)}$ are related to the $H_j^{(+)}$ by an $n \times n$ unitary matrix

$$\phi_j^{(+)} = \sum_{k=1}^{n} Y_{jk}^{-1} H_k^{(+)} = \sum_{k=1}^{n} Y_{kj}^* H_k^{(+)} . \tag{61}$$

One of the fields $\phi_j^{(+)}$ is the Goldstone boson associated with spontaneous breakdown of SU(2) $\times$ U(1) to U(1)$_{e.m.}$ It can be taken to be $\phi_1^{(+)}$ so that

$$Y_{k1}^* = g_2 v_k / \sqrt{2} M_W . \tag{62}$$

Combining eqs (58) - (62) gives

$$\mathcal{L} = \sum_{k=2}^{n} \frac{g_2 \phi_k^{(+)}}{2\sqrt{2} M_W} \left\{ -\left(\frac{Y_{1k}}{Y_{11}}\right) (\bar{u}, \bar{c}, \bar{t}) M_U V (1-\gamma_5) \begin{pmatrix} d \\ s \\ b \end{pmatrix} \right.$$

$$\left. + \left(\frac{Y_{2k}}{Y_{21}}\right) (\bar{u}, \bar{c}, \bar{t}) V M_D (1+\gamma_5) \begin{pmatrix} d \\ s \\ b \end{pmatrix} \right\} + h.c. \ . \tag{63}$$

In unitary gauge $\phi_1^{(+)}$ becomes the longitudinal component of the W-boson so it is omitted from (63). (Its couplings are independent of Y). For $n \geq 3$ the matrix Y contains phases which cannot be removed by field redefinitions and hence are a source of CP violation. A similar situation occurs in the coupling of the (2n-1) neutral scalars. For $n \geq 3$ there are new CP violating phases in these coupling to quarks which arise from the diagonalization of the neutral scalar mass matrix.

The couplings of the charged Higgs to quarks q are proportional to (m_q/M_W). The only quark with a mass comparable to the W-boson mass is the top quark. However, if all other quark masses are neglected then charged Higgs exchange doesn't give rise to CP violation at low energies since in any Feynman diagram the matrix elements of Y will enter as $|Y_{1k}|^2$. One needs a charge $-1/3$ quark mass (and the top quark mass) to get a contribution. This means, for charged scalar masses of order the W-boson mass, the new CP violating phases are not capable of explaining the CP violation in $K^0 - \bar{K}^0$ mixing. They can contribute the ϵ' through the dimension six operator, $m_s \bar{s}_R \sigma^{\mu\nu} G_{\mu\nu} d_L$, in the effective Hamiltonian for $|\Delta s| = 1$ weak nonleptonic decays. However, because this operator transforms as $(3_L, \bar{3}_R)$ under chiral SU(3)$_L\times$ SU(3)$_R$ its $K \to \pi\pi$ matrix elements are order m_s^2 and so its contribution to ϵ' is suppressed by $m_s/\Lambda_{\rm QCD}$.

Similarly the new phases in the charged scalar couplings are not likely to significantly influence the CP violating asymmetries $A_{\psi K_S}(t)$ and $A_{\pi\pi}(t)$ in B decay. The CP violation in charged Higgs couplings can effect the asymmetry $A_{\phi K_S}(t)$ which in the standard model with minimal Higgs sector is given by

$$A_{\phi K_S}(t) = \sin 2\beta \sin \Delta m t \ , \tag{64}$$

assuming that Penguin type diagrams with a top (or charm) quark in the loop dominate. The charged Higgs can replace the W-boson in the loop generating the dimension six operator $m_b \bar{s}_L \sigma^{\mu\nu} G_{\mu\nu} b_R$ in the effective Hamiltonian for $\bar{B}$ meson decay, with a complex coefficient that violates CP. It could significantly alter eq. (64). Experimentally it may be possible to check the assumption that Penguins dominate over the "tree level" $b \to u\bar{u}s$ process. The penguins are purely $I = 0$ and if their effects dominate then there are isospin relations: $\Gamma(\bar{B}^0 \to \rho^+ K^-) = \Gamma(B^- \to \bar{K}^0 \rho^0) = 2\Gamma(B^- \to K^- \rho^0)$. These would be destroyed by the $I = 1$ part of the effective Hamiltonian arising from the "tree level" $b \to u\bar{u}s$ process [44].

The CP violation in the Higgs sector dramatically alters our expectations for the electric dipole moments of the neutron and the electron. If we neglect the small weak mixing angles $\theta_1, \theta_2, \theta_3$ and the small light quark masses m_u, m_d and m_s then the effective Hamiltonian for the electric dipole moment of the neutron has a very simple form. Integrating out the scalars, W-bosons and top quark yields an effective Hamiltonian for the electric dipole moment of the neutron of the form [45,46]:

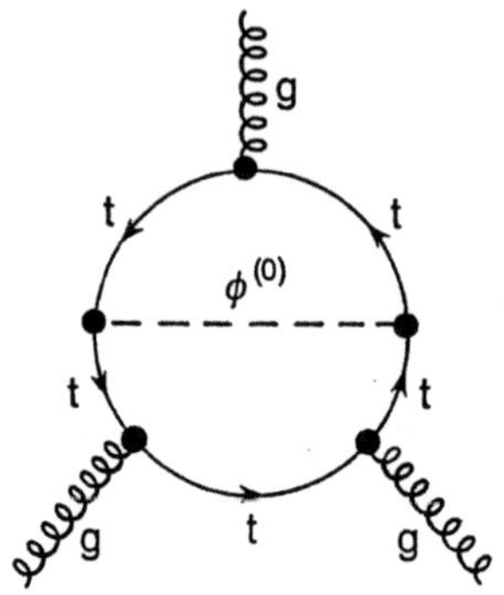

*Fig.*4 Type of Feynman diagram that determines $C_1(M_W)$

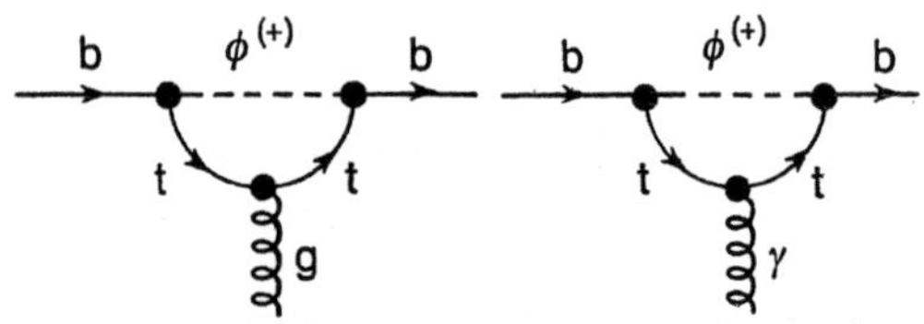

*Fig.*5 Type of Feynman diagrams that determine $C_2(M_W)$ and $C_3(M_W)$

$$H_{\text{eff}} = \frac{G_F}{\sqrt{2}} \sum_{i=1}^{3} C_i(\mu) O_i(\mu) \ , \tag{65}$$

where

$$O_1 = ig^3 Tr[G_{\mu\rho} G^{\rho}_{\nu} G_{\lambda\sigma}] \epsilon^{\mu\nu\lambda\sigma} \ , \tag{66a}$$

$$O_2 = g m_b \bar{b} T^a \sigma_{\mu\nu} b G^a_{\lambda\sigma} \epsilon^{\mu\nu\lambda\sigma} \ , \tag{66b}$$

$$O_3 = e m_b \bar{b} \sigma_{\mu\nu} b F_{\lambda\sigma} \epsilon^{\mu\nu\lambda\sigma} \ . \tag{66c}$$

The dimensionless coefficients $C_i(M_W)$ have a perturbative expansion in $\alpha_s(M_W)$. In leading order $C_1(M_W)$ arises from two-loop graphs like Fig. 4 involving neutral scalar exchange while $C_2(M_W)$ and $C_3(M_W)$ arise one loop graphs like those in Fig. 5 involving charged scalar exchange.

The coefficients of O_1 and O_2, at a subtraction point $M_W > \mu > m_b$, are related to those at $\mu = M_W$ by [47,48]

$$C_1(\mu) = \left[\frac{\alpha_s(M_W)}{\alpha_s(\mu)}\right]^{\frac{54}{23}} C_1(M_W) \ , \tag{67a}$$

$$C_2(\mu) = \left[\frac{\alpha_s(M_W)}{\alpha_s(\mu)}\right]^{\frac{14}{23}} C_2(M_W) \ . \tag{67b}$$

At the scale $\mu = m_b$ it is appropriate to go over to an effective four quark theory. In this theory only O_1 survives (neglecting operators with dimension greater than six). However, because the coefficient $C_1(M_W)$ arises at two loops it is important to include the contribution to the coefficient of O_1 (in the effective four quark theory) that comes from the one loop matrix element of O_2. Doing this gives, for $m_b > \mu$,

$$C_1(\mu) = \left[\frac{\alpha_s(M_W)}{\alpha_s(m_b)}\right]^{\frac{54}{23}} \left[\frac{\alpha_s(m_b)}{\alpha_s(m_c)}\right]^{\frac{54}{25}} \left[\frac{\alpha_s(m_c)}{\alpha_s(\mu)}\right]^{\frac{54}{27}}$$

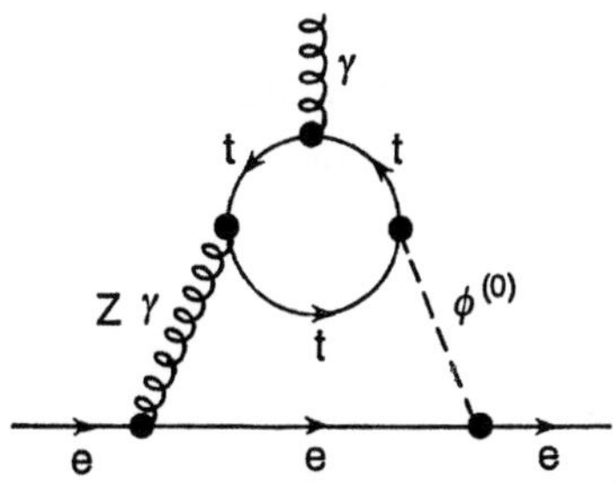

*Fig.*6 Two-loop diagram that generates an electric dipole moment for the electron of order that in eq. (69).

$$\cdot \left\{ C_1(M_W) + \frac{1}{12\pi^2} \left[\frac{\alpha_s(M_W)}{\alpha_s(m_b)} \right]^{\frac{-40}{23}} C_2(M_W) \right\} . \tag{68}$$

It is difficult to estimate the contribution of $H_{\text{eff}} = (G_F/\sqrt{2})C_1(\mu)O_1(\mu)$ to the electric dipole moment of the neutron because nonperturbative strong interaction physics plays an important role in the matrix element of O_1. However, it is generally believed, that for scalar masses of order the W-boson mass and CP violating phases in the scalar couplings of order unity the electric dipole moment is of order $10^{-26} e - cm$ (which is near the current experimental limit).

The electric dipole moment of the electron gets contributions, in the multi Higgs models we are considering, from two loop Feynman diagrams like that in Fig. 6. For scalar masses of order M_W and CP violating phases of order unity the resulting electric dipole moment of the electron is of order [49].

$$d_e \sim \left(\frac{g_2^2}{16\pi^2} \right)^2 \left(\frac{e m_e}{M_W^2} \right) \tag{69}$$

which is near the current experimental limit of $10^{-26} e - cm$ [50].

There are also similar two loop diagrams which generate dipole moments for the light quarks q. Although their contribution to the electric dipole moment of the neutron is suppressed by m_q/Λ_{QCD} compared with the operator O_1 (which we considered previously) it is actually comparable with that of O_1 because O_1 is strongly suppressed by QCD renormalization (see eq. 68).

If an electric dipole moment for the neutron near the current experimental limit was measured, in the future, the minimal standard model is not ruled out. This could be accommodated by the somewhat unattractive possibility of a nonzero (but small) value of θ_{QCD}. However, θ_{QCD} could not be responsible for an electric dipole moment for the electron at roughly the same level. If that were seen it would point very strongly to new sources of CP violation in the Higgs sector.

References

1. Particle Data Group (Review of Particle Properties), Phys. Lett. **B239** (1990) 1.
2. G 't Hooft, Phys. Rev. Lett. **37** (1976) 8; Phys. Rev. **D14** (1976) 3432;
   C.G. Callan, et.al., Phys. Lett. **B63** (1976) 334;
   R. Jackiw and C. Rebbi, Phys. Rev. Lett. **37** (1976) 177.
3. V. Baluni, Phys. Rev. **D19** (1979) 2227;
   R.J. Crewther, et.al., Phys. Lett. **B88** (1979) 123.
4. M. Kobayashi and T. Maskawa, Prog. Theor. Phys. **49** (1973) 652.
5. S. Nussinov and W. Wetzel, Phys. Rev. **D36** (1987) 130;
   M.B. Voloshin and M.A. Shifman, Sov. J. Nucl. Phys. **47** (1988) 511;
   E. Eichten, Nucl. Phys. **B4** (Proc. Suppl.) (1988) 170;
   G. Lepage and B.A. Thacker, Nucl. Phys. **B4** (Proc. Suppl.) (1988) 199;
   E. Eichten and B. Hill, Phys. Lett. **B234** (1990) 511;
   H. Georgi, Phys. Lett. **B240** (1990) 447;
   B. Grinstein, Nucl. Phys. **B339** (1990) 253;
   J.D. Bjorken, SLAC-PUB-5278, invited talk at Les Recontre de Physique de la Vallee d'Acoste, La Thuile, Italy (1990) unpublished.
6. N. Isgur and M.B. Wise, Phys. Lett. **B232** (1989) 113; Phys. Lett. **B237** (1990) 527.
7. M.B. Volosin and M.A. Shifman, Sov. J. Nucl. Phys. **45** (1987) 292;
   H.D. Politzer and M.B. Wise, Phys. Lett. **B206** (1988) 681; **B208** (1988) 504.
8. A.F. Falk, et.al., Nucl. Phys. **B343** (1990) 1.
9. A. Falk and B. Grinstein, Phys. Lett. **B247** (1990) 406.
10. M. Luke, HUTP-90/A051 (1990) unpublished.
11. N. Isgur and M.B. Wise, Nucl. Phys. **B348** (1991) 278;
    H. Georgi, Nucl. Phys. **B348** (1991) 293.
12. H. Georgi, et.al., HUTP-90/A052 (1990) unpublished.
13. J.D. Bjorken Nucl. Phys. **B11** (Proc. Suppl.) (1989) 325;
    M.J. Dugan and B. Grinstein, HUTP-90/A071 (1990) unpublished.
    J. Rosner, Phys. Rev. **D42** (1990) 3732;
    D. Bortoletto and S. Stone, Phys. Rev. Lett **65** (1990) 2951.
14. N. Isgur, et.al., Phys. Rev. **D39** (1989) 799.
15. R. Fulton, et.al., (CLEO Collaboration) Phys. Rev. Lett. **64** (1990) 16.
16. M.B. Wise, in Particles and Fields 3, proceedings of the Banff Summer Institute, Banff Alberta 1988, edited by A.N. Kamal and F. Khanna (World Scientific, Singapore, 1989) 124;
    N. Isgur and M.B. Wise, Phys. Rev. **D42** (1990) 2388.

17. Z. Bai, et.al., (Mark III Collaboration), SLAC-PUB-5341 (1990) unpublished;

J.C. Anjos, et.al., (Tagged Photon Collaboration) Phys. Rev. Lett. **62** (1989) 722; **62** (1989) 1587.

18. T. Inami and C.S. Lim, Prog. Theor. Phys. **65** (1981) 297.

19. Addler, et.al., (Mark III Collaboration), Phys. Rev. Lett. **60** (1988) 1375.

20. C.R. Allton, et.al., Shep 89/90-11 (1990) unpublished.

21. S. Glashow and S. Weinberg, Phys. Rev. **D15** (1977) 1958.

22. L.F. Abbott, et.al., Phys. Rev. **D21** 1980) 1393;

G. Athanasiu and F. Gilman, Phys. Lett. **B153** (1985) 274;

G. Athanasiu, et.al., Phys. Rev. **D32** (1985) 3010.

23. C. Geng and J. Ng, TR 1-PP-90-73, Presented at 25th Int. Conference on HEP Singapore (1990) unpublished.

24. C. Dib, et.al., SLAC-PUB-4840 (1989) unpublished;

J. Ellis and J. Hagelin, Nucl. Phys. **B217** (1983) 189.

25. F.J. Gilman and M.B. Wise, Phys. Rev. **D27** (1983) 1128.

26. A.J. Buras, et.al., Nucl. Phys. **B347** (1990) 491.

27. J. Donoghue, et.al., Phys. Lett. **B119** (1982) 412.

28. J. Bijnens, et.al., Phys. Rev. Lett. **53** (1984) 2367.

29. B.D. Gaiser, et.al., Annls. of Phys. **132** (1981) 66.

30. S. Sharpe, Nucl. Phys. B (Proc. Suppl.) **17** (1990) 146;

C. Bernard and A. Soni, Nucl. Phys. B (Proc. Suppl.) **17** (1990) 495;

G. Kilcup, NSF-ITP 90-233, talk presented at Lattice 1990, Talahasee (1990) unpublished.

31. L. Wolfenstein, Phys. Rev. Lett. **13** (1964) 562.

32. J. Ellis, et.al., Nucl. Phys. **B109** (1976) 213.

33. F.J. Gilman and M.B. Wise, Phys. Lett. **83B** (1979) 83; Phys. Rev. **D20** (1979) 2392.

34. B. Holstein, Phys. Rev. **D20** (1979) 1187.

35. J. Bijnens and M.B. Wise, Phys. Lett. **B137** (1984) 245.

36. J. F. Donoghue, et.al., Phys. Lett. **B179** (1986) 361; **188B** (1987) 511 (E);

W.A. Bardeen, et.al., Phys. Lett. **B108** (1986) 133;

R.S. Chivukala, et.al., Phys. Lett. **B171** (1986) 453;

A.J. Buras and J.M. Gerard, Nucl. Phys. **B264** (1986) 371.

37. J.M. Flynn and L. Randall, Phys. Lett. **B224** (1989) 221; **B235** (1990) 412 (E).

38. For a discussion on the present experimental situation see B. Winstein's Lecture.

39. A.B. Carter and A.I. Sanda, Phys. Rev. **D23** (1981) 1567;

I.I. Bigi and A.I. Sanda, Nucl. Phys. **B281** (1987) 41.

40. M. Gronau and D. London, DESY 90-106 (1990) unpublished.

41. R.D. Peccei and H.R. Quinn, Phys. Rev. Lett. **38** (1977) 1440;
J.E. Kim, Phys. Rev. Lett. **43** (1979) 103;
M. Dine, et.al., Phys. Lett. **B104** (1981) 199.
42. H.B. Nielsen and M. Ninomiya, Phys. Rev. Lett. **62** (1989) 1429;
K. Choi and R. Holman, Phys. Rev. Lett. **62** (1989) 2575;
J. Preskill, et.al., Phys. Lett. **B223** (1989) 26.
43. S. Weinberg Phys. Rev. Lett. **37** (1976) 657.
44. M. Savage and M.B. Wise, Phys. Rev. **D39** (1989) 3346.
45. S. Weinberg, Phys. Rev. Lett. **63** (1989) 2333.
46. G. Boyd, et.al., Phys. Lett. **B241** (1990) 584.
47. J. Dai and H. Dykstra, Phys. Lett. **B237** (1990) 256;
E. Braaten, et.al., Phys. Rev. Lett. **64** (1990) 1709.
48. M.A. Shifman, et.al., Phys. Rev. **D18** (1978) 2583.
49. S.M. Barr and A. Zee, Phys. Rev. Lett. **65** (1990) 21.
50. K. Abdullah, et.al., Phys. Rev. Lett. **65** (1990) 2347.

D- and B-Physics from Lattice QCD

G. Martinelli

Dipartimento di Fisica, Università di Roma “La Sapienza”,
P.le A. Moro 2, I-00185 Roma, Sezione di Roma INFN, Italy

1. Introduction

The study of the hadronic matrix elements which are relevant in weak decays of strange, charm and beauty mesons was started in lattice QCD a few years ago [1-6] and has progressed continuously since its beginning. The results of these calculations for kaon decays has been reported elsewhere [7]. In my talk I will review some progresses which has been done last year in the calculation of D-semi-leptonic decay amplitudes and in the investigation of the properties of mesons constituted by a heavy and a light quark [8-9]. The latter studies are important for their implications on the scaling laws and relations valid in the limit in which the mass of the heavy quark goes to infinity [10]. In this short discussion I will not be able to review other interesting calculations of the electromagnetic and weak properties of the hadrons. I only report in table 1 a list of matrix elements whose calculation has been already attempted, more or less successfully, on the lattice. Among the others, let me mention a very interesting study, the first with lattice techniques, of the gluon component of the proton spin [11].

The plan of this paper is the following. In section 2 I will briefly recall the essential features of the lattice technique and briefly report the results for the pseudoscalar meson decay constants and the kaon B-parameter. In section 3 I will summarize the lattice predictions for D semi-leptonic decays and confront these results with the experimental data and other theoretical approaches. In Section 4 I will present the lattice calculation of the B-meson decay constant and its implications for the scaling laws discussed extensively in the papers of ref.[10]. Finally, in the Conclusion, I will try to give an outlook on future improvements and new calculations which can be done with numerical methods and are of interest in the physics of the Standard Model.

TABLE 1

ELECTRO-MAGNETIC FORM FACTORS	$\langle N\vert J_\mu(0)\vert N'\rangle$ [12,13]
STRUCTURE FUNCTIONS	$\langle N\vert\bar{q}D^\mu\gamma^\nu q\vert N\rangle$ [13]
SIGMA TERM	$\langle N\vert\bar{q}q\vert N\rangle$ [14]
PROTON SPIN	$\langle N\vert\bar{q}\gamma^\nu\gamma^5 q\vert N\rangle$ [11,15]
EDM OF THE NEUTRON	$\langle N\vert\vec{d}\vert N\rangle$ [16]
WEAK DECAYS OF KAONS	$\epsilon, \epsilon'/\epsilon$, $\Delta I=1/2$, $\Delta I=3/2$, ELECTRO-PENGUINS [1,2,3,17,18,19]
D SEMI-LEPTONIC DECAYS	$\langle D\vert J_\mu^W\vert K\rangle$ $\langle D\vert J_\mu^W\vert K^*\rangle$ [4,5,20]
D NON-LEPTONIC DECAYS	$\langle D\vert H_W\vert PP\rangle$ $\langle D\vert H_W\vert VV\rangle$ [21,22]
B-PHYSICS	$\langle B\vert A_0\vert 0\rangle = i\, M_B f_B$ [23,24,25,26]

Springer Proceedings in Physics, Vol. 65 **Present and Future of High-Energy Physics**
Editors: K.-I. Aoki and M. Kobayashi

2. A Short Introduction to Lattice Field Theory

2.1. *General Considerations*

In a given theory, any physical amplitude, from which we may compute the mass spectrum, the decay constants, the scattering amplitudes and so on, can be obtained as a functional integral over the fields of the theory:

$$A = \langle O(\phi) \rangle = \frac{\int d[\phi]\, e^{-S(\phi)} O(\phi)}{\int d[\phi]\, e^{-S(\phi)}} = \frac{1}{Z} \int d[\phi]\, e^{-S(\phi)} O(\phi) \,. \tag{1}$$

In eq. (1) a Wick rotation ($t \to it$) to Euclidean space-time has been applied. In perturbation theory the action $S(\phi)$ has the form:

$$S(\phi) = \sum_{iJ} \phi_i \Delta_{iJ} \phi_J + S_I(\phi) \;, \tag{2}$$

where the interaction, $S_I(\phi)$, can be considered a "small" correction to the kinetic term which is quadratic in the fields. In this case we can expand the functional integral in eq. (1) in powers of $S_I(\phi)$:

$$A \sim \frac{1}{Z} \int d[\phi]\, e^{-\phi\Delta\phi}\, O(\phi)[1 - S_I(\phi) + ..] \,. \tag{3}$$

We are then able to perfom the gaussian integral appearing in eq. (3) and compute A order by order in $S_I(\phi)$. In general, the formal expansion of eq. (3) has also to be substantiated by renormalizing the infinities of the theory.

Beyond perturbation theory we are not able to perfom the functional integral of eq. (1) and we need a non-perturbative method. In the lattice approach one substitutes the continum space-time by a mesh of discrete lattice points as shown in figure 1 and defines the fields only on the vertices of the lattice. The space-time discretization ensures the regularization of the ultraviolet infinities of the theory because the maximum momentum allowed on a lattice is $p_{MAX} = \pi/a$, where a is the lattice spacing. The fundamental approximation made in the numerical simulations of the theory is the replacement of the infinite volume space-time by a lattice with a finite number of points. The accuracy of the results is thus dictated by the limitations imposed on the size of the lattice by computer memory and speed. The situation is illustrated in figs. 2 for the case of QCD. At large values of g_o, the lattice strong compling constant, the size of a hadron is of the order of the lattice spacing and large effects due to a too small ultra-violet cut-off are present, fig. 2a. At small g_o, with a limited number of lattice points, the hadron becomes as large as the whole lattice and feels strong finite size effects, fig. 2b. The optimal situation is shown in fig. 2c: the hadron is much larger than the lattice spacing but much smaller than the lattice volume. This requires the use of huge lattices.

The physical amplitude of eq.(1) can be computed by integrating over the fields:

$$A \sim \frac{1}{Z} \int \prod_i d\phi_i e^{-S(\phi_i)} O(\phi_i) \,. \tag{4}$$

To avoid finite size effects,. the number of points must be very large. In actual lattice calculations the number of variables (i.e. the fields ϕ_i) on which we must integrate is of the order 10^6–10^7!! The integration is done with the method of important sampling, by generating the fields with a probability distribution $P(\phi) = e^{-S(\phi_i)}$.

In spite of the fact that the integration over 10^6 variables seems to be a formidable task, the availability of ever more powerful computers has allowed in the last decade an enormous progress. To illustrate this point I report in table 2 the volumes used in QCD lattice simulations starting from the pioneering works of M.Creutz in 1978 on a 4^4(sic!) lattice. In the last column

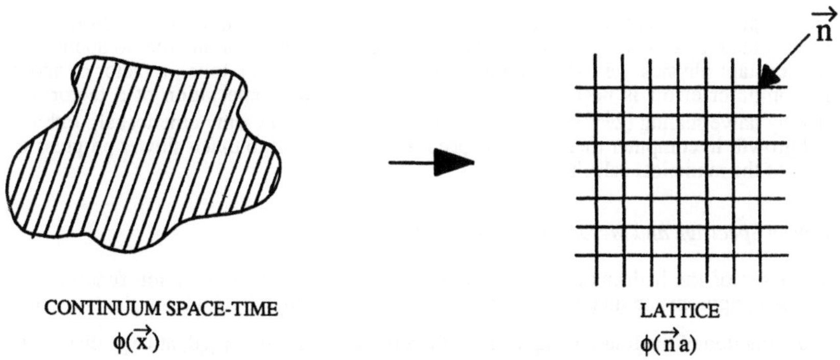

Fig.1: The continuum space-time is replaced by a lattice. The fields are defined only on the vertices of the lattice: $\phi(\vec{x}) = \phi(\vec{x}=\vec{n}a)$*, where a is the lattice spacing.*

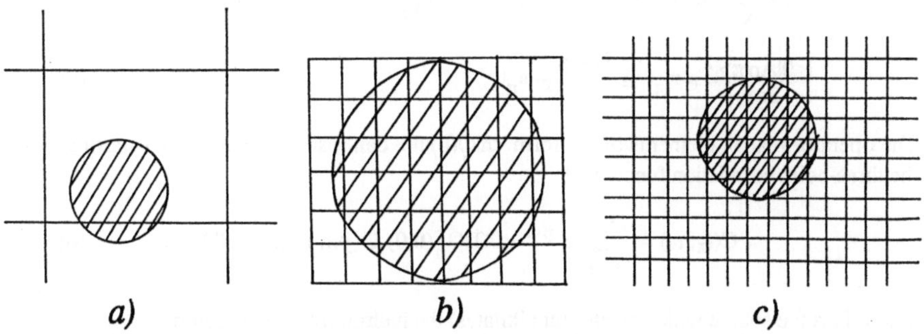

Fig.2: 2a) Hadron on the lattice at g_0 *large: the hadron is smaller than the lattice spacing (effects of order* $\Lambda_{QCD}a$*); 2b) when* $g_0 \to 0$ *the hadron becomes as large as the whole lattice (finite size effects); 2c) the hadron size is much smaller than the lattice and much larger than the lattice spacing.*

TABLE 2

Author	Year	V	#
M.Creutz[27]	78	4^4	1
H.Lipps et al.[28]	83	10^3x20	78
S.Itoh et al.[29]	87	16^3x48	770
Super Ape[30]	92	~50^3x150	7x10^4

the computer resources needed for these calculations, normalized to the original volume 4^4 used by M.Creutz, are also reported. One can see that we have already gained a factor of about 800 in the last 10 years and more is to be expected in the near future. As a reference I have also reported in the last row the characteristic volume on which it will be possible to work, in the quenched approximation, in the next two years, by using the super Ape machine [30]. On this machine it will be also possible to make a full, unquenched, QCD calculation on the same size of lattices (~16^3x48) which are currently used for quenched calculations.

I want also to mention other dedicated computers which are becoming (or will become) operational soon: the Columbia machine [31], the Tsukuba machine [32] and the Fermilab machine [33]. It has been estimated that the rate of growth in computer power is about a factor of a hundred every four years for dedicated machines and of a factor about ten every four years for commercial machines [34].

Is this impressing increase in computer resources enough? Certainly it will allow us to improve the accuracy of the results. However it has been estimated that, in order to double the lattice at constant physics (i.e. with a total volume fixed in physical units), a factor of about 3000 in computer effort is needed, to be contrasted with the expectation of a factor 16, suggested by naïve scaling ($2^{d=4}=16$). This implies that, with the present numerical algorithms, the rapid growth in computer memory and speed will correspond to much slower progress in the quality of the numerical calculations.

2.2. *Hadron Spectrum and Meson Decay Constants*

The masses of the hadrons are obtained from the study of the correlation functions of operators with appropriate quantum numbers. Thus, for example, by studying the two point correlation function of the fourth component of the axial current, $A_0=\bar{u}\gamma_0\gamma_5 d$, at large distances, it is possible to compute the pion mass and decay constant:

$$G(t)=\sum_{\mathbf{x}} \langle A_0(\mathbf{x},t)A_0^+(0,0)\rangle = \sum_n \frac{\langle 0|A_0|n\rangle\langle n|A_0^+|0\rangle}{2m_n} e^{-m_n t} \underset{t\to\infty}{\to}$$

$$\frac{|\langle 0|A_0|\pi\rangle|^2}{2m_\pi} e^{-m_\pi t} = \frac{f_\pi^2 m_\pi}{2} e^{-m_\pi t} . \qquad (5)$$

The operator matrix elements of the form $\langle h|O(0)|h'\rangle$ can instead be calculated starting from the three-point correlation function(*) :

$$C(t_x, t_y) = \sum_{\mathbf{x},\mathbf{y}} e^{i\,\mathbf{p}\cdot\mathbf{x}} \langle 0|T|[J_h^+(\mathbf{0},0)\, O(\mathbf{y},t_y)\, J_h(\mathbf{x},t_x)]|0\rangle \quad , \qquad (6)$$

where J_h is the source which creates (annihilates) the hadron from the vacuum.

Inserting complete sets of states between the operators in eq.(6), taking $t_x > t_y > 0$ and using translation invariance, we have, for sufficiently large t_y and $t_x - t_y$

$$C(t_x, t_y) = \frac{e^{-Et_x}}{4E^2} |\langle h|J_h(0)|0\rangle|^2 \langle h|O(0)|h\rangle \quad . \qquad (7)$$

The matrix element $|\langle h|J_h(0)|0\rangle|$ and the energy (mass) of the hadron are known from the study of the two-point function (5). Therefore, from eqs.(6,7), it is possible to extract the matrix element $\langle h|O|h\rangle$.

The main sources of systematic effects in the hadron spectrum and matrix element calculations are:

i) the use of the quenched approximation;

ii) the effects of corrections of order $\Lambda_{QCD}\, a$, where a is the lattice spacing;

iii) finite size effects.

Finite size effects become more and more severe for small quark masses. For this reason, on the lattices presently used in numerical work, the smallest quark mass is of the order of the strange quark mass. The results are then extrapolated to the values corresponding to the up/down quarks. A certain systematic error is entailed in this extrapolation and the extrapolation is good only for quantities which have a smooth dependence on the quark masses.

The effects of order Λ_{QCD} a tend to disappear in the continuum limit, i.e. $a \to 0$, and are expected to be of the order of 10% for $a^{-1} \sim$ 2-3 GeV. They seem to be more important in the calculation of the hadronic matrix elements, where they can be reduced following the suggestion of Symanzik [35,36].

(*) For the purposes of illustration I consider here the calculation of forward matrix elements. The general case can be easily derived.

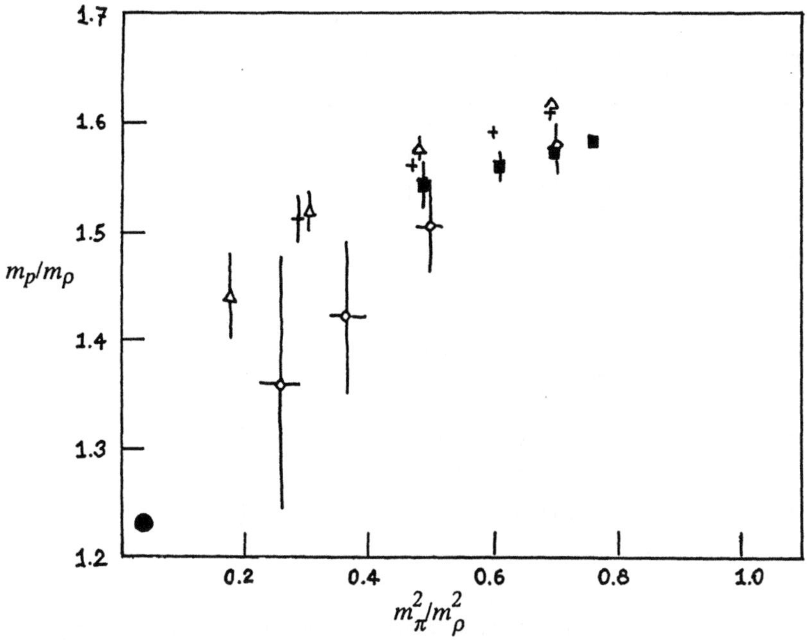

+ β=5.7 12^3x32 , △ β=5.7 24^3x32 , ■β=6.0 18^3x32 , ◊ β=6.0 24^3x32

Fig. 3: The ratio m_p/m_ρ is reported as a funtion of m_π^2/m_ρ^2 from the data of ref. 38. The • indicates the experimental value.

The quenched approximation corresponds to the calculation of the hadronic properties in the limit in which one neglects all the effects due to the presence of extra quark loops, the "sea" quarks in the parton language. The systematic error induced by this approximation depends on the quantity that one has to compute and it is difficult to estimate it a priori. It is known, independently of any lattice calculation, that the quenched approximation is quite inaccurate in the calculation of the nucleon σ-term [14] and of the polarized proton structure function [11,37] and it will certainly give the wrong result for the decay width of the ρ into two pions. For the hadron spectrum, all the lattice calculations give a strong indication that, in the quenched approximation, one obtains values of the proton to ρ mass ratio which are about 20% higher than the experimental one. Although a small part of the effect can still be due to terms of order a, the general impression is that the main effect comes from the quenched approximation. The best results for the quenched case have been obtained by the APE group [38], by combining high statistics with a very effective technique, the "smearing" of the source operators, to reduce the statistical error to a negligible amount. In fig. 3 I report their results obtained at β=5.7 on a 12^3x32 and 24^3x32 lattice and at β=6 on a 18^3x32 and 24^3x32 lattice. From the comparison of the results at the same value of β on two different lattices, it is clear that finite size effects are negligible in the range of quark masses explored; from the comparison of the results at two different values of the lattice spacing, corresponding at β=5.7 and β=6, it seems very difficult to explain the discrepancy between the lattice results and the experimental number, reported as a black dot in the figure, with effects of order Λ_{QCD} a. Notice that in the figure the statistical errors are often smaller than the size of the points.

In spite of the limitations of the method, which I have mentioned above, lattice QCD is superior to any other non perturbative approach for the following reasons:

TABLE 3 Pseudoscalar decay constants from lattice calculations (second column), and their experimental values.

Meson	$f_{lattice}$		$f_{experiment}$
π	[140±20±20(syst)] MeV	[17]	132 MeV
K	(158±13) MeV (173±83) MeV	[17]	160 MeV
D	(180±30) MeV (215±60) MeV (174±52) MeV	[39] [40] [41]	< 290 MeV
D_s	(218±30) MeV ~251 MeV (234±70) MeV	[39] [40] [41]	
B_d	~ 120 MeV (105±34) MeV	[39] [40]	
B_s	~ 150 MeV (160±30) MeV	[39] [40]	

i) it is based on the fundamental theory of strong interactions (plus renormalization group) with *no extra assumption*,
ii) there are no free parameters besides the fundamental scale of strong interactions (for example Λ_{QCD}) and the masses of the quarks (up, down, strange, etc.),
iii) the accuracy of the results is *systematically improvable* and the systematic errors (terms of order Λ_{QCD} a, finite size effects, quenched approximation) are decreasing in time. This is not the case for other non-perturbative methods ($1/N_c$, quark models, chiral models, etc.) as will become clearer in the discussion of D semileptonic decays.

As shown in eq. (5), from the study of the two-point correlation function it is also possible to compute the meson decay constants. The results obtained on the lattice[17,39-42] are reported, together with their experimental values (when known), in table 3(*). The systematic error quoted for f_π in the table comes from the calibration of the lattice spacing. This error is absent in the other cases where one can compute the meson decay constant from the ratio f_M/f_π and then multiply the ratio by the experimental value of f_π. It is interesting to notice the good agreement of f_K with its experimental value; $f_{D,Ds}$ agree well with the results from QCD sum rules [43]. The B-meson decay constant in the table has been obtained from f_D by extrapolating with the non-relativistic formula

$$f\sqrt{M} = \text{const} , \tag{8}$$

where M is the heavy meson mass. It is not clear if this formula, which certainly becomes true as $M \to \infty$ [6] can already be used in the range between the charm and bottom quark masses. So far only the authors of ref. 40 have claimed that the scaling law of eq.(8) is already well satisfied at values of the quark masses slightly above the charm quark mass. These results are at variance with those obtained more recently in refs. 24-26 and 42. I will discuss at length the scaling problem in section 4.

Unfortunately it is not possible to compute directly the B-meson decay constant, because the b-quark is heavier than the presently used ultraviolet cut-offs, a^{-1}~2-3 GeV. As an alternative to a direct calculation of the B-meson properties, at the Seillac lattice conference it was proposed to study heavy-light ($\bar{Q}$,q) systems by systematically expanding in $1/m_Q$, where m_Q is the mass

(*) The results of ref.42 will be discussed more in detail below.

TABLE 4 Kaon B-parameter

Method	$B_{K\bar{K}}$
PCAC+SU(3)[44]	0.33±0.33
QCD Sum Rules[45]	0.33±0.09
QCD Sum Rules[46]	0.50±0.10±0.20
$1/N_c$[47]	0.70±0.07
Lattice QCD Wilson Fermions[7]	0.85±0.20 B(μ=2GeV)=0.65±0.15
Lattice QCD Staggered Fermions[7,49]	0.91±0.03 [B(μ=2GeV)=0.70±0.01]
Lattice QCD Wilson Fermions[48]	1.00±0.05±0.03 [B(μ=2GeV)=0.76±0.04±0.02]

of the heavy quark [6]. The first attempts to use this technique to compute the B-meson properties were not very successful [23]. Recently important progress has been made and will be reported in Section 4 [24,25,26].

As an example of the accuracy which has been already attained in the calculation of the hadronic matrix elements, I report in table 4 a compilation of lattice results for the kaon B-parameter, obtained on the lattice by different groups. I also give in the table other theoretical determinations of the same quantity. The $1/N_c$- expansion and the lattice calculations tend to give a value close to 3/4; the determination of ref.44 using PCAC and of ref.45 with QCD sum rules cluster around a value of about 1/3. It should be noticed, however, that the value of $B_{K\bar{K}}$=0.33 found in ref.44 can be pushed to 0.49 simply by taking f_K instead of f_π in their final formula. This is perfectly legal in the SU(3) symmetric limit, which is used to estimate the B-parameter. On the other hand the error on the determination of the B-parameter quoted in ref.45 could have been underestimated [50] and a different calculation, using QCD sum rules, tends to give a value much closer to the lattice determination.

Notice that the statistical error of the newest lattice determination of this quantity [49], computed with staggered fermions, is less than 3%. A very preliminary result has been presented in ref. 49 with, a smaller lattice spacing, gives a lower value for the B-parameter with staggered fermions, B ~ 0.72. The only relevant source of error remains the use of the quenched approximation(*). Some results, obtained using the Columbia machine by N.Christ et al. give as "unquenched" result for the B-parameter ~ 0.87, very close to the quenched result.

3. Semi-Leptonic Decays of D-Mesons

Semi-leptonic decays of D and B mesons play a crucial role in our understanding of the Cabibbo-Kobayashi-Maskawa (C-K-M) mixing matrix and of the interplay between strong and weak interactions. In particular semi-leptonic decays of neutral D-mesons into pseudoscalar light mesons, $D^0 \to K^- e^+ \nu_e$ and $D^0 \to \pi^- e^+ \nu_e$, are used to measure directly the corresponding C-K-M matrix elements, V_{cs} and V_{cd}, respectively [51,52].

From the theoretical standpoint $D^0 \to K^- e^+ \nu_e$ and $D^0 \to \pi^- e^+ \nu_e$ are particularly simple. The reason is that they proceed via the spectator process in which a charm quark decays into a light quark (s or d) by emitting a W boson, which materializes into a positron and a neutrino, as shown in fig. 4. With only two quarks in the final state, there are no interfering diagrams or

(*) A two loop calculation necessary to a better control of the renormalization of the lattice operator also has yet to be done.

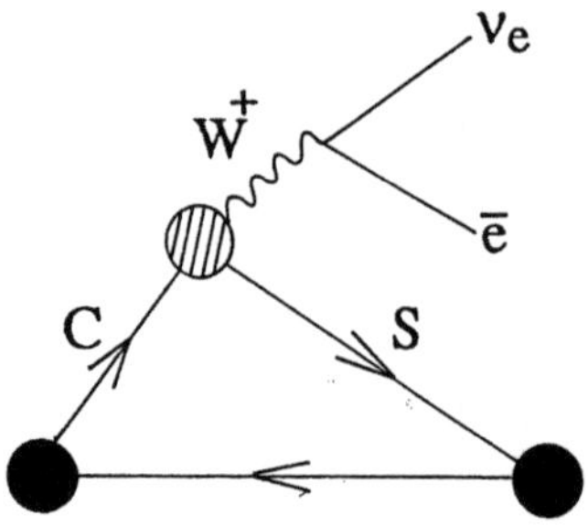

Fig. 4: Feynman diagram relevant in semi-leptonic $D \to K,K^*$ decays

final state interactions to take into account, contrary to what happens in non-leptonic decays. Moreover, in order to extract V_{cs} and V_{cd} from the data, it is necessary to know one form-factor only, $f^+(q^2)$, where q is the momentum of the lepton pair. This form factor is generally taken from some theoretical model.

The $D^+ \to \bar{K}^{*0} e^+ \nu_e$ decay is more complicated than the semileptonic decays of D^0 into pseudoscalar mesons. The relevant quark diagram remains that shown in fig.4. However, because the K^* is a particle of spin 1, there are now three different form factors (which we will denote by $V(q^2)$, $A_1(q^2)$ and $A_2(q^2)$ in the following) involved in the decay. The vector decay channel is very important in itself since it constitutes the only independent test of the theoretical models, which are used to extract the information on the C-K-M angles from the pseudoscalar decay modes of D and, in the future, B mesons.

In $D \to K,\pi$ decays the relevant form factor can be extracted from a lattice calculation of the matrix elements of the weak current sandwiched between the initial and final mesons, $\langle D | J^W_\mu | K,\pi \rangle$.

These matrix elements can be parametrized in terms of two form factors. In the helicity basis:

$$\langle K^- | J_\mu | D^0 \rangle = \left(p_D + p_K - \frac{m_D^2 - m_K^2}{q^2}\right) q_\mu \; f^+(q^2) + \left(\frac{m_D^2 - m_K^2}{q^2}\right) q_\mu \; f^0(q^2) \;, \tag{9}$$

where q is the momentum transfer, p_D-p_K. One only needs to compute the matrix elements of the vector current since the matrix elements of the axial current are zero between two pseudoscalar states. At zero momentum transfer $f^+(0)=f^0(0)$.

In refs. 4 and 53 the two form factors $f^+(q^2)$ and $f^0(q^2)$ were computed at several values of the momentum transfer. By assuming the validity of the vector dominance model:

$$\frac{f^+(q^2)}{f^+(0)} = \frac{m_{D^*}^2}{q^2 - m_{D^*}^2} \tag{10}$$

and using the masses of the vector meson found at the corresponding values of the quark masses, they found:

$$f_\pi^+(0) = 0.57 \pm 0.09 \,,$$
$$f_K^+(0) = 0.66 \pm 0.07 \,. \tag{11}$$

These results are extremely satisfying. The statistical errors are reasonably small and $f_K^+(0)$ is in good agreement with recent experimental data from the Tagged Photon Spectrometer Collaboration [51]. This collaboration measures the relevant branching ratio to be $B(D^0 \to K^-$

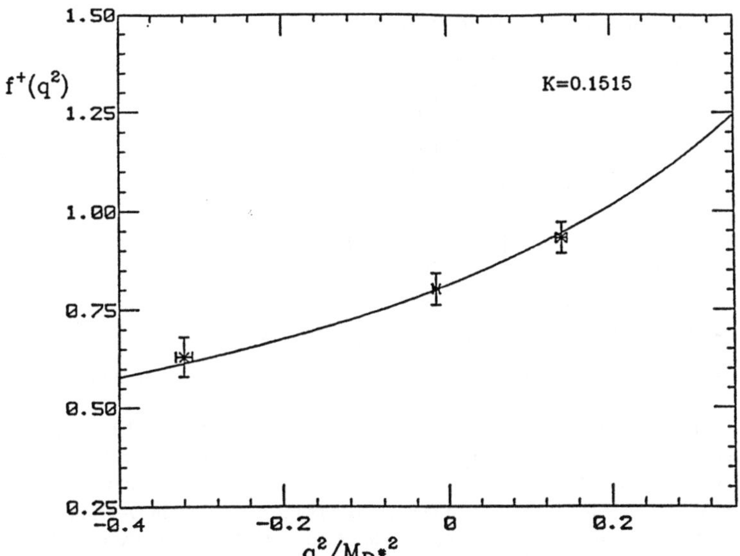

Fig. 5: The lattice results for $f^+(q^2)$ as a function of q^2/m^2_{D*} are reported and compared with the vector meson dominance expectation full curve [4,20].

$e^+\nu_e$) = (3.8±0.5±0.6)%. From this value and the D^0 lifetime, and assuming that the q^2 behaviour of the form factor is given by eq. (10), they obtain the result, $|V_{cs}|^2|f^+(0)|^2$= 0.50±0.07. Assuming three generations and imposing unitarity on the C-K-M matrix (i.e, taking $|V_{cs}|$ = 0.975) this result corresponds to $f^+(0)$ = 0.73±0.05±0.07. The MARK III [52] collaboration obtains an almost identical result for the branching ratio, $B(D^0 \to K^- e^+ \nu_e)$ = (3.9±0.6±0.6)% and hence the deduced value of $f^+(0)$ is also very similar. Both experiments have studied the behaviour of $f^+(q^2)$ with q^2 and found it compatible with vector meson dominance. In figure 5 the lattice results for the form factor $f^+(0)$ as a function of q^2/m^2_{D*} are reported. For comparison with the vector dominance model the curve obtained by using eq.(10) is also plotted. Notice that E653 has found a smaller value for f_K, $f^+_K(0)$=0.59±0.06, which averaged with the other experimental results, gives 0.69±0.04 in excellent agreement with the value given in eq.(11).

The value of $f_K(0)$ reported in ref.48 is $f_K(0)$ = 0.90±0.08±0.21 compatible with the results given in eq.(11).

As can be seen from the figure, the vector meson dominance model gives a good description of the q^2 dependence of the form factor from the lattice calculation.

For $D \to K$, ρ decays one has to compute the matrix elements of the weak current between the initial D and the final vector meson. By varying the current component and the vector meson polarization and momentum it is possible to disentangle the four form factors, V, $A_{1,2}$ and A which are defined, in the helicity basis, by the following relation:

$$<K^*_r|J_\mu|D> = e^\beta_r \left[\frac{2V(q^2)}{m_D+m_{K^*}} \varepsilon_{\mu\gamma\delta\beta} p^\gamma_D p^\delta_{K^*} + i(m_D+m_{K^*}) A_1(q^2) g_{\mu\beta} \right.$$

$$\left. - i \frac{A_2(q^2)}{m_D+m_{K^*}} P_\mu q_\beta + i \frac{A(q^2)}{q^2} 2m_{K^*} q_\mu P_\beta \right] , \qquad (12)$$

TABLE 5 Values of the relevant form factors at zero momentum transfer for $D \to \rho$ and K^*

Process	Form Factors	Lattice result
$D \to \rho$	V	0.77 ± 0.09
	A_1	0.47 ± 0.07
	A_2	-0.08 ± 0.42
$D \to K^*$	V	0.85 ± 0.08
	A_1	0.52 ± 0.07
	A_2	0.05 ± 0.35

where q is the momentum transfer, $q = p_D - p_{K^*}$, $P = p_D + p_{K^*}$ and e_r^β is the polarization vector of the K^*. A (q^2) gives a negligible contribution to the decay rate and will not be considered in the following.

The three relevant form factors V, A_1 and A_2 were recently computed on the lattice in ref. 20 at several values of q^2. Very preliminary results have been also presented by the UCLA group at the Lattice Conference in Tallahassee [48]. The pole dominance gives a good description of the q^2 dependence of these form factors, as was the case for f^+ in $D \to K$ decays. The results of the form factors at zero momentum transfer from ref.20 are reported in table 5. Notice that for V and A_1 the statistical error is reasonably small and that A_2 is zero within (still large) statistical uncertainties.

I also report in table 6 the results of a recent experimental analysis by the E691 Collaboration [54]. The experimental results are confronted in the table with various predictions found by using different theoretical models. I have added to the table a column with the results from the lattice calculation of ref.20. Notice that all the phenomenological models badly fail for A_2, for which they all give a value close to one. Only the lattice seems to be able to reproduce the correct pattern of values, $V \sim 0.9$, $A_1 \sim 0.5$, $A_2 \sim 0$.

TABLE 6 Values measured by E691, compared to predictions of various models (IS,BW,GS and KS and LMS)

	E691[54]	IS[55]	BW[56]	GS	KS[57]	LMS[20]
$A_1(0)$	$0.46 \pm 0.05 \pm 0.05$	0.8	0.9	0.8	1.0	0.52 ± 0.07
$A_2(0)$	$0.0 \pm 0.02 \pm 0.01$	0.8	1.2	0.6	1.0	0.05 ± 0.35
$V(0)$	$0.9 \pm 0.3 \pm 0.1$	1.1	1.3	1.5	1.0	0.85 ± 0.08
$\frac{\Gamma_L}{\Gamma_T}$	$1.8^{+0.6}_{-0.4} \pm 0.3$	1.1	0.9	1.2	1.2	1.7 ± 0.6

The agreement between the experimental data and the lattice predictions is exceedingly good, given the theoretical uncertainties certainly present in the lattice calculation. Work is still in progress in order to reduce the theoretical error. From the form factors of table 5 and using $V_{cs} = 0.975$, one can compute the total rates:

	LATTICE	EXPERIMENT
$\Gamma(D \to K^* e\nu)$	$(5.2 \pm 1.9)\, 10^{10}\, s^{-1}$	$(4.1 \pm 0.7 \pm 0.5) 10^{10}\, s^{-1}$
$\Gamma(D \to \rho)$	$(0.45 \pm 0.23) 10^{10}\, s^{-1}$	------

Before closing this section I wish to compare the technique used to compute the form factors on the lattice with some other theoretical approaches and stress again that the lattice method has the least number of assumptions and the best chances to improve the accuracy of the results with respect to any other method. As an example I will confront the lattice method with the quark model of ref.58:

	LATTICE	ISGW (CONSTITUENT QUARK MODEL)
ASSUMPTIONS	———	$V(r) = -\frac{4}{3}\frac{\alpha_s}{r} + c + br$ SHAPE OF THE POTENTIAL
FREE PARAMETERS	Λ_{QCD} AND CURRENT QUARK MASSES	$\alpha_s \equiv \Lambda_{QCD}$ c, b AND CONSTITUENT QUARK MASSES
q^2-DEPENDENCE	AUTOMATIC	K-TEMPERING FACTOR TO REPRODUCE $F_\pi(q^2)$ ASSUMED TO BE EQUAL FOR ALL FORM FACTORS
SYSTEMATIC EFFECTS	m_s=m_d AND $O(\Lambda_{QCD}\, a)$ (EASILY IMPROVABLE) QUENCHING	? QUENCHING (?)

TABLE 7

	ELC[17,39]	QUARK MODELS[59]	ISGW[58]	EXPERIMENT
f_π	(140±20±20) MeV	(60 - 206) MeV	144 MeV	132 MeV
f_K	(158±13) MeV	(92 - 290) MeV	232 MeV	165 MeV
f_D	(180±30) MeV	(89 - 227) MeV	227 MeV	< 290 MeV
f_{D_s}	(218±30) MeV	(96 - 252) MeV	---	---

The quark model assumes the form of the potential and has two extra parameters, b and c, which have to be fitted, together with α_s and the constituent quark masses, from meson spectroscopy. In the quark model the q^2 dependence of the form factor is not known and it is difficult to evaluate the systematic effects. One possibility is to compare different quark models, which reproduce the meson spectroscopy, and check the stability of the results among the different models. This study has not been done yet for semileptonic decays. There is however a recent study by Capstick and Godfrey [59], where they have compared the predictions from different models for the pseudoscalar decay constants of π, K and D mesons. The allowed intervals, from different quark models, the results from the ISGW model [58], the lattice results from ELC and the experimental values are reported in table 7, from which is possible to deduce the stability and accuracy of quark models compared with the lattice approach. Notice that the uncertainties in evaluating the form factors in $B \to \pi$ and ρ decays, due to the small overlap of the wave functions, are expected to be even larger than for D-decays.

4. B-Meson Physics

As discussed above, among the quantities which have been computed on the lattice are the decay constants of the pseudoscalar mesons, π, K, and D [17,39-42]. In these calculations, the value of the inverse lattice spacing (a^{-1}) is generally in the range between 1 GeV and 3 GeV. Thus it is not possible to simulate hadrons containing heavy quarks, such as the b-quark, on these lattices, since their Compton wavelengths are smaller than the lattice spacing. An interesting technique for the study of heavy quark physics on the lattice has recently been proposed by Eichten [6]. This technique is based on the expansion of the heavy quark propagator in inverse powers of the quark mass.

At the leading order in the $1/m_b$ expansion, where m_b is the mass of the b-quark, the b-quark propagator is given by:

$$S_b(x,y) = P_{\mathbf{x}}(x^0, y^0)\, \delta(\mathbf{x} - \mathbf{y})\, \{\theta(x^0 - y^0)\, e^{-m_b(x^0-y^0)} \left(\frac{1+\gamma^0}{2}\right) + \theta(y^0 - x^0)\, e^{-m_b(y^0-x^0)} \left(\frac{1-\gamma^0}{2}\right)\}\,, \tag{13}$$

where using the lattice formulation of QCD, the path ordered exponential, $P_{\mathbf{x}}(x^0, y^0)$, is written as

$$P_{\mathbf{x}}(x^0, y^0) \equiv U_0(\mathbf{x}, x^0)\, U_0(\mathbf{x}, x^0 + a) \ldots U_0(\mathbf{x}, y^0 - 2a)\, U_0(\mathbf{x}, y^0 - a)$$

for $y^0 > x^0$ and (14)

$$P_{\mathbf{x}}(x^0, y^0) \equiv U_0^{\dagger}(\mathbf{x}, x^0 - a)\, U_0^{\dagger}(\mathbf{x}, x^0 - 2a) \ldots U_0^{\dagger}(\mathbf{x}, y^0 + a)\, U_0^{\dagger}(\mathbf{x}, y^0)\ .$$

for $x^0 > y^0$.

The meson decay constant and mass can be derived, as explained in Section 2, from the two-point correlation function of the axial current, defined by:

$$A^L_\mu(x) = \bar{b}(x)\, \gamma_\mu \gamma_5\, q(x)\ , \tag{15}$$

where b and q represent the fields of the b-quark and a light quark respectively and the label L stands for local.

Using the leading term of the expansion for $S_b(x,y)$, eq.(13), one obtains:

$$G(t) = \sum_{\mathbf{x}} \langle A_0(\mathbf{x},t)\, A_0^{\dagger}(\mathbf{0},0)\rangle = \langle \mathrm{tr}\left[\frac{(1+\gamma^0)}{2}\, P_0(t,0)\, S_q(0,t)\right]\rangle\, e^{-m_b t}\ , \tag{16}$$

where $S_q(0,t) = S_q(\,\mathbf{0},0 \mid \mathbf{x}{=}\mathbf{0},t\,)$ is the light quark propagator between the points $(\mathbf{x}{=}\mathbf{0},t)$ and $(\mathbf{0},0)$, and $\langle\ldots\rangle$ represents the average over the gluon field configurations.

At large times we expect:

$$G(t) \to f_B^2\, \frac{M_B}{2}\, e^{-M_B t}\ , \tag{17}$$

where f_B is defined by the relation $\langle 0|A_0|B(\mathbf{p}{=}0)\rangle = i\, M_B\, f_B$ and M_B is the mass of the heavy meson. On the other hand we expect the following time dependence of the trace in eq.(16):

$$\langle \mathrm{tr}\left[\frac{(1+\gamma^0)}{2}\, P_0(t,0)\, S_q(0,t)\right]\rangle = Z\, e^{-\Delta E t}\ . \tag{18}$$

Consequently $Z = f_B^2\, M_B/2$ and $M_B = \Delta E + m_b$. ΔE can be interpreted as the "binding" energy of the heavy-light quark system; it is however linearly divergent in a^{-1} and requires a non-perturbative renormalization. Since Z is a constant, independent of M_B, we find

$$f_B^2\, M_B = \text{const.} \qquad \text{or} \qquad f_B \sim \frac{1}{\sqrt{M_B}} \tag{19}$$

up to logarithmic corrections which were discussed in refs.60, 61.

Furthermore, because the same correlation, eq.(16), describes the propagation of pseudoscalar and vector states, in this limit we have $M_B = M_{B^*}$, where M_{B^*} is the mass of the vector $\bar{b}$-q heavy meson. Other interesting scaling laws valid for different form factors are discussed in refs. 10.

In ref.23 it has been shown that with 30 gauge field configurations, and using the "local" axial-vector current $A^L_\mu(x)$, it is not possible to isolate the lightest pseudoscalar B-meson in the two-point correlation function, (this result has been confirmed in ref.24). More recently, a new calculation of the leading term of the $1/m_b$ expansion has been done using a "smeared" definition of the axial current, as suggested by the APE Collaboration [38]. Using the "smearing" technique the lightest pseudo-scalar state has been clearly isolated and the asymptotic value of $f_B\sqrt{M_B}$ has been computed with a small statistical error [25]. The results of ref. 25 have been subsequently confirmed in ref. 26 where a different smearing technique has been used and also by the results of ref. 48. Similar results with much larger statistical errors were previously obtained in ref.24.

Including the leading logarithmic corrections, one can derive the following mass dependence of the pseudoscalar decay constant f_P (cf. eq.(19)):

$$f_P = \frac{A}{\sqrt{M_P}} \, (\alpha_S(M_P))^{-2/\beta_0} \quad , \tag{20}$$

where M_P is the mass of P, A is a constant and β_0 is the coefficient of the lowest order term of the β-function.

A comparison of the result for f_B obtained keeping only the leading term in the $1/m_b$ expansion:

$$f_B = (260 \pm 25 + 100) \text{ MeV} \qquad \text{for } a^{-1}(\beta{=}6.0){=}1.8 \text{ GeV} \tag{21}$$

with earlier lattice measurements of f_D, $f_D \simeq (180 \pm 25 \pm 30)$ MeV, leads to the conclusion that there are still large corrections to the scaling law of eq.(20), in the region of masses from 2-5 GeV(*).

The result of f_B reported above (~260 MeV) is indeed larger than the values used in phenomenological models, and than those obtained using QCD sum rules [43]. For example, in their studies of $B^\circ - \overline{B}^\circ$ mixing, the authors of ref.62 take $f_B = 140 \pm 40$ MeV, whereas in ref.63 the author takes (150 ± 50) MeV for this quantity. The above values are based largely on the spread of results obtained in the QCD sum rule calculations for f_B [43].

The value of f_B in eq.(21) is also larger than that for the decay constant of the D-meson, which has been combined recently to give $f_D \simeq (180{\pm}25{\pm}30)$ MeV [64], where the first error is statistical and the second is systematic. Thus there is no evidence of the expected scaling behaviour of the decay constant with the mass of the pseudoscalar meson given by eq.(20) in the range of masses between 1.5 and 5.3 GeV. An extrapolation of the result in eq.(21) to the D-meson (using eq.(20)) would give a decay constant $f_D \simeq 440$ MeV. This implies that the scaling laws, which can be derived in the infinite mass limit, are not valid for charmed mesons. A similar conclusion applies also to the scaling behaviour of other quantities such as the semi-leptonic form factors, and is consistent with the measured ratio of the lifetimes of the charged and neutral D-mesons.

I illustrate this point further with fig.6, in which I plot the values of $f_P \sqrt{M_P}$ as a function of $1/M_P$. The values in this figure were obtained in lattice simulations using full propagators for both the quark and anti-quark [17,39-42]. From fig.6 we see that the quantity $f_P \sqrt{M_P}$ increases with mass in the range 0 - 2 GeV. Only in ref. 40 $f_P \sqrt{M_P}$ appears to level off at a value of about 0.2 $GeV^{3/2}$ probably because the calculation was done at values of the lattice spacing such that $m_b a \geq 1$. We have repeated this calculation at smaller values of the lattice spacing and shown that $f_P \sqrt{M_P}$ is still increasing in the same range of meson masses [42]. In the same figure I present a band, bounded by the two dotted lines, which corresponds to the results obtained by keeping only the scaling term, (i.e. the leading term in the $1/m_b$ expansion). The mild dependence of this band on M_P is due to the logarithmic factor on the right hand side of eq.(20). The width of the band corresponds to the statistical error in eq.(21), (i.e. 25 MeV).

It can be seen from this figure that even the points around the bottom quark mass are far below the scaling result.

To try and quantify the above discussion one can parametrise the deviations from the scaling behaviour by

$$f_P \sqrt{M_P} = A\left(1 - \frac{C}{M_P}\right) , \tag{22}$$

where A and C are constants. Using dimensional arguments it would be natural to assume that

---

(*) In eq.(21) the first error is statistical, obtained by averaging the result over 30 gauge field configurations and the second is systematic, from the calibration of the lattice spacing.

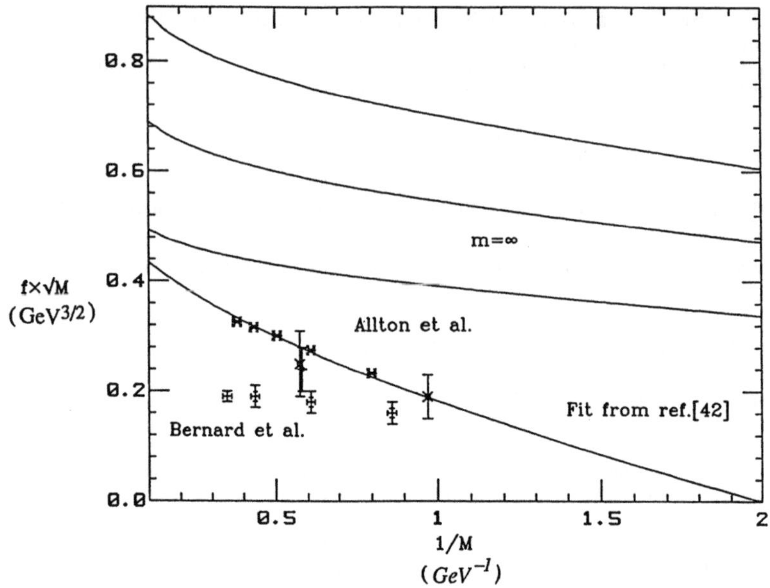

Fig. 6: Values of $f_P\sqrt{M_P}$ as a function of $1/M_P$, obtained from several lattice simulations. The region between the dotted lines corresponds to the results of ref.25 (for a^{-1} = 1.8 GeV at β=6), obtained by keeping only the leading term in the $1/m_b$ expansion.

$C \simeq \Lambda_{QCD}$, however in that case the scaling behaviour would set in at masses lower than that of the charm quark. In order for the result in eq.(21) to be compatible with $f_D \simeq 180$ MeV and with the result obtained in the $m_b \to \infty$ limit we need C to be larger than Λ_{QCD}, or to be more precise one finds C $\simeq$ 1.2 GeV (A ~ 0.6 $GeV^{3/2}$), which says that the correction is of the order of $2\pi\ f_P/M_P$. Using eq.(22), one concludes that eq.(21) overestimates f_B by approximately 25%. For the D-meson the second term in the paretheses on the right hand side of eq.(22) is about 65%. A fit to the points obtained with propagating heavy quarks from ref. 42 (denoted by Allton et al. in fig.6) would give instead A ~ 0.4 and C ~ 0.5 GeV sensibly smaller than those obtained in the static limit.

I want to recall here that a large value of f_B, with a large value for m_{top}, as suggested by the present experimental limits, implies that sin 2β, the angle with controls CP-violation in $B \to K_S\,\psi$ decays, is large (sin 2$\beta \gtrsim 0.6$) [66]. This is certainly a wellcome new for B-factories.

The results presented above raise a number of interesting questions. The static quark method gives a surprisingly large result for the decay constant; considerably larger than an extrapolation of the results obtained using the conventional approach. This indicates that there are still significant systematic errors in one or both of these calculations. Among there are:

i) the errors due to the finiteness of the lattice spacing. These might be expected to be particularly significant in the conventional calculation where they would include correction factors of the form 1 + const. $m_b a$. We estimate that the values of the heavy quark mass which we use range roughly from $m_b a$ ~ 0.2 to 0.75, and so it may not be surprising to find that these corrections may be significant. It will be very interesting to observe how the results change when the "improvement" procedure, designed to reduce these errors, is implemented [36].

ii) Errors due to perturbative corrections. The values of the renormalisation constants which relate the lattice matrix elements of the operators to the corresponding continuum ones. For the axial current which is relevant in the calculation of f_B, these corrections were found reasonably small, ~20%; we cannot exclude however that higher order corrections are important. It is clear that a two-loop calculation is needed to verify if this is the case.

TABLE 8

Measured values of the B-parameter. The superscripts S and D denote whether the singly or doubly smeared axial currents were used as interpolating operators[25]. Similar results were found in refs.24,26. K is the Wilson parameter which regulates the light quark masses.

K	B^S	B^D
0.1515	0.99 ± 0.08	0.98 ± 0.08
0.1530	1.00 ± 0.09	1.03 ± 0.09
0.1545	1.02 ± 0.09	1.01 ± 0.10

It should be noted that both methods agree on the sign of the $1/m_b$ corrections and that they are substantial at the mass of the charm quark, (the conventional method gives about 30% for these corrections at the charm quark mass and the static quark method gives about 60%). This is the main conclusions of the work of refs. 25 and 42.

In order to make progress in understanding the difference between the numerical results a number of possible computations can be performed. First of all by working with a smaller lattice spacing it will be possible to go to large quark masses in the conventional approach before the $O(m_b a)$ corrections become large. In addition by implementing the improvement procedure these corrections can be reduced [65]. In the "static" method the next term in the $1/m_b$ expansion of the heavy quark propagator can be computed explicitly to check how large these corrections are explicitly. These calculations are currently in progress.

The results for the B-parameter, reported in table 8, are consistent with 1, the expected value in the heavy quark limit [25]. It is however important to study the renormalisation of the operator $O_{\Delta B=2}$ in the effective theory, to check if there are any significant corrections to the lattice result in the determination of the physical value of B. Very recently B.Hill and H.Flynn have reported that this calculation has been just finished.

5. Conclusions

In this short review on lattice calculations of B- and D-matrix elements, I have tried to give a flavour of the present state of the art in this field. I have chosen to discuss only two examples, D semi-leptonic decays and the pseudoscalar meson decay constant, f_B, because they are relevant in our understanding of the Standard Model and show the advantages and limitations of the lattice approach.

I have reported the lattice predictions for $D \rightarrow K$ and $D \rightarrow K^*$ decays and shown that they agree fairly well with the experimental results, with a small theoretical uncertainty. In the case of $D \rightarrow K^*$ the agreement is even more striking if one compares the lattice results with the predictions of several quark models. In particular all quark models are in disagreement with the results of the E691 Collaboration, which finds $A_2(0) \sim 0$. The lattice results are even more valuable because no extra assumptions or arbitrary parameters are present in the calculation. A sensible improvement of the accuracy of the lattice calculation will soon be obtained by studying SU(3) breaking effects and terms of order Λ_{QCD} a.

The main information that came from the lattice calculation of the pseudoscalar meson decay constant a là Eichten, i.e. in the limit of an infinitely heavy quark, is that corrections to the asymptotic behaviour, due to $1/m_b$ terms, are between 30-65% for charmed mesons and ~20% for B mesons. This is larger than what was expected ($\Lambda_{QCD}/m_B \sim 5\%$) and seems to indicate that pre-asymptotic effects can be as large as $2\pi f_B/M_B$. It is then important to study the terms of order $1/m_b$ of the Eichten expansion and to investigate the mass dependence of the different decay constants and form factors in the region above the charm quark mass, with propagating quarks and at a larger value of the inverse lattice spacing [42].

Before closing my talk I will give some suggestions on new lattice calculations which may give relevant phenomenological implications in B-physics.

i) The mass of the Λ_b baryon. Suppose that we define a baryon operator with the u-d quarks in a iso-singlet state, which corresponds to the Λ_b. In the correlation function of this operator I expand the b-quark propagator a la Eichten, Then the correlation will go as:

$$\sum_{\vec{x}} < \Lambda_b(\vec{x},t)\, \Lambda_b^+(0,0) > \sim \text{const.} \exp(-\Delta E_\Lambda t) \ . \qquad (23)$$

ΔE_Λ is linearly divergent, but the divergent term is universal, i.e. equal to the corresponding term in the B-meson propagator, ΔE_B. Then one can write the relation:

$$R = \Delta E_\Lambda - \Delta E_B = M_\Lambda - M_B + O(1/M_B) \ . \qquad (24)$$

We can predict the value of M_Λ as:

$$M_\Lambda = M_B^{exp} + R + O(1/M_B) \ , \qquad (25)$$

where R is taken from the lattice calculation of $\Delta E_\Lambda - \Delta E_B$.

ii) Semi-leptonic B decays a la Eichten. A new problem arises is if one tries to compute $B \to D,D^*$ and $\Lambda_b \to \Lambda_c$ semi-leptonic decays, with both the b- and c- quark propagators taken in the infinite mass limit. In fact, on the lattice, the quark, in the infinite mass limit, can only fo at infinite or zero velocity, and only in average we can let it go to a fixed four velocity v. To treat this case we have to use the Georgi formalism, which is able to describe the propagator of a particle moving at costant velocity v[10]. Let us take the configuration in which the b- quark is at rest and the c- quark has a speed v. The c- propagator can be obtained from the static case by a Lorentz transformation, which, on an Euclidean lattice, is described by:

$$\begin{aligned} x' &= x \cos\theta - t\sin\theta \ , \\ t' &= x \sin\theta + t\cos\theta \ . \end{aligned} \qquad (26)$$

Where θ is related to the speed of the c-quark; the four velocity, $v = (v_o, \vec{v})$, which has to satisfy the condition $v_o^2 + |\vec{v}|^2 = 1$, has components $v_o = \cos\theta$ and $|\vec{v}| = \sin\theta$. I define the quantity $H(\vec{x},t)$ as follows:

$$S_H(\vec{x},t) = \frac{(1+\not{v})}{2}\, H(\vec{x},t) \ . \qquad (27)$$

By taking the static solution, i.e. $v = (1,0,0,0)$, and applying the trasformation given in eq.(26) to eq.(13), one obtains:

$$H(\vec{x},t) = \left(\exp(ig_o \int_0^{t/v_o} A(z) \cdot v dz)\right)_{\vec{z}=\vec{v}/v_o z_o} \delta^3(\vec{x} - t\frac{\vec{v}}{v_o})\, \Theta(t) \frac{1}{v_o} \ . \qquad (28)$$

The function H satisfies the differential equation:

$$(v \cdot D)\, H(\vec{x},t) = \delta^3(x) \qquad (29)$$

which is the correct one [10].

Now we have to traslate the above formulae on the lattice. In order to do so, let me define the following P-lines:

$$\begin{aligned} P_t(x=0, x_o=0) = (&U_0^+(1,0,0,t-1)\, U_0^+(1,0,0,t-2)\, U_0^+(1,0,0,0)\, U_1^+(0,0,0,0) + \\ &2U_0^+(1,0,0,t-1)\, U_0^+(1,0,0,t-2)\, U_0^+(1,0,0,1)\, U_1^+(0,0,0,1)\, U_0^+(0,0,0,0) + \dots \end{aligned}$$

$$U_1^+(0,0,0,t)\; U_0^+(0,0,0,t-1)\ldots U_0^+(0,0,0,0))\,/(2t)/\sqrt{(1+t^2)}\;, \qquad (30)$$

where the links position and direction is defined as $U_\mu = U_\mu(x,y,z,x_0)$.

It is easy to show that, provided $ta \ll 1$ the above equation corresponds to:

$$[1 - ig_0(tA_0(x=1/2,\, t/2) + A_1(x=1/2,\, t/2))]\,/\sqrt{(1+t^2)} + O(a^2) \quad . \qquad (31)$$

Now by calling $|\vec{v}| = \sin\theta = 1/\sqrt{(1+t^2)}$ and $v_0 = \cos\theta = t/\sqrt{(1+t^2)}$, I can define the following differential (discrete) equation:

$$H(\vec{x},x_0)/\sqrt{(1+t^2)} - P_t(\vec{x},x_0)\, H(\vec{x}-\hat{i},\, x_0 - t) = \delta(x) \qquad (32)$$

which expanding in a becomes:

$$[v_0\,(\partial_0 + ig_0A_0) + v_1(\partial_1 + ig_0A_1)]\; H(\vec{x},x_0) = \delta(x) \qquad (33)$$

which coincides with the continuum eq.(29) for $v=(v_0,v_1,0,0)$. For a velocity which has only the x- and t- components, this is the expected equation. It is trivial to find the solution of the lattice eq.(32).

Let me define the velocity of the particle as $|\vec{u}| = |\vec{v}|/v_0 = \Delta x/\Delta t$. For t=1 in eq.(13), $|\vec{u}| = 1$ = c, the speed of light; for t=2, $|\vec{u}|$ =1/2, for t=3, $|\vec{u}|$ =1/3 etc. Certainly we can do also $|\vec{u}|$ =2/3, 4/5, ... or any rational you want, but in that case the number of path increases enormously. In $B\rightarrow D$, the maximum momentum is $p_D^{max} = (M_B^2 - M_D^2)/(2M_B)$, which corresponds to $|\vec{u}| \simeq 0.77$, not too far from 0.5, which is what we can very easily do.

With our present techniques, using the solution of eq.(32), we can study the following decays:

a) B(static) $\rightarrow$D,D* (static) + $\ell\nu$,
b) B(static) $\rightarrow$D,D* (non-static) + $\ell\nu$,
c) Λ_b(static) $\rightarrow\Lambda_c$ (static) + $\ell\nu$,
d) Λ_b(static) $\rightarrow\Lambda_c$ (non-static) + $\ell\nu$,
e) D(static) $\rightarrow$ (K,K*,$\pi\rho$) + $\ell\nu$,
f) D(non-static) $\rightarrow$ K,K*,π,ρ + $\ell\nu$[53] ,

where static means infinite mass limit.

A comparison of a) with b) (c with d) or e) with f) will measure the size of $1/m_b$ corrections in these channels. More over a), b), e) and f) can be directly compared with the experimental results. The study of $D\rightarrow\pi\rho$ with a static or non static source will also test our capability of computing $B\rightarrow\pi\rho$ which will become in the future one the best methods to measure (V_{bu}).

The results reported in this talk and many others that I could only list in the introduction, will certainly improve in accuracy in the near future. They are however already at the stage to be able to provide us very useful information in the phenomenology of particle physics.

Acknowledgements

I wish to thank the Organizing Committee and particularly Proff.Aoki and Kobayashi for their kind invitation to this Conference and my colleagues C.Allton, V.Lubicz, L.Maiani and C.T.Sachrajda for many useful discussions on the subjects reported here.

References

1) N.Cabibbo, G.Martinelli and R.Petronzio, Nucl. Phys. B244 (1984) 381.
2) R.C.Brower et al., Phys. Rev. Lett. 53 (1984) 1318.
3) C.Bernard in Gauge Theory on a Lattice: 1984, C.Zachos et al., eds. National Technical Information service, Springfield, VA, 1984.
4) M.Crisafulli, G.Martinelli and C.T.Sachrajda, Phys. Lett. B223 (1989) 90.
5) C.Bernard, A.El-Khandra and A.Soni, Nucl. Phys. (Proc. Suppl.) 9 (1989) 186.
6) E.Eichten, Nucl. Phys. (Proc. Supp.) 4 (1988) 170 and refs. therein.
7) G.Kilcup at LATTICE '90, Tallahassee, Florida, to appear in the proceedings.
8) E.Eichten at LATTICE '90, Tallahassee, Florida, to appear in the proceedings.
9) Proceedings of Lattice '89, Capri, Sept. 89, N.Cabibbo et al. eds. Nucl. Phys. B (Proc. Suppl.) 17.
10) N.Isgur and M.Wise, Phys.Lett. B232 (1990) 113; Phys.Lett. B237 (1990) 527, UTPT-90-03- CALT-68-1696 (1990), CALT-68-1625-UTPT-90-02 ;
H.Georgi and M.B.Wise, Phys.Lett. B243 (1990) 279;
A.F.Falk et al., Nucl.Phys. B343 (1990) 1;
H.Georgi, B.Grinstein and M.Wise, HUPT-90/A052 - CALT-68-1664 (1990);
J.Chay and H.Georgi, HUTP-90/A035 (1990).
11) J.E.Mandula, PRINT-90-0356 (DOE) 1990 and refs. therein.
12) W.Wilcox and R.M.Woloshyn, Phys.Rev.Lett. 54 (1985) 2653. R.M.Woloshyn, Phys.Rev. D34 (1986) 605; R.M.Woloshyn and A.M.Kobos, Phys.Rev. D33 (1986) 222; T.Draper, R.M. Woloshyn and K.F. Lin, Nucl.Phys.B (Proc.Suppl.) 9 (1989) 175, TRI-PP-89-61 (7/89); Wilcox in ref. 9.
13) G.Martinelli and C.T.Sachrajda, Phys.Lett. B196 (1987) 184; Nucl. Phys. B (Proc. Suppl) 9, (1989) 175; Nucl.Phys. B306 (1988) 365; Nucl.Phys. B316 (1989) 305.
14) L.Maiani et al., Nucl.Phys. B293 (1987) 420; S.Güsken et al.,Nucl.Phys.B327 (89) 763; The APE Collaboration, P.Bacilieri et al., Nucl. Phys. B317 (1989) 509 and refs. therein.
15) S.Güsken et al., Phys. Lett. 227B (1989) 266.
16) S.Aoki and A.Gocksch, Phys. Rev. Lett. 63 (1989) 1125.
17) G.Martinelli, Phys.Lett. 141B (1984) 395; M.Bochicchio et al., Nucl.Phys. B262 (1985) 331; L.Maiani et al., Phys.Lett. 176B (1986)445; Nucl.Phys. B289 (1987) 505; L.Maiani and G.Martinelli, Phys.Lett. B181 (1986) 344; M.B.Gavela et al., Nucl.Phys. B306 (1988) 677; Phys.Lett. 211B (1988) 1139; E.Franco et al., Nucl.Phys. B317 (1989) 63; the ELC Collaboration, presented by G.Martinelli in ref.9.
18) C.Bernard et al., Phys Rev. D32 (1985) 2343; Phys. Rev. Lett. 55 (1985) 2770; C.Bernard, T.Draper and A.Soni, Phys. Rev. D36 (1987) 3224; C.Bernard in ref.9.
19) S.Sharpe in Lattice Gauge Theory '86, H.Satz, I.Harity and J.Potrin eds., Plenum, New York, 1987; G.W.Kilcup and S.R.Sharpe, Nucl. Phys. B283 (1987) 493, R.Gupta et al., Nucl. Phys. B286 (1987) 253; R.Gupta et al., Phys. Lett. 192B (1987) 149; S.R.Sharpe in ref. 9.
20) V.Lubicz, G.Martinelli and C.T.Sachrajda, SHEP 89/90-13 and Rome prep. n.748, to appear in Nucl. Phys. B.
21) C.T.Sachrajda in ref.9.
22) J.Simone in ref. 9 and refs. therein.
23) Ph.Bouchaud et al., Phys.Lett. 220B (1989) 219.
24) E.Eichten, G.Ockney and H.B.Thacker, Nucl.Phys. B (Proc.Suppl.) 17 (1990) 529.
25) C.R.Allton et al., SHEP 89/90-11 (1990), to appear in Nucl. Phys. B.
26) C.Alexandrou et al., PSI-PR-28/WU B90-18, (1990).
27) M.Creutz, Phys.Rev.Lett. 43 (1979) 553; Phys.Rev. D21 (1980) 2308.
28) H.Lipps et al., Phys.Lett.126B (1983) 250.
29) S.Itoh, Y.Iwasaki and T.Yoshie, Phys.Lett. 183B (1987) 351.
30) M.Albanese et al., Proc. of the Asilonar Conference, in Comp.Phys.Comm. 45 (1989) 449.
31) N.H.Christ and A.E.Terrano, Byte Magazine, Vol.II, No.4 (1986) 145.
32) Y.Iwasaki et al., Comp.Phys.Comm. 49 (1988) 449.
33) M.Fischler et al., Third Conf. on Hypercube Concurrent Computers and Applications, Pasadena, CA (1988).
34) R.Tripiccione in ref.9.
35) K.Symanzik, Nucl.Phys. B226 (1983) 187 and 205.

36) G.Heatlie et al., Nucl. Phys. B (Proc. Suppl.) 17 (1990) 607; Nucl.Phys. B352 (1991) 60.
37) The EMC Collaboration, Nucl.Phys. B328 (1989) 1.
38) The APE Collaboration, presented by E.Marinari in ref.9.
39) M.B.Gavela et al., Phys.Lett. B206 (1988) 113.
40) C.Bernard et al., Phys.Rev. D38 (1988) 3540.
41) T.A.De Grand and R.D.Loft, Phys.Rev. D38 (1988) 954.
42) C.R.Allton, D.B.Carpenter, G.Martinelli and C.T.Sachrajda, presented by C.R.Allton at LATTICE '90, Tallahassee, Florida, to appear in the proceedings.
43) E.V.Shuryak, Nucl.Phys. b198 (1982) 83; V.L.Chernyak et al., Sov.J.Nucl.Phys. 38 (1983) 773; A.Aliev et al., Sov.J.Nucl.Phys. 38 (1983) 936; S.Narison, Phys.Lett. 197B (1987) 405; C.A.Dominquez and N.Paver, Phys.Lett. 197B (1987) 423; erratum Phys.Lett. 199B (1987) 596; L.J.Reinders, H.Rubinstein and S.Yazaki, Phys.Reports 127 (1985) 1; V.S.Mathur and T.Yamawak, Phys.Rev. D29 (1984) 2057.
44) J.Donoghue, E.Golowich and B.Holstein, Phys.Lett. 119B (1982) 412.
45) E.De Rafael and A.Pich, Phys.Lett. 158B (1985) 477.
46) N.Bilic, C.A. Diringuer and B.Guberina, Desy-preprint 87-162.
47) A.J.Buras and J.M.Gerard, Nucl.Phys. B264 (1986) 371.
48) C.Bernard and A.Soni, at LATTICE '90, Tallahassee, Florida, to appear in the proceedings.
49) S.Sharpe, at LATTICE '90, Tallahassee, Florida, to appear in the proceedings.
50) E.De Rafael, private communication.
51) C.Anjos et al., The Tagged Photon Spectrometer Collaboration, Phys.Rev.Lett. 62 (1989) 1587.
52) J.Adler et al., The Mark III Collaboration, Phys.Rev.Lett. 62 (1989) 1821.
53) V.Lubicz et al., presented at LATTICE '90, Tallahassee, Florida, to appear in the proceedings.
54) A.Bean at the PANIC Conference, MIT, Boston, June 1990, to appear in the proceedings.
55) N.Isgur and D.Scora, Phys.Rev. D40 (1989) 1491.
56) M.Bauer, B.Stech and M.Wirbel, Z.Phys. C 29 (1985) 637; M.Bauer and M.Wirbel, Z.Phys. C34 (1987) 103; rep. no. HD-THEP-88-82.
57) J.G.Körner and G.A.Shuler, Z.Phys. C38 (1988) 511.
58) B.Grinstein et al., Phys.Rev. D39 (1989) 799.
59) S.Capstick and S.Godfrey, GIPP-89-10; CMU-HEP 89-20.
60) M.B.Voloshin and M.A.Shifman, Sov.J.Nucl.Phys. 45 (1987) 292; Sov.J.Nucl.Phys. 47 (1988) 511.
61) H.D.Politzer and M.B.Wise, Phys.Lett. 206B (1988) 681; Phys.Lett. 208B (1988) 504.
62) G.Altarelli and P.J.Franzini, Z.Phys. C37 (1988) 271.
63) Y.Nir, Nucl.Phys. B306 (1988) 14.
64) S.R.Sharpe in ref.9.
65) G.Heatlie et al., Rome preprint n.737 (1990) to appear in Nucl. Phys. B;
G.Gabrielli et al., Roma preprint n.765 (1990);
G.Martinelli, C.T.Sachrajda and A.Vladikas, Roma preprint n.766 (1990).
66) For a recent discussion see M.Lusignoli, L.Maiani. G.Martinelli, L.Reina, Rome prep.792 (1991).

Concluding Remarks

M. Kobayashi

KEK, National Laboratory for High Energy Physics,
1-1 Oho, Tsukuba, Ibaraki 305, Japan

As was mentioned in the opening address, this symposium marks the fifth year of the Nishinomiya-Yukawa symposium on fundamental physics. This seemed to us a good occasion to review the present status of high energy physics, thereby obtaining a perspective for the future, so we invited best speakers who are actively working in each field. I think that almost all aspects of the present trend of high energy physics have been fully discussed by our speakers at this symposium. First of all, on behalf of the organizing committee, I would like to express my sincere thanks to all the speakers for their excellent talks.

Now let me summarize very briefly what has been discussed in these two days.

Results from TRISTAN, LEP and CDF at Tevatron were discussed by Professors Watanabe, Mashimo, Treille, and Brandenburg. All these results show that the standard model works *too* well and we have no indication of what would exist beyond the standard model.

A remarkable fact revealed by LEP is that the number of light neutrinos is just three and the room for various possible neutral objects is very limited. A new lower bound for the Higgs mass, $m_H > 44\text{GeV}$, was also reported this morning.

Top quark has not been observed yet. CDF results give a new lower bound of 91 GeV for the top quark mass. Precise determination of the Z-boson mass by LEP has enabled us to pin down the mass range of the top through the calculation of radiative corrections, as discussed by Professor Sirlin. From such an analysis, the top mass is estimated to be somewhere between 100 and 180 GeV. If this estimate is correct, a luminosity increase of Tevatron would be the most prompt possibility of finding the top.

In connection to the top quark mass, a very interesting theoretical idea, discussed by Professor Bardeen, is a scenario of the top quark condensation. A minimal scheme of this senario predicts the top mass around 220 GeV. It was argued that this prediction is a very stable one.

Another unsolved issue of the standard model is the problem of CP violation. The situation is still uncertain both experimentally and theo-

Springer Proceedings in Physics, Vol. 65 Present and Future of High-Energy Physics
Editors: K.-I. Aoki and M. Kobayashi

retically. As was discussed by Professor Winstein, experimental results on the ratio of ϵ' to ϵ appear to be in serious conflict between CERN NA31 and Fermilab E731. The theoretical ambiguity partly comes from our ignorance of the top quark mass, but a major problem is the determination of the relevant hadronic matrix elements, in which we are confronted with nonperturbative effects of the strong interaction. Recently much effort has been paid for this problem as described by Professor Wise. One of the frontal attack is the lattice calculation of the matrix elements, which was discussed by Professor Martinelli. Professor Winstein concluded that ϵ'/ϵ is the best place to study CP violation in the coming decade. On the other hand, we can expect that a clean test of the standard model of the CP violation is possible in the B-meson system. In order to do this we need a new type of e^+e^- collider, called an asymmetric B-factory. Construction of B-factories is under discussion at many laboratories in the world, and KEK is seriously considering the construction of this type of B-factory.

Here let us recall future plans for accelerators at the energy frontier. HERA is the first electron-proton collider and experiments are going to start in the next year. LEP II is the energy increase of the present LEP machine by installing superconducting RF cavities. The physics run is scheduled in 1994. Main physics aims will be the three gauge boson coupling and Higgs search. Luminosity upgrade of Tevatron is planned in several steps. New particle searches, in particular, top quark search, will be the most important physics aim there.

Going up to much higher energy, SSC is a 20TeV + 20TeV proton-proton collider at Texas, and LHC is an 8TeV + 8TeV proton-proton collider planned at CERN. Commissioning is expected in 1999 for SSC and 1998 was mentioned for LHC. Besides these hadron colliders, there are many plans of the e^+e^- linear collider in TeV region at various places. As for the physics in this energy region, theorists have been proposing many scenarios. Those were discussed by Professor Peskin in a previous lecture. To find an experimental answer to these theoretical ideas, we have to wait for still many years.

When we look back upon the development of the standard model, we notice that basic theoretical inputs of the standard model are only three ingredients related to non-Abelian gauge theory, which are renormalizability of non-Abelian gauge theory, Higgs mechanism, and the scenario of confinement. With these three findings, we came to posess a new device or a new language describing the nature. This framework is quite flexible and accomodates various possibilities. What we have done under the name of the standard model is just to describe the nature in this language as it appears. We don't know, for example, the deep reason why such and such quark exists with such and such mass. Of course, we have learned many things in these 20 years along the development of the standard model, and this is a great history of understanding the nature. From the viewpoint of the fundamental law of the nature, however, what we have made may be a tiny step. I feel that unfettered imagination of human being is required in order to reveal the secret of the nature.

Index of Contributors

Springer Proceedings in Physics

Managing Editor: H. K. V. Lotsch

Springer Proceedings in Physics

Managing Editor: H. K. V. Lotsch

44 *Optical Fiber Sensors*
Editors: H. J. Arditty, J. P. Dakin, and R. Th. Kersten

45 *Computer Simulation Studies in Condensed Matter Physics II: New Directions*
Editors: D. P. Landau, K. K. Mon, and H.-B. Schüttler

46 *Cellular Automata and Modeling of Complex Physical Systems*
Editors: P. Manneville, N. Boccara, G. Y. Vichniac, and R. Bidaux

47 *Number Theory and Physics*
Editors: J.-M. Luck, P. Moussa, and M. Waldschmidt

48 *Many-Atom Interactions in Solids*
Editors: R .M. Nieminen, M. J. Puska, and M. J. Manninen

49 *Ultrafast Phenomena in Spectroscopy*
Editors: E. Klose and B. Wilhelmi

50 *Magnetic Properties of Low-Dimensional Systems II: New Developments*
Editors: L. M. Falicov, F. Mejía-Lira, and J. L. Morán-López

51 *The Physics and Chemistry of Organic Superconductors*
Editors: G. Saito and S. Kagoshima

52 *Dynamics and Patterns in Complex Fluids: New Aspects of the Physics–Chemistry Interface*
Editors: A. Onuki and K. Kawasaki

53 *Computer Simulation Studies in Condensed Matter Physics III*
Editors: D. P. Landau, K. K. Mon, and H.-B. Schüttler

54 *Polycrystalline Semiconductors II*
Editors: J. H. Werner and H. P. Strunk

55 *Nonlinear Dynamics and Quantum Phenomena in Optical Systems*
Editors: R. Vilaseca and R. Corbalán

56 *Amorphous and Crystalline Silicon Carbide III, and Other Group IV – IV Materials*
Editors: G. L. Harris, M. G. Spencer, and C. Y.-W. Yang

57 *Evolutionary Trends in the Physical Sciences*
Editors: M. Suzuki and R. Kubo

58 *New Trends in Nuclear Collective Dynamics*
Editors: Y. Abe, H. Horiuchi, and K. Matsuyanagi

59 *Exotic Atoms in Condensed Matter*
Editors: G. Benedek and H. Schneuwly

60 *The Physics and Chemistry of Oxide Superconductors*
Editors: Y. Iye and H. Yasuoka

61 *Surface X-Ray and Neutron Scattering*
Editors: H. Zabel and I. K. Robinson

62 *Surface Science: Lectures on Basic Concepts and Applications*
Editors: F. A. Ponce and M. Cardona

63 *Coherent Raman Spectroscopy: Recent Advances*
Editors: G. Marowsky and V. V. Smirnov

64 *Superconducting Devices and Their Applications*
Editors: H. Koch and H. Lübbig

65 *Present and Future of High-Energy Physics*
Editors: K.-I. Aoki and M. Kobayashi

Volumes 1–43 are listed on the back inside cover